CW00868789

Great Powers, Small Wars

Great Powers, Small Wars:

ASYMMETRIC CONFLICT SINCE 1945

Larisa Deriglazova

Woodrow Wilson Center Press
WASHINGTON, D.C.

Johns Hopkins University Press
BALTIMORE

EDITORIAL OFFICES

Woodrow Wilson Center Press
One Woodrow Wilson Plaza
1300 Pennsylvania Avenue, N.W.
Washington, DC 20004-3027
Telephone: 202-691-4029
www.wilsoncenter.org

ORDER FROM

Johns Hopkins University Press
Hopkins Fulfillment Services
P.O. Box 50370
Baltimore, MD 21211-4370
Telephone: 1-800-537-5487
www.press.jhu.edu/books/

Printed in the United States of America

Library of Congress Cataloging-in-Publication Data

Deriglazova, L.V. (Larisa Valer'evna)
Great powers, small wars : asymmetric conflict since 1945 / Larisa Deriglazova.
p. cm.
ISBN-13: 978-1-4214-1412-6
ISBN-10: 1-4214-1412-0
1. Asymmetric warfare—History—20th century. 2. Asymmetric warfare—History—21st century. 3. Great powers—History, Military—20th century. 4. Great powers—History, Military—21st century. 5. Great Britain—Colonies—History, Military—20th century. 6. Iraq War, 2003–2011. I. Title.
U163.D476 2014
355.4'2—dc23

2013042522

The Wilson Center, chartered by Congress as the official memorial to President Woodrow Wilson, is the nation's key nonpartisan policy forum for tackling global issues through independent research and open dialogue to inform actionable ideas for Congress, the Administration, and the broader policy community.

Conclusions or opinions expressed in Center publications and programs are those of the authors and speakers and do not necessarily reflect the views of the Center staff, fellows, trustees, advisory groups, or any individuals or organizations that provide financial support to the Center.

Please visit us online at www.wilsoncenter.org.

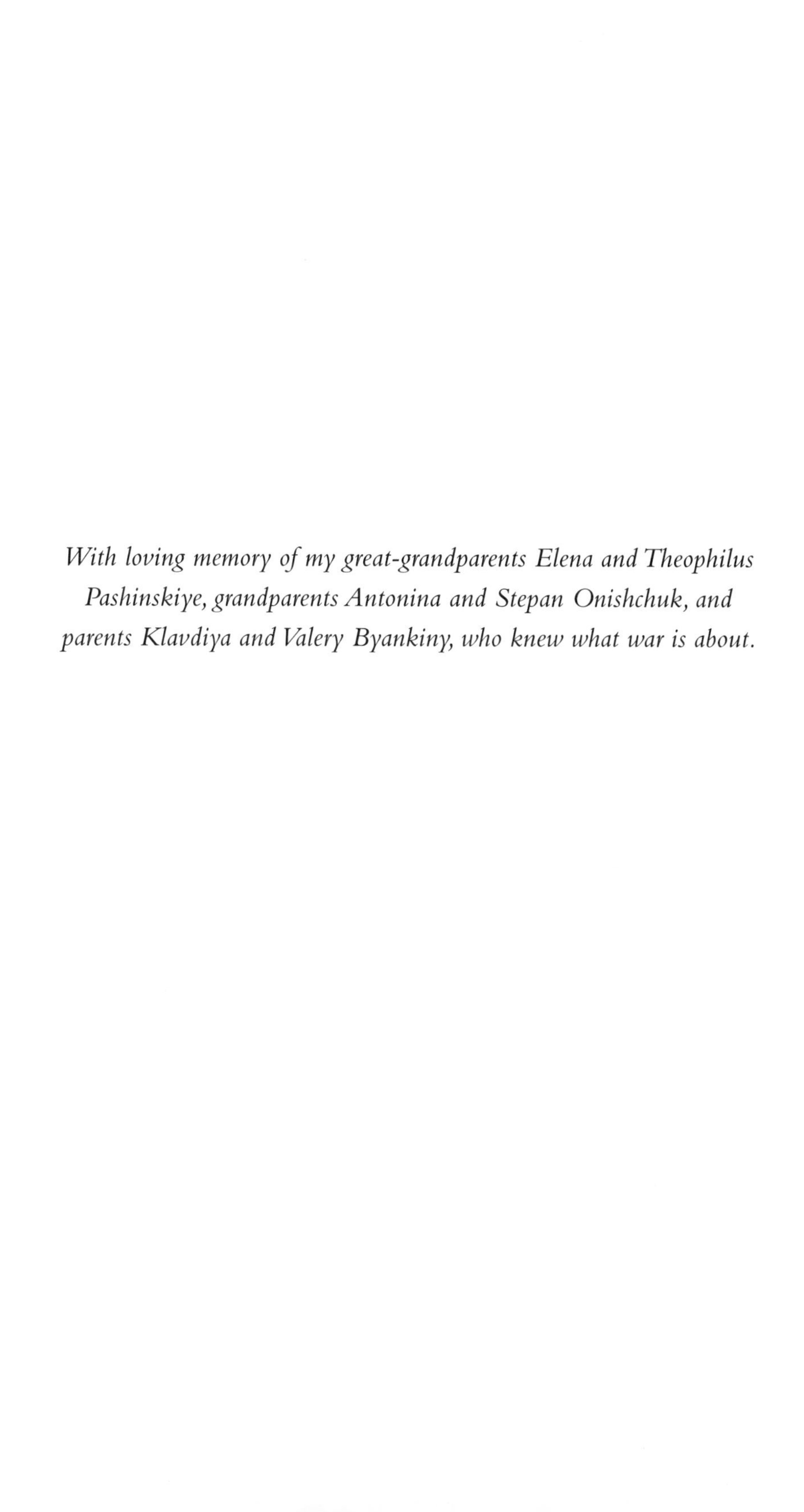

With loving memory of my great-grandparents Elena and Theophilus Pashinskiye, grandparents Antonina and Stepan Onishchuk, and parents Klavdiya and Valery Byankiny, who knew what war is about.

Contents

List of Tables

List of Figures

PREFACE

Asymmetric Conflicts— An Equation with Many Unknowns

The study of conflicts involving great powers has long been at the core of international relations. Post–World War II history demonstrated that great powers could refrain from direct military clashes with one another. However, in no single year in the postwar period did the military forces of the great powers rest entirely. Scholars of the post–World War II period are unanimous in assigning the status of great power to the five permanent members of the UN Security Council, namely, the United States, the USSR/Russia, the United Kingdom, France, and China. Statistical data on post–World War II armed conflicts also indicate that these states were the most active in waging war during and after the Cold War. The collapse of the bipolar system in the late 1980s and early 1990s inspired hope for a lasting peace, but the following two decades saw this optimistic forecast dismissed.

The theory of asymmetric conflict aims to identify recurring patterns in the post–World War II armed conflicts that involved the great powers. It was originally developed to understand the trends of limited armed conflicts involving great powers, but its use has since been extended beyond this initial theoretical basis. Although the idea of asymmetric conflict appears frequently in the research literature and political essays of the past decade, the term is neither fully defined nor widely accepted. The following tentative definition can be offered: an asymmetric conflict is an armed struggle between adversaries that are unequal in power, resources, and status, in which the weaker party relies on strategies designed to compensate for its weakness and tries to exhaust the will of the stronger party and force it to stop fighting. The emergence of asymmetric conflict theory in the mid-1970s marked an attempt to identify the reasons for the counterintuitive defeat of the great powers in a number of armed conflicts with weaker adversaries, including national liberation movements and other nonstate actors. The strengthening of this trend in the second half of the twentieth century and its culmination in the early twenty-first century undermine the logic of developed states' domination of the periphery and semiperiphery of the world system as a result of their military, technical, economic, and political strengths.

This book largely focuses on one particular type of asymmetric conflict, namely, armed conflict between a great power and a weaker adversary that results in political defeat for the great power. Narrowing the problem in this way aids the analysis because such situations represent a "pure" phenomenon. They served as a foundation for the development of the theory and vividly underscore the counterintuitive nature of asymmetric conflict, making it possible to identify the most important recurring elements of such conflicts.

The factors accounting for the defeat of developed states (including the Soviet Union beginning in the 1950s, that is, after its postwar economic recovery) in asymmetric conflicts can be divided into three groups. The first group contains factors that for research purposes can be classified as endogenous, or arising from the internal characteristics of the belligerent parties. This group includes both available resources and political, economic, and moral characteristics specific to the chosen parties. The second group comprises exogenous factors, or those externalities that have an impact on the participants in armed conflicts or the development of the conflict itself. This group may include aspects such as the effect of negative international opinion on the conduct of the stronger power, or the availability of overt or covert support for a weaker party from another great power. The third group includes factors related to the tactics and strategy of the struggle, such as the use of guerrilla warfare or terrorist activity contrasted with the presence of fixed fighting units. According to the hypothesis put forward in this book, commonalities shared by different states that have engaged in asymmetric conflicts can be discovered in all three groups of factors.

Despite the diverse historical contexts for such conflicts, it is possible to discern similarities both in the underlying situations and in the dynamics and outcome of the struggle, which leads to the suggestion of recurrent patterns that hold across asymmetric conflicts. Connections can be identified among the reasons for a great power's defeat, derived from the three different groupings presented above, that are also maintained across conflicts. Therefore, one can assume that certain basic (essential) asymmetries exist that give rise to the whole chain of asymmetric characteristics.

The book's hypothesis also holds that the basic asymmetry in the adversaries' relationship becomes an obstacle to managing their interstate relations within the existing system. At least one party strives to eliminate this asymmetry and establish a new, more symmetric kind of interaction. From this perspective, defeat of the stronger party (the great power) in the conflict results in destruction of the hierarchical system in which the stronger party dominated, and the emergence of a more egalitarian relationship between the two opponents. This perspective on asymmetric conflicts offers a way to resolve the problem associated with one of their numerous illogical characteristics, the discrepancy between the military superiority of the stronger party and the political victory of its weaker adversary.

The unpredictable outcome of armed conflicts in the post–World War II period became key to the great powers' reconsidering their attitude toward the use of force. Owing to the unpredictability of the short- and long-term consequences of armed conflict, the great powers more and more often regard war as a means of last resort, not as a normal instrument of foreign policy.

The novelty of the research reported in this book is reflected in the fact that for the first time, an attempt is made to systematically study asymmetric conflicts using quantitative methods to explore this phenomenon in the post–World War II era, based on in-depth historical research for individual cases. The first part of the book, devoted to quantitative manifestations of asymmetric conflicts, reinforces the argument about the need to correct certain simplified perceptions concerning the nature of the interactions between the great powers during the Cold War. A critical rethinking of the Cold War period is taking place, as evidenced in recent scholarly publications by historians studying certain aspects of postwar history, but some stereotypes still need to be dispelled. A persistent belief in the unyielding antagonism between the West and the Soviet Union prevents many scholars from understanding the complex dynamics of relations within the Western bloc and accepting that the positions of states in opposing blocs can sometimes coincide. Such instances of co-occurring views should not be considered exceptional. Rather, they should prompt an in-depth and comprehensive investigation of postwar history to counterbalance its often simplistic and schematic interpretation.

The dissolution of the British Empire and the US war in Iraq of 2003–2011 were selected as case studies for in-depth analysis. The United Kingdom and the United States have at certain times been the undoubted global leaders. The peak of the United Kingdom's global influence came in the early twentieth century and terminated with the collapse of the empire in the post–World War II years and its transformation into the independent states of the British Commonwealth of Nations. The United States started to test its power beyond its regional influence and position itself with increasing confidence as a global leader in the postwar period, especially in light of the United Kingdom's decreased willingness and ability to resist the pressures of national liberation movements and the growing influence of Soviet ideas in its former possessions. After the end of the Cold War, the United States attained a position of world superiority, and its ability to influence global politics seemed unlimited. However, the launch of the war on terrorism in 2001 revealed the vulnerability of the global giant. Despite significant differences between the two countries, this study shows that it is possible to identify consistent reasons for the political defeat of these two world powers in armed conflicts against considerably weaker adversaries.

The choice of the British Empire as a research subject was conditioned by existing assertions as to the peaceful and voluntary dissolution of the empire. Surprisingly, this opinion dominates both the Russian and more broadly the Western research literature. However, an attentive historian would notice that the dissolution of the British Empire was neither peaceful nor voluntary. In fact, the United Kingdom waged military operations in many of its colonies and dependent territories until the mid-1960s. As a result of these actions, called "small wars" or "emergencies," the British Empire experienced a swift disintegration in the first two postwar decades. The United Kingdom used its experience accumulated in the course of these small wars to "consult" with the United States during its war in

Vietnam, to organize peacekeeping operations in the third world countries, and to render support to the governments of the Commonwealth countries in their fight against local opposition movements.

An additional point in support of using the experience of the United Kingdom as a case study is the appearance, beginning in the late 1990s, of numerous official documents on the dissolution of the British Empire. The new body of sources has stimulated the emergence of historical studies on the British Empire and its dissolution. This book attempts to provide a critical analysis of the process of empire disintegration that takes into account the newly published documents, as well as opinions expressed in research publications by authors on different sides of the political spectrum.

The war in Iraq in 2003–2011 represents a recent example of an armed conflict in which the most powerful country in the world, the United States, in a coalition with other developed countries, failed to achieve its objectives in a war against an adversary vastly incommensurate in terms of power. Analysis of the war dynamics and of attempts to bring the war to a close offers an opportunity to apply the model of asymmetric conflict as a way to understand better the reasons for the political, if not military, failure of the United States in Iraq. In the American research literature, the model of asymmetric conflict has been used to analyze a limited aspect of the war operations, specifically guerrilla and terrorist actions, but no complex or comprehensive study of the Iraq War as an example of asymmetric conflict has appeared.

The research presented in this book is of practical value, as the model of asymmetric conflict can be used to analyze both historical and contemporary events and contexts. It serves to identify the most significant components of the armed conflicts between parties incommensurate in resource and power capabilities and helps shed light on the reasons for the political defeat of the stronger parties and on the strategy choices of their weaker opponents.

Understanding the logic of armed clashes between asymmetric antagonists can be useful in applied military and strategic analysis, as well as in political decision making with regard to the possible use of force and how it is used. Such an understanding may also be useful in setting conditions and restrictions on the use of force in international and domestic conflicts.

The book consists of four sections, a conclusion, and an appendix. The first section looks at the origins and development of the asymmetric conflict concept. It examines the relevant literature and critical approaches to the concept of asymmetric conflict, and proposes an analytical model to investigate the phenomenon. The second section identifies the asymmetric factors in armed conflicts, looking at the structural and dynamic characteristics of such conflicts and assessing the impact of asymmetry on great-power conflicts. To do so, it draws on statistical data from two main sources: the database of armed conflicts compiled by the Uppsala Conflict Data Program, Sweden, in cooperation with the International Peace Research Institute, Oslo, and the database on national and international conflicts compiled by the COSIMO project at the University of Heidelberg, Germany, under the leadership of Professor Frank R. Pfetsch. The third section focuses on the first case study of asymmetric conflict: the dissolution of the British Empire and the resulting conflicts of the post-decolonization period. The fourth section focuses on the second case study of asymmetric conflict: the US war in Iraq from 2003 to 2011, and the asymmetric conflict model approach to the war's outcomes and international discussion of those outcomes. The synopsis summarizes the results of the book's analysis and presents the author's concluding thoughts. The appendix provides statistical and other quantitative information from the COSIMO database on armed conflicts to support the analysis and conclusions of the study.

ACKNOWLEDGMENTS

This book would not be possible without the many special people who taught, encouraged, supported, and challenged me. Sometimes, it was essential that they simply stood by my side. My studies at the Kroc Institute for International Peace Studies at the University of Notre Dame opened the topic of asymmetric conflicts to me in the mid-1990s. My dear teachers and colleagues Savely Wolfson, Vasily Zinoviev, Alexei Timoshenko, Inna Karataeva, Valentina Sumarokova, and Anna Korchuganova from Tomsk State University deserve credit for making this book happen. I am immensely thankful to Professor Alexei Bogaturov for his encouragement and life investment into the development of international relations studies in Russia's vast territory beyond Moscow. Richard Sakwa, Judith and David Marquand, and Eric Remacle provided intellectual guidance and models to follow, and their warm friendship crossed borders. I owe a lot to Professor Sir Hew Strachan, Chichele Professor of the History of War at All Souls College, Oxford University, who generously provided me with exceptional insight into British military literature while being critical of the very concept of asymmetric conflict as such. The greater part of this manuscript was written during my time as a Fulbright scholar at the Kennan Institute of the Woodrow Wilson International Center for Scholars in Washington, DC. This "academic paradise" gave me a refuge for productive intellectual work with excellent library and logistic support, as well as a rich and diverse landscape of research projects that challenged my ability to concentrate on my own project. I want to express my deep appreciation to the Kennan Institute staff, and especially to Blair Ruble, for their support and the feeling of belonging to this very special institution devoted to Russian studies in the United States, with the belief that Russia is a country that deserves to be known better. For the English edition of the book, translator Nina Rozhanovskaya and editor Marjorie Pannell provided important revisions that helped to bring this project to a strong conclusion. Finally, my family—Denis, Vasily, Lena, Sergey, and Asya—have never doubted my work, and they are my biggest source of energy and reward.

CHAPTER 1

Origin and Development of the Asymmetric Conflict Concept

POSTWAR PEACE: FROM TOTAL WAR TO ASYMMETRIC CONFLICTS

The two world wars of the twentieth century marked the culmination of the development of international relations involving the great powers and the peak of their interstate aggression. The numbers of the dead tell part of the tale. More than 10 million people perished in World War I, and more than 17 million military personnel and 34 million civilians died during World War II. World War I gave rise to the hope that war on such a scale would never again happen, leading it to be dubbed "the war to end all wars"—only for such hopes to be belied by the horrors and immense resource consumption of World War II. The desire to end all wars has been at least partially fulfilled, as since 1945, the great powers have not gone to war with one another. Competition among Western countries was transferred to the economic realm, while the West and the East have developed certain "rules of the game" intended to prevent military clashes.

The concept of asymmetric conflict emerged because the nature of armed conflicts changed significantly in the post–World War II era. The direct military confrontation of major powers gave way to indirect participation in armed conflicts on the periphery of the international system.

Several new terms were introduced to emphasize the distinctive features of these conflicts, among them small wars, low-intensity conflicts, local wars, limited wars, counterinsurgency operations, and antiterrorist campaigns. Notably, a significant proportion of the armed conflicts involving the great powers were not classic interstate wars, or wars between parties of roughly equal status and capabilities.

The French philosopher, historian, and political scientist Raymond Aron (1905–1983) was one of the first to point out that the character of war had changed with the emergence of nuclear weapons, and that "nations on the periphery [of the United States and the Soviet Union] acceded to the first rank." In his 1948 book *The Great Schism* (*Le Grand schisme*), Aron coined the popular postwar saying "impossible peace, improbable war."[1] Later, in his 1962 *Peace and War: A Theory of International Relations* (*Paix et guerre entre les nations*), this aphorism was elaborated in another memorable statement: "Inter-state relations present one original feature which distinguishes them from all other social relations: they take place within the shadow of war, or, to use a more rigorous expression, relations among states involve, in essence, the alternatives of war and peace."[2]

Aron believed that the main trend in the evolution of wars was their limited character, resulting from the need to avoid total war between two opposing camps, as it "cannot not be atomic." In his 1951 book *Chain Wars* (*Les Guerres en chaîne*) and in the articles "A Half-Century of Limited War?" (1956) and "On War without Victory" ("De la guerre sans victoire," 1951), Aron further developed the idea of the transformation of war from a military-strategic phenomenon into a political one. In his memoirs, he wrote that "the history of war could only be understood in the context of the history of political relations," and that the Korean War (1950–1953) seemed to him to be a turning point, when "for the first time in its history, the United States gave up an annihilating victory. After a half century of total wars, there began the half century of limited war."[3] Furthermore, Aron predicted the binary impact of nuclear weapons in terms of states able to conduct total war versus limited war: "A hierarchy of the regions of the world becomes apparent, with certain regions protected by thermonuclear weapons, and certain objectives worth, in the eyes of both sides, the risk of mutual suicide. Elsewhere, the rivalry will be pursued in traditional ways, with or without the use of conventional military techniques (with guerrilla warfare playing an increasingly important role)."[4]

Hans Morgenthau (1904–1980), a prominent twentieth-century student of international politics and law, pointed to two mechanisms restraining the bellicose aspirations of nations and contributing to the preservation of peace: first, the balance of power as a spontaneous mechanism of self-regulation of states' behavior in the international arena as states strive to maximize their power; and second, normative restrictions on the struggle for power imposed by international law, international morality, and global public opinion.[5] Furthermore, the balance of power as competition and struggle between autonomous forces or unions of states in the postwar period becomes global, and the existence of a dominant system of the balance of power between superpowers and their allies subordinates and determines the local balance of power systems.[6]

Analyzing the elements of the international system that constrain major powers (international law, international morality, and global public opinion), Morgenthau pointed out that they were primarily relevant for the conduct of "total war." In his view, modern war had become total "in four different respects: (1) with regard to the fraction of the population engaged in activities essential for the conduct of the war; (2) with regard to the fraction of the population affected by the conduct of the war; (3) with respect to the fraction of the population completely identified in its convictions and emotions with the conduct of the war; (4) with respect to the objectives of the war." Morgenthau also stressed that "Warsaw and Rotterdam, London and Coventry, Cologne and Nuremberg, Hiroshima and Nagasaki are stepping stones, not only in the development of the modern technology of war, but also in the development of the modern morality of warfare. The Indochina war for all practical purposes obliterated the distinction between combatants and civilian population."[7]

The prominent American political scientist and foreign relations expert Kenneth Waltz (1924–2013) wrote in 1967 that "the striking characteristics of world politics since the war have been: peace among the powerful; their occasional use of force against others; war at times within and among the weak; the failure of such forces as have been used to lead to wider wars at higher levels of violence." He also noted that "never in this century have so many years gone by without the great powers fighting a general war," and that "small wars have been numerous, but somehow violence has been controlled and limited," and "despite dreadful dangers, a relative peace has prevailed." "But," Waltz emphasized,

> by the size of stakes and the force of the struggle, ideology was subordinated to interest in the policies of America and Russia, who behaved more like traditional great powers than like leaders of messianic movements. In a world in which two states united in their mutual antagonism far overshadow any other, the incentives to a calculated response stand out most clearly, and the sanctions against irresponsible behavior achieve their greatest force. Thus two states, isolationist by tradition, untutored in the ways of international relations, and famed for impulsive behavior, soon showed themselves—not always and everywhere, but always in crucial cases—to be wary, alert, cautious, flexible, and forbearing.[8]

The posture of forbearance that Waltz singled out is further supported by the potential for economic loss from warmaking, a matter of concern even to wealthy developed countries. Jack S. Levy, professor of political science at Rutgers University and a specialist in wars between the great powers, has pointed out that the main reason for the decreased frequency of such wars was that the potential gain from war was much reduced compared to the economic and human costs associated with it. Military actions are accompanied by numerous casualties and human suffering, the destruction of infrastructure, increasing expenditure on modern armaments, and a growing demand for manpower. Such wars expand by involving new participants, which further increases the costs and reduces the potential gain from the war: either one's costs increase because of the need to wage war against yet another adversary or one is obliged to share the gains with a larger number of allies.[9]

Thus, the great powers, when creating the postwar world order, had to elaborate and adhere to mechanisms of nonviolent conflict resolution, no matter how acute the disagreements dividing them might be. The Australian-born Hedley Bull (1932–1985), a leading international relations theorist of the twentieth century and a dominant voice for the so-called English school, noted that "great powers manage their relations with one

Table 1.1. Participation of Great Powers in Armed Conflicts with Casualties of at Least 25 People, COSIMO Database, 1945–1999

State	Direct participant	Indirect participant	Initiator	Aggressor	Mediator in the conflict settlement
United Kingdom	27	16	14	1	10
USSR/ Russia	17	55	8	3	7
France	17	20	1	3	8
China	16	19	11	1	0
United States	11	78	4	3	33

Note: States are listed in order of their direct participation in armed conflict, from most (United Kingdom) to least (United States), for the period 1945–1999.

Source: Calculations were performed using a database of national and international conflicts from 1945 to 1999 created as part of the COSIMO project at the University of Heidelberg, Germany, under the direction of Professor Frank Pfetch, Cosimo (http://www.hiik.de/en/kosimo/data/). The complete record of data from which table 1.1 was extracted is given in the appendix to this book.

another in the interests of international order by (i) preserving the general balance of power, (ii) seeking to avoid or control crises in their relations with one another, and (iii) seeking to limit or contain wars among one another. They exploit their preponderance in relation to the rest of international society by (iv) unilaterally exploiting their local preponderance, (v) agreeing to respect one another's spheres of influence, and (vi) joint action, as is implied by the idea of a great power concert or condominium."[10]

However, the self-imposed moderation with regard to direct confrontation did not mean that great powers would completely forgo the use of force. Table 1.1 provides data on the participation of states in armed conflicts resulting in casualties of at least 25 people. The first five positions among the most belligerent nations belong to the great powers. They are all permanent members of the UN Security Council.

These figures highlight the fact that the United Kingdom was the absolute leader in terms of direct participation in armed conflicts during the latter half of the twentieth century. Its leading position can be explained by the rise of national liberation movements in the British Empire's vast colonial possessions and mandate territories after World War II. France and the Soviet Union (Russia) are the two other clear leaders in terms of direct participation in armed conflicts, with 17 instances each. They are followed by China (16 instances) and the United States (11 instances).

Statistics on indirect participation in conflicts, however, offer a different perspective, confirming the widespread assertion that the superpowers actively struggled for influence through indirect participation in armed conflicts in the third world, with the United States engaged in 78 instances and the Soviet Union/Russia in 55. The indirect participation of France and China exceeded the degree of these states' direct involvement in armed conflicts. For France, 20 instances of indirect involvement were recorded, compared with 17 instances of direct participation; for China, these figures are 19 and 16. The United Kingdom is distinguished by the domination of its direct involvement (27 instances) over indirect participation (16 instances).

Patterns of conflict initiation and mediation also speak to the superpowers' further participation in armed conflicts as both instigators and peacemakers. The United Kingdom led among conflict initiators (14 instances), followed by China (11) and the Soviet Union/Russia (8). The United States was the most active mediator (33 instances), followed by the United Kingdom (10), France (8), and the Soviet Union/Russia (7).

Postwar peace turned out to be an illusion for the great powers, but even more surprising was that their power and ability to win wars against weaker opponents also appeared to be illusory. In a number of cases, the great powers suffered political defeat at the hands of adversaries that had considerably inferior power and resources. Major examples include the defeat of France in Indochina and North Africa, the dissolution of the British Empire, and conflicts in which Belgium, the Netherlands, and Portugal lost their various colonial possessions in Africa and Southeast Asia. The outcome of such conflicts demonstrated that military domination did not always lead to victory, and military success did not ensure political triumph. According to calculations by the American researcher Ivan Arreguín-Toft, an international security expert and specialist in asymmetric warfare, in 1800–2003 the strong won 71.5 percent of asymmetric conflicts, with the proportion steadily falling over the next century: in 1900–1949 the strong won 65.1 percent of asymmetric conflicts, but in 1950–1999 won only 48.8 percent.[11] The political victory of weaker adversaries was reflected in the fact that the great powers were forced to enter into agreements contrary to their interests, which basically neutralized their superiority in the balance of power.

It is possible to identify several distinctive features of armed conflicts in the postwar period that were characterized by ascending asymmetry. In particular, the number of internal conflicts kept growing throughout the entire postwar period. These conflicts took place within the borders of a single state, as opposed to "classic" interstate conflicts. Such internal or national-level conflicts are asymmetric in that they represent a struggle between the whole and its parts (the center and the periphery). Many internal struggles, however, became international with the active involvement of external players or the international community in conflict settlement.[12]

The third world was the scene of most asymmetric conflicts in the postwar period. Internal conflicts in Africa, Asia, and Latin America often took the form of civil wars in which guerrilla tactics and terrorist strategies were used. Guerrilla warfare is not a novelty; its origins can be traced to far-distant historical periods. In the second half of the twentieth century, however, favorable conditions emerged for the broad use of this form of armed struggle as an efficient means to defeat a superior adversary.[13] The great powers frequently became enmeshed in these protracted domestic wars, providing military, technical, economic, and other assistance to one or another of the belligerent parties. However, this assistance did not guarantee victory to the recipients.

The developed countries faced their own internal conflicts during this time and were not always able to resolve the issues without resorting to force. Examples include ongoing actions by the separatist Basque movement in Spain, the activities of the Irish Republican Army in Northern Ireland, and conflicts in Russia surrounding the separatist movement in the Chechen Republic. Such conflicts require long-term efforts to find a political solution, isolate the radical elements, and "win the hearts and minds" of the separatists.[14] Sometimes, too, national conflicts lead to complications in relations with other countries, such as when migrant communities provide financial support to radical movements in their country of origin.[15]

The end of the Cold War and the collapse of the East-West division put at risk the existing balance of power, the system of great power relations, and precipitated a dangerous chaos in international relations. As Kenneth Waltz wrote in 1967, "I am tempted to predict, perversely, that in the coming years students of politics will look back on the era of the Cold War, if indeed it has ended, with the nostalgia that diplomatic historians have long felt for nineteenth-century Europe."[16] The superpower confrontation gave way to new and no less grave security threats.

★ ★ ★

Beginning in the late 1990s, mentions of asymmetric threats or challenges started to appear in political analyses. Around the same time, US security and defense agencies undertook a thorough examination of the concept of asymmetric threats.[17] Asymmetric threats were defined as the capability of an opponent weaker in terms of resources and power to strike a significant blow or inflict considerable damage on a superior adversary and thus influence the outcome of the conflict. If asymmetric conflict is understood as conflict characterized by an asymmetry in the power, resources, status, and interests of the parties to the conflict, then asymmetric threats and challenges represent primarily tactics aimed at finding a stronger adversary's vulnerabilities and striking against them. The actions of terrorist groups and the proliferation of weapons of mass destruction (WMDs) on the part of economically and politically weaker states serve as examples of asymmetric strategies or threats. The acquisition of WMDs is one of the sought-after ways for some developing countries to gain authority and power.

Asymmetric threats force powerful international actors to reconsider their military doctrines and systems of military personnel training, and to pay more attention to the early detection and prevention of possible dangers. Actors that are weak in respect to power and resources come to rely on asymmetric strategies more and more frequently. At present, most analysts refer to guerrilla tactics and terrorist strategies as asymmetric. However, this assessment largely reflects the evolution of these tactics and strategies in conflicts between unequal adversaries in the postwar period, rather than the belief that guerrilla warfare and terrorist activities are inherent characteristics of asymmetric conflicts.

The distinctive features of asymmetric conflicts were readily apparent in the wars of the early twenty-first century. The war on terror initiated by the United States in response to the September 11, 2001, attacks materialized in two protracted wars, in Afghanistan and Iraq. The dynamics of the military operation in Afghanistan, launched in 2001, demonstrate that the United States and the NATO countries are far from achieving unequivocal victory over the Taliban and local al-Qaeda cells, despite eliminating their primary terrorist adversary. The Iraq War, launched in 2003, also demonstrates that the strategy of relying on superior power is unproductive. After Saddam Hussein's regime was overthrown, the most powerful states in the world failed to ensure the necessary level of security in the country or to implement a plan for postconflict reconciliation and nation-building. What was supposed to be a war of liberation for the people of Iraq turned into a guerrilla fight against Western occupation forces. This fight relied on terrorist methods of struggle against the occupation forces and local collaborationists.

The volatile Middle East offers other examples of asymmetric conflicts of recent vintage. One such conflict was Israel's 2006 war against Hezbollah in Lebanon, which precipitated a feeling of defeat and a deep political crisis in Israeli society. Though the 2006 Israel-Hezbollah conflict is far from resolved, it already exhibits features that would allow one to characterize it as asymmetric, and to make certain predictions about its outcome based on the asymmetric conflict model. Unrest in North Africa in 2011 and a military operation by Western countries against the authoritarian regime of Muammar Gaddafi and in support of the Libyan opposition forces are other recent examples of great power involvement in domestic conflicts, with the great power here backing the nonstate actors.

Thus, in armed conflicts of the twenty-first century, asymmetry has been evident across the board. Moreover, the manifestations of asymmetry are increasingly associated with a nonlinear sequence of events—that is, sporadic skirmishes in which neither side makes lasting military gains—and a break with the simplistic logic of military power superiority. For this reason, it is important to pay special attention to the phenomenon and try to evaluate the impact of asymmetry in individual cases of armed conflicts involving the great powers and their weaker adversaries.

Some experts believe that it is not worthwhile analyzing wars according to asymmetry principles, because in almost all armed conflicts, the power and the resources of the adversaries are a priori not identical. Besides, it is held, rooting out an adversary's vulnerabilities so as to inflict maximum damage and minimize one's own losses and costs—in other words, mounting an asymmetric response—is the task of any military commander. From that perspective, the notion of asymmetry is indeed not useful or relevant to an analysis. Analysis within the framework of the asymmetric conflict concept presumes greatly incommensurate power between adversaries. Developed, industrialized countries have state-of-the-art high-tech weapons systems, advanced economies, educated personnel, and advanced transportation and communication networks, while their third world adversaries are disproportionately weaker in military, technical, and economic strength. It is the conflicts (and their outcomes) between two such clearly delineated adversaries that will prove most helpful in developing a robust model of asymmetric warfare, the goal of this book.

The phenomenon of asymmetric conflict and its development in the post–World War II era can be briefly summarized in the following points.

1. In turning to the asymmetric conflict phenomenon, researchers seek to identify recurring patterns in contemporary armed conflicts that cannot be explained from the perspective of existing international relations theories or strategic analysis. The asymmetries reflect chiefly a qualitative rather than quantitative disparity between the belligerent parties. To quickly summarize the argument: Great powers and superpowers possessing enormous military power were forced to constrain themselves in its use after World War II, which effectively ended direct interstate conflicts between the great powers. Thereafter, the international system imposed restrictions on the behavior of the great powers through formal and informal associations, international law, economic cooperation, and the global economy. Nonmilitary factors have played a crucial role in the outcome of post–World War II armed conflicts, for it was after that war that public opinion began to play a meaningful role in the foreign affairs

of democratic countries and the influence of the mass media increased greatly, especially with the widespread availability of television.

2. The direct military confrontation of great powers has given way to indirect forms of armed conflict as the great powers participate in wars on the periphery of the international system. The nature of the warfare, strategy, and tactics; the degree of involvement of a nation's armed forces; and the resources available to the great powers cannot be categorized within the conventional framework of war. The great powers face forms of warfare with which their large armies are not familiar, and must develop special strategies and train special military forces to engage in such wars.

3. The participation of the great powers in armed conflicts in the third world is an important factor contributing to the dynamic character and outcome of such conflicts. Ideological confrontation within an international system consisting of great powers at the center and weaker states on the periphery has enabled relatively weak actors to manipulate the interests of the great powers, drawing the latter into protracted wars and attracting the resources of developed countries to achieve the objectives of the lesser states. Such an intertwining of interests, manipulation, and the cynical use of ideology has created a special environment in which the core and the periphery of the international system interact. The rhetoric of a struggle for ideals and justice has become an integral part of warfare.

4. Most armed conflicts in the post–World War II era do not conform to our customary understanding of war between states, that is, between parties symmetric in terms of status and capabilities. Moreover, symmetry in this context does not imply parity, a simple equivalence of antagonists' forces and resources, which would be a rather basic and uninformed interpretation of equality in international relations. As Brantly Womack, professor of political science at the University of Virginia, puts it, "symmetry does not require absolute equality, but it does imply potential reciprocity in the interaction: what A could do to B, B might likewise do to A." Such an understanding of the relationship between the two parties allows us to take into account the possibility of influence exerted by a stronger party through "soft power" without resorting to

"hard power," while still relying on the existence of hard power and the veiled threat of using it.[18] Thus, conventional wars have been replaced by various armed conflicts characterized by numerous asymmetries.

5. The post–World War II era can be divided into three periods defined by the distinctive features of great power participation in armed conflicts on the world system periphery and of their interaction in the international relations system. The first period, from 1945 until the mid-1960s, was characterized primarily by armed conflicts in colonies struggling for independence against the European great powers, especially the United Kingdom and France. The second period, from the early 1960s to the 1990s, witnessed the intensification of US and Soviet participation in armed conflicts in the third world. The collapse of the Eastern bloc and the dissolution of the Soviet Union marked the beginning of the third period, which has been characterized by deepening economic and political competition between developed states, even as new factors of asymmetry arise from actors on the periphery of the world system. Powerful extrasystemic actors, such as al-Qaeda and other international terrorist groups, have challenged the stability of the post–Cold War period and in doing so have once again raised the profile of asymmetric conflict.

CRITICAL APPROACHES TO THE CONCEPT OF ASYMMETRIC CONFLICT

The literature on asymmetric conflicts is of three general kinds. In the first group are studies investigating specific historical cases that could be regarded as examples of asymmetric conflict. This group constitutes an abundant research literature that analyzes postwar decolonization and includes the memoirs of eyewitnesses, among them military personnel, politicians, and diplomats. The second group includes studies written by political scientists and theorists of international relations and conflict studies; these studies often proceed from specific case studies to generalizations. In the third group are publications that address the problem of asymmetric conflict as one of tactics and strategy, as a need to adapt military strategies and the structure of military forces to wage small wars, as well as efficient counterinsurgency and antiterrorist campaigns. The writings of politicians, strategists, and revolutionaries from developing countries who elaborate the tactics and strategies of victorious wars against imperialist countries are allied with this group in terms of subject matter.

The term "asymmetric conflict" was introduced by the international relations scholar Andrew Mack in a 1975 article titled "Why Big Nations Lose Small Wars: The Politics of Asymmetric Conflict." His study, which was supported by the Social Science Research Council of the United Kingdom and the Rockefeller Foundation, aimed "to undertake an analysis of several asymmetric international conflicts in which an external power confronts indigenous insurgents." Mack devoted most of his analysis to the US war in Vietnam; however, he also pointed to several other defeats of developed states that corresponded to the concept he was exposing, namely, the conflicts in Indochina, Indonesia, Algeria, Cyprus, Aden, Morocco, and Tunisia. According to Mack, "Local nationalist forces gained their objectives in armed confrontations with industrial powers which possessed an overwhelming superiority in conventional military capability."[19]

Mack wrote that asymmetric conflicts refuted the experience of great power control over the third world. Furthermore, this experience could not be reduced to colonial domination and its deposition. These conflicts destroyed the "once prevalent assumption—that conventional military superiority necessarily prevails in war." In most of these conflicts, according to Mack, the strong states neither won nor lost militarily, but did lose politically as they failed to impose their will on their opponents. Thus, the main justification for the use of force to achieve one's goals and the rationale for entering the war were lost. Mack argued that in every case, "success for the insurgents arose not from a military victory on the ground—though military success may have been a contributory cause—but rather from the progressive attrition of their opponents' *political* capability to wage war. In such asymmetric conflicts, insurgents may gain political victory from a situation of military stalemate or even defeat."[20]

Mack's article offers hypotheses about the reasons for the paradoxical defeat of great powers, though Mack himself characterized his work as offering a "pre-theoretical perspective." He noted that the defeat of great powers was driven by several factors, among them (1) the loss of political will to continue war; (2) a complex of asymmetric relations between adversaries, defined by, among other things, the assumption of total war by the weaker party and limited war by the stronger party; (3) the use of strategies of asymmetric struggle (guerrilla warfare); and (4) the impact of nonmilitary factors—domestic, social, and international—on the decision to stop fighting. Following Mack's argument, we can define the essence of asymmetric conflict as the political defeat of a great power in a war against an a priori weaker adversary, under circumstances in which military superiority does not guarantee victory and might even be counterproductive.

Andrew Mack deserves credit for having brought together seemingly isolated facts into a single conceptual model and for offering a capacious and succinct definition of the phenomenon as asymmetric conflict. He applied the term "asymmetric" to both the structural and the dynamic elements of conflict—to resources, status, interests, the ability to mobilize, strategies, conflict outcomes—a move that allows the qualitative changes in a conflict to be explored once the quantitative disparities have been identified. At the same time, he saw the need to wrap the concept of asymmetric conflict in a holistic cover, and invoked an axiom from Aristotle to make his point: "The asymmetries described in this paper—in the interests perceived to be at stake, in mobilization, in intervention capability, in 'resource power,' and so forth—are abstracted from their context for the sake of analytical clarity. But the whole remains greater than the sum of its parts, and it is the conflict *as a whole* which must be studied in order to understand its evolution and outcome."[21] The title of his article contains the key words that Mack's followers often use: *small wars* of great powers, and *why* big nations *lose* small wars. Surprisingly, after the publication of this article, which laid the foundation for the asymmetric conflict concept, Mack seems not to have developed the idea further.

Andrew Mack's biography reflects a uniquely diverse set of life experiences. His pre-academic career included six years in the Australian Royal Air Force, two and a half years in Antarctica as a meteorologist and deputy base commander, a year as a diamond prospector in Sierra Leone, and two years with the BBC's World Service, writing and broadcasting news commentaries and producing the *Current Affairs* program. Later, he studied at the University of Essex, worked at the Copenhagen Peace Research Institute and the London School of Economics, became research director of the Richardson Institute for Peace and Conflict Research in London, and taught at leading US and Asian universities. In 1998–2001, he was director of the Strategic Planning Office in the Executive Office of UN Secretary-General Kofi Annan. He then took up an appointment as head of the Human Security Centre at the University of British Columbia, Canada. Since 2007, he has directed the Human Security Report Project at the School for International Studies, Simon Fraser University, British Columbia.[22] Professor Mack presented the Human Security Report 2009/2010 on January 20, 2010, at the United Nations in New York. Some of the conclusions of that report refute received opinion about the dynamic of current armed conflicts, the number of casualties, and the results of peacekeeping efforts.[23]

⋆ ⋆ ⋆

Before the 1990s, the concept of asymmetric conflict was not on the charts, though the issues that Mack addressed in his 1975 article—especially the sources of power and influence in international relations and how nonmilitary factors might condition victory and defeat—had attracted the attention of other researchers. The changing character of warfare was a crucial component of the scholarly debate: the transition from direct military confrontation, or conventional warfare, to indirect forms of struggle (guerrilla strategies, civilian participation, terrorism), and the spread of small wars involving great powers on the periphery of the world system (e.g., low-intensity conflicts, limited wars, local wars, proxy wars, counterinsurgency operations, peacekeeping operations) as opposed to big wars between great powers. Thus, it seems that asymmetric conflict, parsed as the paradoxical defeat of a great power by a vastly weaker adversary, was investigated by many scholars who, for one reason or another, did not fully apply Andrew Mack's concept.

In reflecting on the US political defeat in Vietnam, for example, many analysts sought the reasons for such an outcome and tried to draw lessons from it. Such a pragmatic approach led to the conclusions of these studies being applied in political analysis, in decision making, and in military planning. Such well-known scholars as Hans Morgenthau; British historian Michael Howard; Jeffrey Hart, speechwriter for President Richard Nixon; and the researchers James Lee Ray, Ayse Vural, Zeev Maoz, and many others were among those who debated the lessons of

Vietnam.[24] These debates frequently referenced the biblical battle between David and Goliath,[25] which underscores the scholarly interest in conflicts between unequal adversaries, even if Mack's full conceptualization had not yet come to the fore.

★ ★ ★

Research using asymmetric conflict as a guiding concept has been pursued in a variety of genres. A vast research literature has been devoted to the US war in Vietnam,[26] for example, including books by Henry Kissinger and Robert McNamara detailing the reasons for the US defeat and the lessons of the Vietnam War.[27] The distinctive features of limited conflicts involving superpowers were actively researched in the 1970s and 1980s,[28] and the topic was still drawing scholarly attention in the mid-1990s.[29]

Other notable case studies of this phenomenon have dealt with the history of British engagement in small wars and the Soviet war in Afghanistan.[30] US, coalition forces, and NATO military operations in Afghanistan and Iraq in the 2000s reminded analysts of the war in Vietnam and the Soviet defeat in Afghanistan, and encouraged comparisons.[31] The war in Lebanon in 2006, the military campaigns in the Chechen republic of the Russian Federation, and other conflicts became the subject of case studies and theoretical analyses in the late twentieth and early twenty-first century.[32]

Military tactics and strategy represent a somewhat different area of research, and publications in this field have addressed guerrilla warfare, counterinsurgency operations, the problem of army efficiency in small wars, and the ways in which the great powers revised their military strategies in light of the changing international system.[33] In surveying such works, it becomes clear that the organization of "small wars" and guerrilla and sabotage groups in the adversary's rear, or the same sort of operation undertaken by regular forces in an occupied territory, holds a prominent place historically in the development of the military strategy of many countries. Organizing guerrilla warfare is described in the writings of the French, Russian, and British military strategists of the eighteenth and nineteenth centuries, for example.[34]

The importance of training military personnel to organize and wage guerrilla warfare or small war varies with the time period, but such training continued to be an integral part of military theory and practice during the twentieth century. At the same time, guerrilla warfare tactics were studied by the leaders of developing countries, who deemed them the most efficient in the fight against "international imperialism." Prominent among such leaders is Mao Zedong, author of the Chinese revolution, though he preferred discussing the advantages of a protracted engagement rather than guerrilla warfare.[35] Equally famous are the ideas of the legendary Cuban revolutionary Ernesto "Che" Guevara, who believed that revolution could be exported and who perished in Bolivia in 1967 while trying to foment a revolution there.[36] Another example of this genre is provided by the writings of the British scholar and army officer Colonel T. E. Lawrence, the famous Lawrence of Arabia, who wrote an article on the theory and practice of guerrilla warfare for *Encyclopaedia Britannica* based on his experience in such wars in the Arab East.[37] He died in 1935, but books devoted to his legacy are still being published, and he has been called many things, from a genius of friendship and military strategy to "a prince of our disorder."[38]

For the past decade, international terrorism has been the most dangerous adversary of the leading world powers, particularly the sort of terrorism embodied in the al-Qaeda movement and its former leader, Osama bin Laden. Some analysts even began considering international terrorism as an example illustrating the evolution of asymmetric strategies in the era of confrontation between the developed global north and the poor and oppressed global south. Earlier studies of guerrilla movements also highlighted instances of the use of terrorist tactics, but in those cases such tactics were not the key mode of struggle for national liberation.[39]

In the 2000s, the ideologists of the international terrorist movement attempted to equate terrorist actions with the actions of guerrilla groups and conventional forms of armed struggle in order to obtain public approval for such actions and to try to position terrorist acts within the compass of international law. This view may seem absurd, if one overlooks the fact that in the 1970s the actions of guerrilla groups were de facto equated with the actions of regular troops under international humanitarian law, which regulates the conduct of participants in armed conflicts. The 1976 Additional Protocols I and II to the Geneva Conventions of 1949 included provisions according to which participants in guerrilla groups were covered by international humanitarian law. The main idea is that guerrilla fighters are to follow the rules of military actions and wear insignia distinguishing them from civilians. The issue of whether guerrilla groups should follow these rules is the subject of lively debate among lawyers, politicians, and ideologists of radical movements, as well as their opponents.[40]

★ ★ ★

In Soviet science, the victory of relatively weak parties over stronger states was viewed through the prism of research on national liberation and anticolonial movements, and an understanding of such conflicts was rooted in a class-based approach. Nevertheless, although Soviet and Russian researchers did not propose analytical schemes similar to those propounded in the West (such as a theory of asymmetric conflict, an analytical approach to the problem of defeat of democracies in small wars, the effect of restrictions on military actions in democratic countries, or a complex understanding of power and influence in international relations), they did identify similar patterns that could explain the defeat of developed countries in such wars.[41] Soviet historians often pointed to crucial reasons for the defeat of developed countries in anticolonial wars that Western scholars and politicians did not openly discuss. For instance, Soviet historians paid much more attention than the West to "interimperialist contradictions" with regard to colonies, postwar arrangements, the global monetary system, and the struggle for influence in the third world.

Another reason for defeat, one noted in both the Russian literature and Western studies, though with a certain amount of bias, was of ideological competition and the attractiveness of socialist rhetoric to anticolonial movements. In rereading the writings of Russian and overseas historians of the 1950s to the 1970s, it seems clear that ideological biases influenced the way in which historical events were assessed, often resulting in a deliberate veil of silence or poor coverage, but the level of analysis and argumentation lends undeniable credibility to many papers. Fortunately, researchers today have an opportunity to compare and reconcile the writings of Russian and foreign authors, thus uncovering both persistency and missing links in the historiography.

One aspect commonly mentioned is the superpowers' self-imposed restriction on the use of force during the confrontation era of the 1960s to the

1980s. This voluntary restraint was intended to prevent a direct armed conflict that could escalate into full-scale global war. Western and Soviet research literature are also united by their fealty to the notion of a just war, which touches on the issue of morality. The Soviet interpretation of just war was peculiar, as it secularized this theological concept, which was first suggested by Saint Augustine. Soviet historians instead relied on the understanding of just war promulgated by Vladimir Lenin. The cornerstone of his interpretation was the popular legitimization of violence for the sake of national liberation and revolutionary transformation. By way of comparison, Western literature acknowledged the immorality of war within the pacifist and liberal traditions, whereas Soviet literature condemned the "crimes of imperialism."

From the standpoint of Russian military doctrine, the topic of small wars and guerrilla warfare was viewed primarily as part of military personnel training. Furthermore, the importance assigned to this component of professional military training varied. There were periods of heightened attention to the topic, such as after the Patriotic War against Napoleon in 1812, again in the second half of the nineteenth century, when the Russian army participated in the Balkan wars, and during World War II. However, there is no indication in the open sources that Russian military science would develop strategies to put down insurgencies, though at present that is a crucial task for the armies of all developed countries.

It is difficult to assess how deeply Russian military science investigated counterinsurgency tactics, as the military domain by tradition remains almost entirely closed, so that any existing practical and methodological studies that might have been used to train military personnel participating in local conflicts in the twentieth century are unavailable, though one occasionally comes across references to those writings. It is also true that the writings of military strategists on small wars and guerrilla strategies published in the pre-Soviet period can be found in library collections.[42] The book by Vladimir V. Kvachkov is an exception to the wall of silence surrounding Soviet and Russian military training methods, for it is available both on the Internet and in print. Kvachkov studies the evolution of small war strategies, including special operations and guerrilla warfare, in Russian and Soviet military science.[43]

Until the early 1990s, there were no indications in the Russian literature—at least in the publicly available literature—of the possibility of using the experience of guerrilla movement organization in military doctrines or that the Soviet army had used such experience in third world countries. However, in the Western literature one finds papers investigating the participation of the Soviet Union and developed countries in organizing guerrilla and liberation movements.[44] Information on such operations was usually restricted, and only after the collapse of the Soviet Union did a significant amount of new memoir and research literature emerge on the extent to which the Soviet Union and its military specialists participated in organizing guerrilla movements overseas.[45] There are also bibliographic materials on Western concepts of limited wars, prepared at the request of the Soviet General Staff.[46]

In the late 1990s to early 2000s, a number of publications devoted to the Soviet experience in Afghanistan appeared. They included books written by participants and eyewitnesses, chiefly military personnel and correspondents; the memoirs of politicians; and the recollections of and interviews with ordinary war participants.[47] At present it is of research interest to compare the Soviet and US experiences in Afghanistan. Although Russian authors tend to believe that the United States repeated the Soviet Union's mistakes, American analysts are not so sure.[48]

In conversation with Professor Tatyana Alexeeva, head of the political theory department at the Moscow State Institute of International Relations (MGIMO), I learned that Soviet analysts were familiar with Andrew Mack's seminal article on asymmetric warfare. Indeed, Professor Alexeeva, along with other MGIMO faculty members, had prepared a special executive summary for the country's leadership outlining the key ideas of the article. However, until the 2000s, asymmetric conflict was not a specific research topic in Russia. The collapse of the Soviet system also meant an opening of the country to free interaction with the global research community, and many topics that once were taboo came under the spotlight. Conflict studies and political theory were on the upswing, and asymmetric conflict drew the attention of scholars familiar with Western research in this field.

The concept of asymmetry is used in the Russian sciences to study ethnopolitical conflicts, gender relations, legal and economic relations, and European integration.[49] It is also used in strategic analysis. In this regard, the Russian historian and ethnologist Airan R. Aklaev has articulated a distinction between symmetric and asymmetric conflicts based on political-administrative differences of the parties to the conflict. Vertical or hierarchical conflicts, or conflicts between actors at different levels of a political hierarchy, such as between a state and an ethnic group,[50] are defined to be asymmetric, in contradistinction to horizontal conflicts, which are symmetric.[51] Aklaev provides examples further adumbrating this understanding of symmetric versus asymmetric relations:

> Horizontal conflicts involve equal-status actors and/or power holders of the same order, such as groups within the ruling elite, moderates and radicals, nonruling parties, or factions of the same political movement. Horizontal conflicts are those occurring between two ethnic groups or political/administrative units when none of the ethnic groups controls the central government (e.g., conflict between Armenia and Azerbaijan for Nagorno-Karabakh in the Soviet Union). In vertical conflicts the parties are not of equal status, and there are differences in power, that is, the parties are involved in relations of hierarchy, of domination and subordination, as happens in instances of conflict between an ethnic group and a state, such as when there is an attempt to secede or a unilateral push for autonomy (e.g., Transdnistria and right-bank Moldova, Georgia and South Ossetia, Quebec and Canada, the Basque Country and Spain, the Kurds in the Middle East).[52]

Aklaev's formulation of asymmetric conflict thus invokes a status difference between parties in a hierarchical relationship characterized by relations of subordination and inequality of the conflicting parties. It is worth noting here that unequal status relations between entities already are recognized as part of the political and legal discourse at national and international levels. For instance, an "asymmetric federation," as discussed in the political organization literature, means that units that differ in their legal authority are united under a federal system.[53]

In characterizing ethnopolitical conflicts in the modern world, Aklaev argues that asymmetric armed conflict, the predominant form that such conflicts take, is a complex phenomenon that can be identified "based not only on the participants involved (not two sovereign states but a state and a rebellious section of politicized identity population groups) but also on other qualitative characteristics ([such as] decentralized decision making by fragmented authorities, armed violence [conducted] not by regular but by paramilitary gunman groups, [or] the widespread use of terrorist methods and guerrilla warfare)."[54]

In Russian strategic analysis, the term "asymmetric" has a specific meaning: it describes the Soviet (Russian) strategies that were developed in response to new military programs and systems of the United States. In particular, the Soviet concept of an asymmetric response evolved in answer to the US Strategic Defense Initiative (SDI) of 1983. The concept implied looking for the most vulnerable spots in the antiballistic missile defense system of the United States and developing ways to counter those system components so as to make the SDI redundant.[55] This approach, in its essence, does not contradict the understanding of asymmetric response as a way to compensate for an inequality in power capabilities when it is impossible to achieve superiority over an adversary in an arena dominated by the adversary. Precisely this approach—neutralizing a stronger adversary by seeking out and attacking its weak points—is integral to debates over military strategy and tactics and foreign policy doctrines.

In recent years, several dissertations relying on the concepts of asymmetry and asymmetric conflict have been defended in Russia.[56] Articles on this topic have been published in analytical and military journals,[57] and there are even discussions of asymmetry in college textbooks.[58] At this point, it is hard to predict how valuable and independent the contribution of Russian researchers to this topic will be. Their writings are likely to continue the debate over the changing nature of armed conflicts and the sources of power and influence in the contemporary world system. It is imperative, however, that Russian analysts move beyond a stance of noninvolvement. Further development of asymmetric conflict theory by Russian political scientists should result in identifying both the positive and negative asymmetries within Russian power structures, state institutions, and society. To date, however, statements mentioning asymmetries are predominantly valorizations of Russian power structures that have been victorious in asymmetric confrontations, rather than a critical analysis of the country's internal mistakes and weaknesses. It is possible that in accordance with the Russian tradition, which is dictated to a certain extent by the Soviet legacy, any discussion of the "weaknesses of strength" will remain an embargoed topic.

★ ★ ★

The research literature on asymmetric conflicts as instances of the political defeat of great powers in wars against weaker adversaries appeared almost simultaneously with the articulation of this concept in the mid-1970s. However, as usually happens, a more systemic investigation of the phenomenon, generalization from individual cases, and theoretical work had to wait. Therefore, the phenomenon bears further scrutiny, even if it has already been studied within other conceptual frameworks such as small wars, guerrilla and antiguerrilla strategies, and national liberation and anticolonial movements. Debates as to the relevance of this concept and its analytical significance do not undermine its heuristic contribution to the study of post–World War II armed conflicts. The asymmetric conflict concept brings together in one explanatory construct recurring patterns of behavior exhibited by the great powers and by new actors on the world political stage. When working with this concept, it is important to analyze it in connection with other analytical constructs that allow the identification of interconnected events taking place in different parts of the postwar world.

DEFINING ASYMMETRIC CONFLICT

The concept of asymmetric conflict developed chiefly along one of two paths: the further development of Mack's hypotheses, or a rethinking of existing perceptions about international relations that also took Mack's hypotheses into account. Two major trends in analysis have also emerged: asymmetric conflict is studied either as a tactical and strategic phenomenon[59] or as a sociopolitical phenomenon. The first approach is concerned with strategies of warfare that might account for victory or defeat. The second approach sees war as subordinate to politics, and pays significant attention to the process of foreign policy–making in developed countries, the participation of society in foreign policy decisions, the ways in which the media present the war, and other national, international, and economic factors. Regardless of their scholarly predilection, however, all authors examining the concept of asymmetric conflict proceed from the foundation of Andrew Mack's pioneering work.

The US literature on military strategic analysis is an example of the consistent development and application of the concept of asymmetric conflict. In the 1990s, the concept gained popularity because of the obvious superiority of the United States after the end of the Cold War and the vanishing possibility of any military conflict based on the symmetric scenario; that is, conflict with an adversary equal in military strength. The concept of asymmetry answered perfectly to the need to understand the US position in the new world system and to assess potential threats. Robert M. Cassidy, a US Army major who holds a PhD in international security, has observed that "asymmetric" became a term du jour in the mid-1990s.[60] Robert H. Scales, commander of the US Army War College, has also noted that "asymmetric warfare" became a Pentagon buzzword in the 1990s.[61] Several different meanings of the term have been codified in military doctrinal documents, and furthermore, changes in the term's meaning can be seen to have tracked changes in the global political environment.

In the early 1990s the term "asymmetric" was used to characterize US Army strategies. To quote from "A Doctrinal Statement of Selected Joint Operational Concepts" (1992), by Colin L. Powell: "When required to employ force, Joint Force Commanders seek combinations of forces and actions to achieve concentration in various dimensions, all culminating in attaining the assigned objective(s) in the shortest time and with minimal casualties. Joint Force Commanders arrange symmetrical and asymmetrical actions to take advantage of friendly strengths and enemy vulnerabilities and to preserve freedom of action for future operations." Another interpretation of the term in the same document turns on the type of force used: "Engagements with the enemy may be thought of as symmetrical, if our force and the enemy force are similar (land versus land, etc.) or asymmetric, if forces are dissimilar (air versus sea, sea versus land, etc.)."[62] The term was used to conjure an indistinct threat in the "U.S. Joint Doctrine" and "National Military Strategy" documents of 1997 ("While we no longer face the threat of a rival superpower, there are states and other actors who can challenge us and our allies conventionally and by asymmetric means such as terrorism and weapons of mass destruction").[63] The following definition of the term appeared in the *Joint Doctrine Encyclopedia* of 1997, under the authorship of the Joint Chiefs of Staff: "Asymmetrical actions that pit joint force strengths against enemy weaknesses and maneuver in time and space can provide decisive advantage."[64] Similarly, the *Doctrine for Joint Interdiction Operations* of 1997 reiterates the need to "arrange symmetrical and asymmetrical actions to take advantage of friendly strengths and enemy vulnerability." It notes that "Joint Forces Commanders must aggressively seek opportunities to apply asymmetric force against an enemy in as vulnerable aspect as possible—air attacks against enemy ground formation in convoy (e.g., the air and special operations forces [SOF] interdiction operations against German attempts to reinforce its forces in Normandy), naval air attacks against troop transports (e.g., US air attacks against Japanese surface reinforcement of Guadalcanal), and land operations against enemy naval, air, or missile bases (e.g., allied maneuver[s] in Europe in 1944 to reduce German submarine bases and V-1 and V-2 launching sites)."[65] This interpretation of the term asymmetric is traditional in strategic analysis and characterizes the choice of an efficient strategy that allows maximizing success and minimizing one's costs and casualties. The essence of asymmetric strategy is to find smart solutions to fighting, and to overcome an adversary's advantages in armaments or strategic position.

This gradual shift in emphasis and interpretation of symmetric and asymmetric was reflected in the new notion of "asymmetric threats." This is evident in the May 1997 *Quadrennial Defense Review*, a six-month analysis of the threats to US national security that also reviewed US defense strategy and programs, including force structure, infrastructure, and readiness. The report was prepared under the leadership of US secretary of defense William S. Cohen and states, in part:

> U.S. dominance in the conventional military arena may encourage adversaries to use such *asymmetric* means to attack. ... That is, they are likely to seek advantage over the United States by using unconventional approaches to *circumvent* or *undermine* our strengths while *exploiting* our vulnerabilities. Strategically, an aggressor may seek to avoid direct military confrontation with the United States, using instead means such as terrorism, NBC [nuclear, biological, or chemical] threats, information warfare, or environmental sabotage to

> achieve its goals. If, however, an adversary ultimately faces a conventional war with the United States, it could also employ *asymmetric* means to delay or deny U.S. access to critical facilities; disrupt our command, control, communications, and intelligence networks; deter allies and potential coalition partners from supporting U.S. intervention; or inflict higher than expected U.S. casualties in an attempt to weaken our national resolve.[66]

This review is frequently cited, and later documents repeat almost word for word the definition of asymmetric strategies that could be used by US adversaries. In the *Joint Strategy Review* of 1999, asymmetric approaches were defined as "attempts to circumvent or undermine US strengths while exploiting US weakness using methods that differ significantly from the United States' expected method of operation."[67] The term "asymmetric" was used in the same sense in the February 2010 *Quadrennial Defense Review* presented by the US secretary of defense, Robert Gates.[68] Thus, over time, "asymmetric" came to be used to characterize the strategies and tactics of US adversaries of inferior power capabilities. In accordance with this approach, "asymmetric" came to mean the opposite of "conventional," "ordinary," or "traditional" in describing threats, attacks, and military actions.

The problem of asymmetric threats, strategies, and military operations is actively explored by think tanks, at US military schools and academies,[69] and in the pages of professional journals.[70] *Challenging the United States Symmetrically and Asymmetrically: Can America Be Defeated?* was one of the first publications of this kind, appearing in 1998. It explored US military and technical superiority and the changing character of military actions, as well as terrorism, information warfare, and the ability to wage asymmetric wars.[71] Another report, *Asymmetry and U.S. Military Strategy: Definition, Background, and Strategic Concepts*, prepared in 2001 by the Strategic Studies Institute, part of the US Army War College, suggested two kinds of asymmetries: positive and negative. Positive asymmetry is understood as the use of the strategic advantages of the US Army, while negative asymmetry is understood as situations in which adversaries of the United States attack its vulnerabilities. As the report correctly pointed out, there is nothing new about these interpretations from the standpoint of military art; rather, such situations had simply not been characterized as asymmetric before.[72]

The launch of the US war on terror ramped up interest of military analysts in wars against asymmetric adversaries. In the late 1990s, authors looked at the phenomenon from a more theoretical perspective, but the realities of two wars in Afghanistan and Iraq turned their attention instead to actual instances of asymmetric conflicts. In 2003, Robert M. Cassidy published his monograph, *Russia in Afghanistan and Chechnya: Military Strategic Culture and the Paradoxes of Asymmetric Conflict*, in which he viewed the problem of Soviet/Russian defeat through the prism of Andrew Mack's hypotheses. Cassidy thought that the key reason for the Soviet defeat was its backward strategic culture—its incompatibility with the contemporary world. According to Cassidy and other writers, the strategy of "big war" typical of the great powers is anachronistic for strategic culture since it implies conflict with an equal adversary based on a scenario of symmetry.[73]

The term "asymmetric" became the main characteristic of *irregular warfare* in publications of the Joint Chiefs of Staff, and in the 2006 version of *Joint Publication 3-0* irregular warfare was defined as "a violent struggle among state and nonstate actors for legitimacy and influence over the relevant population. Irregular warfare favors indirect and asymmetric approaches, though it may employ the full range of military and other capacities, in order to erode an adversary's power, influence, and will."[74] The 2011 revision of this document states that "an enemy using irregular methods often will use terrorist tactics to wage protracted operations in an attempt to break the will of their opponent and influence relevant populations. At the same time, terrorists and insurgents also seek to bolster their own legitimacy and credibility with those same populations."[75] Of note, there is an important shift from the 2006 document to its 2011 revision in the definition of participants in military actions, which in the 2011 report includes both state and nonstate actors, and in the objectives of military actions, which moved from inflicting military defeat on an adversary in the earlier report to earning the trust of local civilians, whose interests the adversaries are trying to protect. In the 1960s, this strategy was nicknamed "winning hearts and minds" in the British literature on counterinsurgency,[76] and this expression is now constantly and predictably used by military analysts in many countries to characterize the goals of the stronger party in an asymmetric conflict.

In 2004, the collection of essays *A Nation at War in an Era of Strategic Change* was published. The essays explore different aspects of US participation in the war on terror, and the volume title references the words of President George W. Bush on the military campaigns in Afghanistan and Iraq as a situation of the entire nation being at war. The introduction notes that "the American military needs to think in a more holistic[77] fashion about the conduct of war at the operational level." The volume editor, Williamson Murray, echoes Carl von Clausewitz's nineteenth-century rationale of the use of military force as an alternative method of achieving political goals: "Since war is a political act, the defeating of enemy military forces in combat operations only represents a portion of the far larger mosaic that must include not only the planning stages, but the transition stages from war to peace as well. In fact, as Americans are discovering in Iraq and Afghanistan, the latter may represent as important a component of operational art as the direct battlefield confrontations in securing the political ends for which the United States has waged war. And those political aims are the only conceivable reason that the U.S. military will engage in war."[78]

US military analytics thus evolved from interpreting asymmetry through traditional strategic analysis (that is, as a kind of descriptor for the tactic of producing an efficient nonequivalent response to challenges posed by traditional adversaries), to identifying individual elements of asymmetry in armed conflicts with weaker or nonstate actors (in the sense of asymmetric challenges or military actions), to finally accepting a complex or mosaic of factors as constituting asymmetric conflict and defining a special type of armed confrontation. The special character of the confrontation derives from the need to obtain the support of the population living in the territory where the military actions are taking place, rather than simply to secure a military victory. Thus, American military strategic analysis started by partially accepting elements of the asymmetric conflict concept and then developing a holistic understanding of it in line with Andrew Mack's initial articulation of the concept. It should also be noted that the military services of Israel, Australia,[79] and the United Kingdom regularly use the term "asymmetric,"[80] and these issues are discussed in joint seminars and conferences involving military personnel from different countries.[81]

For military analysts, the problem of asymmetric conflict is dictated by a pragmatic task, the need to understand the phenomenon so as to select the right strategy and tactics. It is not surprising that the military is more interested in understanding armed asymmetric conflicts than in abstract theorizing about asymmetric relations. The military needs instrumental definitions and operational models. This point—the need for clear and functional language—was underscored in an eloquent epigraph to the article "Unorthodox Thoughts about Asymmetric Warfare," by Montgomery C. Meigs, the American general who commanded the NATO Stabilization Force in Bosnia and Herzegovina from October 1998 to October 1999: "Bad terminology is the enemy of good thinking."[82] The pragmatic turn in the American literature on asymmetric conflict may have prompted Russian scholar Ekaterina A. Stepanova to observe that the American literature "excessively militarizes" the nature of such conflicts, to the point of dismissing or at least downplaying other crucial elements of the concept.[83] However, if one sees asymmetric conflict as armed struggle between adversaries incommensurate in power, resources, and status—and this is in fact the starting point of the concept—then the military usually subscribes to a broader, more holistic approach to the concept and relies on researchers' conclusions. Moreover, some military analysts are fascinated by how asymmetry is constructed, as indicated in the title of British Royal Air Force officer J. G. Eaton's "The Beauty of Asymmetry: An Examination of the Context and Practice of Asymmetric and Unconventional Warfare from Western/Centrist Perspective." Eaton's article suggested a practical model of an "asymmogram" that he believed "would reflect a shorthand notation of the balance of negative and positive asymmetry, thus concentrating defensive or offensive planning."[84]

An interesting footnote is that US military officers often hold advanced academic degrees in international security or international relations. For instance, General David H. Petraeus,[85] who was a commanding general of Multi-National Force in Iraq during the "surge" in 2007–2008 and whose name is associated with a turning point in the situation in Iraq, defended his doctoral dissertation in international relations at Princeton University in 1987. His thesis was titled "The American Military and the Lessons of Vietnam: A Study of Military Influence and the Use of Force in the Post-Vietnam Era." In an article that he published based on his dissertation work, he noted that American popular support for protracted wars is limited, and described an optimal scenario of "nasty little wars": "if the United States is to intervene, it should do so in strength, accomplish its objectives rapidly, and withdraw as soon as conditions allow." Moreover, politicians should not interfere in the specifics of the operation after agreeing to the use of force and specifying the objective. Petraeus pointed to the inconstancy of politicians, the unclear war objectives they formulate, and the all too common impossibility of achieving the objectives stipulated by Congress or of solving certain problems by military means, which basically makes military personnel prisoners of the situation and forces them to take the fall for a political defeat. He cited a well-known book, *The Soldier and the State: The Theory and Politics of Civil-Military Relations* (1957), by Samuel Huntington, who described the armed services' strong preference for peace: "The military man tends to see himself as the perennial victim of civilian warmongering. It is the people and the politicians, public opinion and governments, who start wars. It is the military who have to fight them." Petraeus also recalled the powerful image that General William A. Knowlton introduced into public discourse in a graduation speech to the Army War College Class of 1985 in reference to the US engagement in Vietnam: "Those who ordered the meal were not there when the waiter brought the check."[86] Petraeus is the author of multiple papers on counterinsurgency strategies and actively participates in discussions of counterinsurgency measures in congressional hearings and debates sponsored by independent think tanks.

In sum, academic research, doctrinal documents, and the public speeches of American military experts all underwrite the view that armed conflicts involving the US Army with an adversary significantly inferior in military power and resources are regarded as a qualitatively new phenomenon that needs to be taken into account when developing and implementing military operations. This brief overview of the US interpretation of the concept of asymmetric conflict also highlights the short distance between political science research findings and their practical application in the contemporary United States.

★ ★ ★

It took a few years before asymmetric conflict appeared as a distinct field of inquiry; however, the 1990s saw the publication of research papers in which asymmetric conflict was classified as a category separate from other kinds of warfare. Subsequently in the United States, a scientific and professional journal, *Dynamics of Asymmetric Conflict*, started publication in 2008. The mission of the journal is "to contribute to understanding and ameliorating conflict between states and non-state challengers [as] many experts believe that this is the predominant form of conflict in the world today, and will be the predominant source of violent conflict in the twenty-first century."[87]

Unlike military analysts, social scientists have interpreted the notion of asymmetry more broadly and use the term "asymmetric" in nonmilitary as well as military situations. In the 2000s, attempts were made to elaborate a general scholarly approach to the use of the term "asymmetry."[88] On the one hand, the term conjures up the whole complex of asymmetric relations in social interactions; on the other hand, this broad usage creates terminological difficulties—which in turn encourages researchers to try to delimit its meaning more accurately.

Given the traditional understanding of asymmetric conflict as a power imbalance, political scientists focus on instances of aggressive behavior exhibited by weak states, and it is within this framework that T. V. Paul uses the term. Paul, an Indian-born political scientist, is a founding director of the McGill University Centre for International Peace and Security Studies in Montreal, Quebec. In his 1994 book, *Asymmetric Conflicts: War Initiation by Weaker Powers*, he analyzed factors that could account for such seemingly illogical behavior on the part of weak states. Paul examined the Japanese offensive against Russia in 1904, the Japanese attack on Pearl Harbor in 1941, the Chinese intervention in Korea in 1950, the Pakistani offensive in Kashmir in 1965, the capture of Sinai by Egypt in 1973, and the Argentinean invasion of the Falkland Islands/Islas Malvinas in 1982.

According to Paul, the study of asymmetric wars should demonstrate why simple superiority in power and resources is insufficient to deter the aggression of a weaker adversary. Paul emphasized that the investigation of such cases should include an analysis of the domestic and international factors that exert a decisive influence on the weaker adversary's posture. To this end, he proposes four conditions that must be met before war initiation by a weaker state in an asymmetric conflict: (1) there is a serious conflict of interest; (2) the weaker side places greater value than the stronger side on the issue under dispute; (3) the weaker party is dissatisfied with the status quo; and (4) the weaker party fears an unchanged status quo or a worsening from the status quo in the future. The following variables then shape the aggressive behavior of the weaker party: (1) a limited aims strategy that employs military forces in battle to achieve limited goals, such goals not being equivalent to the decisive defeat and surrender of the enemy; (2) an offensive weapons system; (3) defensive support from a great power; and (4) a changing domestic power structure.[89] Of these, a limited aims strategy and an alliance with a great power for defensive purposes appear to be the most important conditions. Paul also noted that history offers examples of the weak party realizing the inevitability of its own military defeat but hoping nevertheless for political gains, in accordance with the axiom "One may lose the battle, but not the war." "For some such states," Paul pointed out, "the prospect of a limited defeat is better than living with an unbearable status quo."[90] The short war between Georgia and Russia in August 2008 is perhaps the best example that can be adduced in support of Paul's assertions about the logic of asymmetric wars. In his recent papers, Paul has analyzed the conflicts between India and Pakistan within the framework of asymmetric warfare theory.[91]

American political scientist Michael Fischerkeller, in his 1998 article "David versus Goliath: Cultural Judgments in Asymmetric Wars," similarly considered the situation of weak states being aggressive toward great powers. According to his calculations, 54 percent of all conflicts in which major powers became involved over the period 1816–1996 started in this manner. Fischerkeller stressed that "the aggressive behavior of weaker powers in asymmetric wars is often incongruous with the basic propositions of balance-of-power theories," and also incongruous with traditional assessments based on quantifiable measures of capability, such as troop numbers and military effectiveness. Fischerkeller argued that "a reliance on objective, quantitative indicators places a theorist in peril of deducing unfounded behavioral propositions because subjective, cultural prejudice can play an equally monumental role in the assessment process." His basic proposition regarding asymmetric wars was that "the weaker state's judgment of the target as culturally inferior results in discounted capability evaluation of the quantitatively superior enemy. Viewing itself as culturally superior to its rival, the weaker state is encouraged to sound the trumpets for war when its quantitative inferiority seems to call for a more cautious policy."[92] This argument is based on an analysis of the beginning of World War II and the cultural assessments conducted by the Axis powers and the future Allies of the opposing side. Fischerkeller formulates two propositions and deduces two hypotheses to explain the logic of perception and the choice of strategy in asymmetric wars:

1. Where a target is reported as possessing superior, similar, or inferior quantitative capability and judged to be equally culturally sophisticated, the quantitative measures will be confirmed as accurate and representative of the overall net assessment of the target. Consequently—and here is the first hypothesis—"the perceiving power is encouraged to adopt a defensive, independent fortress or containment strategy."

2. Where a target is reported as possessing superior, similar, or inferior quantitative capability and judged to be culturally inferior, the quantitative measures will be discounted and represented in a net assessment that paints the target as much weaker than the quantitative measures would suggest. Consequently—the second hypothesis—"the perceiving power is encouraged to adopt an aggressive, imperialist strategy."

The second hypothesis is the focus of Fischerkeller's case studies. He criticized T. V. Paul's research for examining conflicts that are not, in Fischerkeller's view, truly asymmetric "since the weak powers initiated these wars with the understanding that there was a near-equal or preponderant coalition willing to support them if strategic expectations went awry. Through ignoring the contribution of alliance commitments [Paul exhibits] a lack of discrimination between strong and weak powers in his selection which, consequently, confounds the study of symmetric and asymmetric wars." However, Fischerkeller himself examined the Russo-Japanese war of 1904–1905 and Indo-Pakistani war of 1971 as examples of asymmetric conflict in which the warring parties have more comparable political, military, and economic capacities.[93] The thrust of his research is how cultural factors condition aggressive stances. He writes, "Classical realists and other power-determinists have written of such factors as national character and national morale in their conceptual discussion of power. Since these factors are actually derivative of subjective cultural judgments, they should be considered separate from conventional measures of power.

This separation is not merely superficial, it has significant theoretic utility, as the deduced partial explanation for the 'incongruous' weak power behavior in asymmetric wars demonstrates."[94]

Both Paul and Fischerkeller looked at motivations for weak states to initiate aggression against an adversary of unquestioned greater military might and resources. A complementary view, the paradoxical inability of great powers to win small wars, came from the Israeli political scientist Gil Merom in his 2003 monograph, *How Democracies Lose Small Wars: State, Society, and the Failures of France in Algeria, Israel in Lebanon, and the United States in Vietnam*. Merom was educated at Hebrew University, Israel, and Cornell University in the United States, where he earned a PhD. He currently combines teaching positions at Tel Aviv University and the University of Sydney. In his book, Merom explored Andrew Mack's hypothesis that one of the reasons for the defeat of great powers is the inability of democratic societies to wage small wars. Developed countries, Mack posited, are defeated because of a political and moral self-restriction on continuing the war. Merom argued that to understand the defeat of strong democratic powers, it is necessary to take into account the complex nature of relations between society, state, and war, and set aside the simplified perception of an "amorphous collective of society [that] was by and large considered important only in relation to its potential as a source for men and material needed for war."[95] Rather than regarding the outcome of a war as a foregone conclusion based primarily on the participants' respective military capacities, in which society plays only a passive role, modern concepts of warfare must take a broader view of society's ability to shape the conflict and its outcomes. Merom described the "modern power paradox" as "a struggle between two forces on three realms over three issues." The two forces are the state and "[that] part of the educated middle class which is a proxy of society," and the three realms are the three aspects of warfare: instrumental, political, and normative. Finally, the three issues, or three interrelated dilemmas that democracies are unable to resolve, and therefore prompt democracies to fail militarily and politically in small wars, are the following: (1) how to reconcile the humanitarian values of the educated class with the brutal requirements of counterinsurgency warfare; (2) how to find a domestically acceptable trade-off between the brutality and civilian sacrifices; and (3) how to preserve support for the war without undermining the democratic order.[96] Merom's argument is consonant with the conclusions of historians about the reasons for developed countries' misfortunes in specific historical incidents. It is also to a large extent congruent with what is often referred to as the lessons of Vietnam, the subject of numerous papers, with new publications coming out every year.[97]

If the strong lose wars for identifiable reasons, can the determinants of success be similarly narrowed down for weak states that win wars? The American political scientist Ivan Arreguín-Toft published an article (2001) and then a monograph (2005) under the title *How the Weak Win Wars: A Theory of Asymmetric Conflict*.[98] Arreguín-Toft served as electronic warfare/signal intelligence analyst at the US Army Field Station, Augsburg, West Germany. He received his undergraduate degree in political science and Slavic languages and literatures from the University of California, Santa Barbara, and his PhD from the University of Chicago. Arreguín-Toft has developed a novel approach that examines struggle strategies as success determinants, on the basis of which he proposes a theory of strategic interaction. He distills the various forms of struggle into two "ideal-type strategic approaches," direct and indirect. Direct strategies are conventional military actions aimed at neutralizing an antagonist's armed forces and destroying an antagonist's *ability* to continue war. Indirect strategies are designed to destroy an adversary's *will* to continue the struggle while avoiding direct clashes. The novelty of his approach, however, lies in introducing *barbarous forms of struggle* as an indirect strategy typical of both state and nonstate actors. He defines barbarism as "the deliberate or systematic harm of noncombatants (e.g., rape, murder, and torture) in pursuit of a military or political objective." There follows a hypothesis: "When actors employ similar strategic approaches (direct-direct or indirect-indirect), relative power explains the outcome: strong actors will win quickly and decisively. When actors employ opposite strategic approaches (direct-indirect or indirect-direct), weak actors are much more likely to win, even when everything we think we know about power says they shouldn't."[99]

Arreguín-Toft tested his hypothesis on five cases representing five historical periods: the Murid war of Russia in the Caucasus, 1830–1859; the South African war or Boer War of Britain, 1899–1902; the Italo-Ethiopian war, 1935–1940; the Vietnam War of the United States, 1965–1973; and the Soviet war in Afghanistan, 1979–1989. In addition to strategic factors, he considered other reasons for the victory of weaker actors, including those proposed by Andrew Mack, namely, asymmetry of interests, the sociopolitical nature of the actors (including/encompassing here Gil Merom's thesis that democracies are unable to win out of moral hesitation, to which Arreguín-Toft refers as "democratic social squeamishness"), and arguments over arms diffusion, specifically the supply of weapons to weak actors by their external supporters.

According to Arreguín-Toft, "the problem for strong actors is weak actors who pursue an indirect defense strategy, such as a GWS [guerrilla warfare strategy] or terrorism. This presents strong actors with three unpalatable choices: an attrition war lasting perhaps decades; costly bribes or political concessions, perhaps forcing political and economic reforms on repressive allies as well as adversaries; or the deliberate harm of noncombatants in a risky attempt to win the military contest quickly and decisively." He believes that his study has an advantage "as an explanation of all asymmetric conflict outcomes, and in particular as a guide to strategy and policy," and thus can be applied to counterterrorism as well as counterinsurgency strategy, both of which US policy-makers must face in the coming decades.[100] Arreguín-Toft concludes with other important reasons for a strong actor's loss, directly addressing the case of the United States, by significantly enlarging his original arguments and supporting his predecessors in asymmetric conflict studies:

> If the United States wants to win wars it must build two different militaries. If it wants to win the peace—a far more ambitious and useful goal—it must support its resort to arms by eliminating foreign policy double standards and by increasing its capacity and willingness to use methods other than violence to resolve or deter conflicts around the world. . . . The current US government confused military power with state power, and by over-applying the former has actually undermined its interests. If this policy continues and follows the historical pattern of *every previous attempt to accomplish the same ends* (peace) *by the same means* (the overwhelming application of military force unsupported by political, economic, and administrative recourses), the result will be costly quagmires such as Vietnam, Afghanistan (1979 and 2002–), and Iraq (2003–), and a future attack

on the United States or its allies that makes the terror attacks of September 11, 2001 pale by comparison.[101]

The writings of T. V. Paul, Michael Fischerkeller, Gil Merom, and Ivan Arreguín-Toft are among those most cited on asymmetric conflicts, along with Andrew Mack's seminal paper. All show different angles of entry into understanding asymmetric warfare; at a common, meta level of analysis, they attempt to develop the theory of asymmetric conflict and identify its determinants. Unlike their military colleagues, who are largely concerned with practical applications, political scientists narrow their research question to establish correlations between certain variables and be able to test hypotheses based on historical material. These publications show that the authors are familiar with their colleagues' work, though some, it must be said (especially Fischerkeller and Arreguín-Toft), turn a more critical eye on the achievements of other researchers and try to highlight the advantages of their own ideas. In doing so, they risk distorting the information supporting their own arguments and seem not to fully recognize their colleagues' achievements.

Another monograph that bears mention for its nod to technology was published in 2007: *Americans and Asymmetric Conflict: Lebanon, Somalia, and Afghanistan*. The author, Adam B. Lowther, a defense analyst of the Air Force Research Institute, holds a PhD in international relations. At the beginning of the book, Lowther related military art and the concept of asymmetric conflict, then analyzed US military operations from the standpoint of that concept. Drawing on a deep historical knowledge of warfare, he argued that the modern concept of asymmetric conflict is merely a "reinvention of concepts developed decades, centuries, and millennia ago. What is often mistaken for innovation is the rediscovery of well-worn ideas modified by the application of technological innovation." He further observed that asymmetric strategy and tactics developed along distinctly different paths in the East and the West. Thus, the West prefers the direct confrontation of opposing armies, while the East has refined the art of defeating an adversary without a pitched battle. However, these cultural predilections do not preclude the development and application of other forms of struggle, as evidenced by classical texts on military strategy.[102] Lowther seems correct in saying that analysts who see the current situation as something new lack an "understanding of the nature of war, which has changed very little over the past 7,000 years of human history. At the heart of war is the need to overcome an enemy's will to fight. This may be done by destroying an adversary's fighting ability or by overcoming his cost tolerance."[103]

Lowther analyzed specific instances of military actions involving US armed forces that conform to the asymmetric scenario and concludes that the war in Afghanistan is "the most successful major American military operation since the end of the Second World War."[104] In comparing the US and Soviet experiences in Afghanistan, Lowther found that they differed dramatically in a number of ways. "Where the Soviet Union deployed large-scale ground forces ... to support an unpopular regime and establish an economic and social system foreign to Afghans, the United States relied and continues to rely on experienced Mujahideen with the aid of American air power and Special Forces to defeat an unpopular fundamentalist regime supported by an even less popular foreign presence (al-Qaeda). Sensitive to the mistakes made by the Red Army, American military planners sought to mitigate many of the problems incurred by the Soviets by limiting the number of American forces used in ground operations, mollifying clan animosities, and providing tangible aid to Afghans." He believed that the United States succeeded in ensuring the support of loyal political forces and in resolving a host of problems, including the repatriation of displaced persons; the reconciliation of feuding tribes; the training of local security forces; the provision of assistance to Afghans in carrying out reforms in the economy, politics, health care, and education; and overcoming the legacy of the Islamist regime's attitude toward women. Lowther emphasizes that the US success in Afghanistan was made possible by an understanding of the essence of asymmetric military conflict as a need to ensure the security of the local population.[105]

Lowther's conclusions on the US engagement in Afghanistan might seem flawless if one overlooked the current state of affairs in that country and ignored the reasons that prompted the United States to choose an absolutely different strategy for Iraq. The Iraq War brought the United States to the brink of its largest military and political fiasco in post–World War II history. A drastic worsening of the situation in Afghanistan in 2009 coincided with the Obama administration's resolve to end the war in Iraq and led to a forced increase in troop size in Afghanistan and the implementation of full-scale military operations, which were accompanied by numerous civilian casualties. This picture contradicts Lowther's logic of success. Recent developments in Afghanistan, which saw the US administration attempting to engage the Taliban in the peace process, also contradict the initial intentions of American politicians. Such a reversal, however, is in keeping with the logic of asymmetric conflict settlement described in the writings of American experts.

As we turn to the termination of asymmetric conflict, an outstanding voice in the field is that of the Russian political scientist Ekaterina Stepanova. Her recommendations for resolving asymmetric conflicts stem from a specifically structural understanding of asymmetry, which she explores in her book, *Terrorism in Asymmetrical Conflict: Ideological and Structural Aspects.*[106] Published in 2008, the book aims to combine mobilization with a structural approach, drawing on the concept of asymmetry to stress a disparity in the structural arrangement of states and terrorist networks that benefits the latter. She regards terrorism as "the most asymmetrical of all forms of political violence," and tries to explain the great vulnerability of states with regard to nonstate actors. The main argument of the book is that "[even though] within the asymmetrical framework... states and the international community of states are incomparably more powerful in a conventional sense, enjoy a much higher formal status within the existing world system, and remain its key formative units, in the situation of a full-scale conflict of ideologies with violent Islamists they put themselves at a disadvantage." She continues: "It is precisely because of the modernized, moderate, relatively passive nature of the mainstream ideologies of state actors that they cannot compete with a radical quasi-religious ideology. They can offer little to compete with Islamist extremism as a mobilizing force in asymmetrical confrontation at the transnational level. In other words, on the ideological front the state and the international system may be faced with a reverse (negative) asymmetry that favors their radical opponents."[107]

For the international community and world powers that are engaged in a big struggle, Stepanova's research suggests "politicization as a toll for structural transformation." By politicizing radicals she means integrating them into existing political structures, which should gradually destroy the radicals' key networks and other structural advantages.[108] The strategy of co-opting and pacifying radical elements by integrating them into the political process is well established. Such strategies were used in Northern Ireland, in the Chechen Republic, and in the Middle East settlement. Nevertheless, certain problems emerge when this strategy is implemented, among them the unreliability of co-opted radicals, the difficulty of gaining control over the entire radical network, and, frequently, the high cost attached to such strategies. Furthermore, the issue of morality in politics remains relevant for the leaders and citizens of developed countries since it plays a crucial role in the normal functioning of the system as a whole. The moral hazard lies in the necessity to cooperate with people who are willing to resort to criminal actions (such as terrorism) in the interests of their political goals, and to give up possible moral and juridical restitution as a price for reconciliation and lasting peace.

Scholars and practitioners of conflict resolution have long observed that negotiating rather than continuing an armed struggle is the best strategy for conflict termination. As William Zartman pointed out, "[the military] defeat of the rebellion often merely drives the cause underground, to emerge at a later time." He noted further that "negotiation is the best policy for both parties in an internal conflict" but that it is rarely used and is associated with significant subjective and objective difficulties. He argued that "negotiations under conditions of asymmetry (asymmetrical negotiations) are a paradox, because one of the basic findings about the negotiation process is that it functions best under conditions of equality, and indeed only takes place when parties have some forms of mutual veto over outcomes. Asymmetry means that the most propitious conditions for resolving conflict are difficult to obtain." Negotiations also require reciprocal recognition by the parties, and that in itself is a subject of dispute. There is also the problem of representation in negotiations on behalf of the opposition. It is difficult for a government to acknowledge that such representatives have the right to express group interests, and thus difficult to acknowledge the validity of rebels' claims. The interests of parties to asymmetric conflicts are often caught in a zero-sum game: the weaker party strives to change the existing power structures and the stronger party strives to retain it. Zartman wrote, "The government seeks to turn asymmetry into escalation, to destroy the rebellion and break its commitment, and force the rebels to sue for peace. The insurgents usually seek to break out of their asymmetry by linking up with an external host state and neighbor, thus internationalizing the conflict. In so doing, insurgents radically change the structure of the conflict from a doubly asymmetric dyad to a wobbly triad of great complexity."[109] Internationalization of a conflict enlarges the number of actors involved. These actors have their own agendas and visions of how the conflict needs to or could be resolved. For state actors, internationalization of a conflict undermines the authority of their central government and its sovereignty in resolving domestic problems. whereas for nonstate challengers the internationalization of a conflict raises their status and the legitimacy of their case.

Christopher R. Mitchell, an expert in conflict resolution and professor emeritus at George Mason University, has examined and identified many key asymmetries in protracted conflicts and peacemaking strategies: asymmetry in capability (e.g., coercive ability, external support, access, visibility, cost experience, survivability, bargaining ability), in structure (e.g., intraparty cohesion, leadership legitimacy, leadership insecurity, constituent mobilization, elite entrapment), in commitment (e.g., goal salience, constituent commitment, external dependency, commitment to change, expectation of success), in interdependence (e.g., isolated or interdependent status, parallelism, historical justification), in legality or status (e.g., representativeness, existence, legitimacy), in morality (e.g., existential acceptance, issue acceptance, goal acceptance), and in behavior (e.g., violence, coercion, persuasion, conciliation, avoidance). Mitchell noted that the weak party strives to reverse the key asymmetries and create conditions of equality. However, in his opinion, "conflicts are essentially dynamic phenomena," and there is "no straightforward, linear relationship between relative coercive capacities and the probability of conflict reduction." In fact, his analyses of real instances support two opposite arguments for stopping conflict. The first argument proposes eliminating asymmetry to stop conflict, as "equals make peace more readily and more easily than unequals." The second, opposing argument proposes retaining asymmetry, as "[a] very high coercive inequality between parties may lead to avoiding or reducing conflict, on the grounds that the weaker may be more ready to end its efforts at protest and coercion, while the stronger may be willing to consider making a more generous offer to avoid the possibility of trouble later."[110] Thus, there seems no sure exit path from conflicts involving asymmetric parties.

As this section has shown, the writings of military analysts and scholars of asymmetric conflict exhibit both differences and similarities. Military art is overall more concerned with pragmatics. Researchers, by contrast, strive to develop a theory expressed in terms of determinants, correlations, and variables, and this stance sometimes forces them to artificially narrow their scope to hypotheses that can be verified by means of existing methods and means of verification. Scholars choose to take the problem of asymmetric relations outside the framework of military conflict and military praxis to develop a broader, truly conceptual understanding of the nature of asymmetric relations in all their diverse manifestations. As a result, the concept of asymmetric relations has been accepted as a useful tool for exploring problems in the social sciences and humanities, as well as in the more obvious fields of economics and international relations.

DEVELOPING AN ANALYTICAL MODEL OF ASYMMETRIC CONFLICT

An analysis of the literature in which the concept of asymmetric conflict is used shows that the theory of asymmetric conflict is underdeveloped, and that researchers prefer to deal with individual aspects of the phenomenon or with the analyzable factors in a small number of cases, rather than trying to derive a full-fledged theory. I propose that asymmetric conflict as a theory should be considered as a suite of variables in which the weight of each factor is rarely predictable in any given case, and in this section I suggest my own model for analyzing asymmetric conflicts. A brief review of the history of the terms will help set the stage for the theoretical work to follow.

For centuries, symmetry was seen as a sign of harmony, balance, order, and norm in the universe and scientific knowledge, while asymmetry was considered a sign of disorder and anomaly. In the nineteenth century, the French scientist Louis Pasteur proved that asymmetry was a norm rather than an aberration, and was in fact one of the main characteristics of nature. Gradually, the understanding of asymmetry as a particular way to organize the organic and inorganic world started diffusing into the arts and humanities. For instance, the notions of symmetry and asymmetry play a pivotal role in game theory and in negotiations. A special use of asymmetry occurs in logic, where the term indicates two entities that are related but that do not relate to each other in precisely the same way. An example would be the logical expression of a relationship between husband and wife, in which the statement "*Napoleon is the husband of Josephine*" is true but "*Josephine is the husband of Napoleon*" is not true, and the relation "being the husband of" is asymmetrical.[111] There is perhaps a hint of this sense of the word in the structural and cultural differences that divide the participants engaged in asymmetric conflict, where a great power does not have the same relationship with its adversary that it would have with another great power.

In the social sciences, the concept of asymmetry is most often drawn on to study conflicts of various kinds, from confrontations between small groups to global clashes. Christopher R. Mitchell, whose work on conflict resolution was cited in the immediately preceding section of this chapter, emphasizes that asymmetry refers to more than just a power imbalance between parties. He defines asymmetry as "a dynamic as well as multidimensional phenomenon, consisting of a differential distribution of relevant resources and salient characteristics between adversaries in a conflict system."[112] In this way, the concept is extended to cover a panoply of features and affordances, and its inherent elasticity suggests the multiple dimensions for which any theory must account.

Lawyers as well as political and social scientists use the concepts of symmetry and asymmetry to analyze relations between actors in a given system, which might be social, political, or legal. This approach implies that asymmetry is an essential characteristic of relations between participants in interactions in counterpoise: equal versus subordinate, horizontal versus vertical, pluralist versus hierarchical. The struggle is usually initiated by a subordinate party, which seeks to change the situation and achieve symmetry or equality, while the actions of the dominant party are directed toward restoring order and preserving the hierarchical status quo—which is also asymmetric and unequal.

Different disciplines either use the concept to characterize individual elements of conflict[113] or, more holistically, regard the phenomenon as a suite of several asymmetric characteristics subject to differential conditioning. The journal *Dynamics of Asymmetric Conflict,* for example, emphasizes a more complex approach to asymmetric conflict by considering the status and resources of adversaries as determinants of the phenomenon, as well as the psychological aspects of adversaries' behavior. Political and military analyses, in contrast, tend to present a narrower vision of asymmetric conflict, and focus more pragmatically on the tactical and strategic aspects of individual conflicts.

Despite such differences in exactly what is meant when analysts introduce the concept of asymmetric conflict in their work, a handful of consistent applications of the concept in conflict analysis can be identified:

1. To analyze confrontations between adversaries characterized by unequal status within one legal system: the political-legal approach.

2. To characterize adversaries' inequality in power and resource capabilities: the traditional approach.

3. To explain the political defeat of the dominant party in an armed conflict: the paradoxical approach.

4. To characterize tactics and strategies that compensate for adversaries' inequality in power and resources: the tactical and strategic approach.

5. To highlight the incongruity of the parties' interests and their attitude to conflict: the subjectivist approach.

6. To assess parties' motivation to continue the struggle: the mobilizational, psychological, or ideological approach.

7. To identify differences in the organization or structure of adversaries, such as a struggle between state actors and nonstate actors—for example, a terrorist organization—with a network structure: the structural approach.

A common aspect of conflicts that are analyzed according to the asymmetric conflict concept is the incommensurability of the adversaries' resources, power, and status. The "system," which represents the conditions under which conflicting parties interact, may be either a single state or a system of international relations within an existing hierarchy of power and state capabilities. The resource and power inequalities represent basic asymmetries, which often are accompanied by status asymmetry. These main asymmetries give rise to the nonlinear development of events through asymmetric forms of struggle (e.g., guerrilla warfare, protracted wars using terrorist actions). A nonlinear course of events—sporadic encounters, without clear military, economic, or political gains for the stronger party—strengthens asymmetries in relations between fighting parties (that is, their mobilization capabilities) and in their attitude toward the conflict, or will to win. This could result in political rather than military defeat of the stronger party and the victory of its weaker opponent—the paradoxical outcome of asymmetry.

Andrew Mack pointed out the dichotomy in the parties' attitude toward war, whereby the weaker party fights a "total war" (i.e., there is full mobilization of resources for the victory) while the stronger one wages a "limited war." In such a situation, the stronger party is unable or unwilling to mobilize all its available resources to achieve its war objectives. For the United States, the war in Vietnam had limited importance, as it did not affect major US interests and did not seriously threaten national security. For Vietnam, however, the war was total. Under the conditions of a limited war, a large-scale mobilization of resources to achieve victory was impossible for the United States, both politically and logically. The obvious power superiority of the United States made such a use of resources unrealistic.

As Mack observed, the war in Vietnam showed that the theater of military operations is not limited to the actual battlefield and may significantly influence domestic politics and the social situation at home:

> The Vietnam war may be seen as having been fought on two fronts—one bloody and indecisive, in the forests and mountains of Indochina, the other essentially nonviolent—but ultimately more decisive—within the polity and social institutions of the United States. The nature of the relationship between these two conflicts—which are in fact different facets of the same conflict—is critical to an understanding of the outcome of the war. However, the American experience was in no sense unique, except to Americans. In 1954 the Vietminh destroyed the French forces which were mustered at Dien Bien Phu in a classic set piece battle.[114]

In both cases, the French engagement during the 1950s and the US war in Vietnam, the stronger party lacked the political will and popular support to mobilize resources and continue the struggle. A famous moment of such rescission came in 1968, when the United States managed to inflict a decisive defeat on the insurgents, which could have been a turning point in the war but instead became a turning point in a totally different sense. Popular opposition to the war in the United States reached such a pitch that plans to further mobilize resources and deploy additional troops were not approved by the Congress.

In asymmetric conflicts, the perception of the legitimacy of a war and the justification for casualties often change over time. An important requirement for victory is the readiness of the populations of the belligerent countries to accept politically motivated mobilization and possible casualties for the sake of the war cause. The willingness of career military and draftees to sacrifice their lives even if there is no direct military threat to the nation is a crucial condition for continuation of a limited war. Another one is support of war objectives by the population. A society involved in a total war views the issue of national survival in a different way from a nation involved in limited war. The longer the limited war goes on, the greater are its costs. A high casualty count encourages opposition movements, provokes debate over the morality and objectives of war, and visibly divides society along the lines of attitude toward the war. A state involved in a limited war typically finds belief in the war's legitimacy constantly shrinking, and the logic of a protracted war ensues. Henry Kissinger, US secretary of state during the Vietnam War, defined what he called "the basic equation of guerrilla war" this way: "The guerrilla army wins as long as it can keep from losing; the conventional army is bound to lose unless it wins decisively. Stalemate almost never occurs."[115]

One of the most powerful mobilization factors capable of uniting a state is nationalism. Many political scientists pay heed to political mobilization when a nation comes under threat. As a rule, political opponents facing a common external enemy unite and find common national interests. Once the majority perceives that a conflict is a common national danger, the populace becomes ready to mobilize all available resources and to subordinate all other interests to victory over the external enemy. The experience with guerrilla wars also indicates that victory belongs to the party that is supported by its own population.

As Andrew Mack noted, most strategic theorists agree that "in war the ultimate aim must be to affect the will of the enemy. But in practice, and at the risk of oversimplification, it may be noted that it is a prevalent military belief that if an opponent's military capability to wage war can be destroyed, his 'will' to continue the struggle is irrelevant since the means to that end are no longer available."[116] As an analysis of asymmetric conflicts shows, destroying the adversary's will to wage war may turn out to be a more efficient method of inflicting political defeat on an adversary that still has sufficient military power to continue fighting. In this case, if the will of the stronger adversary to continue war is undermined, its military strength does not have a decisive influence. The US defeat in Vietnam came about as a result of the eroded political capability of the United States to mobilize resources and ensure a sense of legitimacy at home for the war's objectives and methods.

Many anticolonial movements are examples of asymmetric conflict in which the parties to the conflict were without question incommensurate in their power capabilities. The key feature of such conflicts is the inability of the stronger participants to impose their will on their weaker adversaries, subdue them, and achieve their war objectives. Thus, one of the most important dicta of politics and military science concerning the strong winning the struggle and imposing its will on an adversary was challenged. But it should be noted that in none of the conflicts in the third world countries were the insurgents able to invade the territory of a major power or wreak significant military or material damage on it. Victory was possible only because the political ability of the major power to continue war was shattered: the stronger party did not have the political wherewithal to continue mobilizing human and material resources and press on with war. It had to cease fighting the weaker opponent and agree to conditions that ran counter to its original objectives of engagement.

Raymond Aron characterized the essence of dissymmetry in a colonial conflict as follows:

> The nationalists who demand the independence of their nation (which has or has not existed in the past, which lives or does not live in the hearts of the people) are more impassioned than the governing powers of the colonial state. At least in our times they believe

> in the sanctity of their cause more than their adversaries believe in the legitimacy of their domination. Sixty years ago the Frenchman no more doubted France's *mission civilisatrice* [civilizing mission] than the Englishman questioned the "white man's burden." Today the Frenchman doubts that he has the moral right to refuse the populations of Africa and Asia a *patrie* (which cannot be France), even if this *patrie* is only dream, even if it should prove to be incapable of any authentic independence.... The inequality of determination among adversaries was still more marked than the inequality of material forces. The dissymmetry of will, of interest, of animosity in the belligerent dialogue of conservers and rebels was the ultimate origin of what French authors call the defeats of the West.

Aron also pointed out an important feature of such conflicts, namely, their absolute character. The nationalists fight an "*absolute enemy*, the one with whom no reconciliation is possible, whose very existence is an aggression, and who consequently must be exterminated."[117] The longer and harder the struggle, the stronger the conviction of its legitimacy and the greater the solidarity of a nation engaged in total war.

Thus, in keeping with the theory of asymmetric conflict, the key reason for the defeat of the strong power is the dilution of its political will to continue war as a result of domestic economic, political, and social processes. International factors may also place pressure on the behavior of the belligerent parties, thereby limiting the aims and methods used to achieve objectives and influencing in particular the political elite, as well as the populace as a whole.

Andrew Mack referred to his work as providing a pretheoretical perspective, and many of his followers have sought to test the relevance of his hypotheses, studying the impact of individual factors on the outcome of armed conflicts between unequal antagonists. Arguably, however, no one has suggested an alternative parsing of the reasons for great power defeat. This book attempts to do just that, by integrating the factors accounting for defeat into an analytical model of asymmetric conflict, all rooted in the work of Mack and his followers.

In the research presented in this book, asymmetric conflict is understood to be a conflict in which a strong actor loses to a weaker actor in an armed struggle. Narrowing the field of candidate conflicts to ones in which wins and losses are easily verified, we will consider only those armed conflicts involving the great powers. In the postwar period only five countries have generally been considered to be great powers, and all are permanent members of the UN Security Council: the United States, the United Kingdom, France, the Soviet Union/Russia, and China. Weak parties, on the other hand, may encompass very diverse actors, including, in addition to weak independent states, a political group or movement, a dependent territory, or an extrasystemic terrorist group. These latter nonstate entities may engage regular state troops in armed confrontation, or they may avoid direct confrontation and use indirect tactics to pursue a certain political objective. As a rule, the goal is to establish an independent political entity, either a state or an autonomous region within an existing state. Contemporary international terrorist organizations undoubtedly pursue other political objectives, such as pushing Western countries out of the developing world and eradicating Western ideas from developing countries.

The concept of asymmetric warfare as described above can be tested by verifying the various hypotheses that have been proposed to account for the defeat of developed countries in wars against weak antagonists. These are as follows:

1. A weak adversary wins because of its unyielding will to win and the more powerful opponent's loss of such will. This hypothesis reflects the dichotomy of the small war waged by the strong adversary versus the total war waged by the weak party. This hypothesis pays attention to differences in the capabilities of the state and political elites to mobilize resources (human, material, and nonmaterial) to achieve victory, which differences in turn depend on the importance of the war to the society.

2. A weak adversary wins as a consequence of the stronger adversary's fatigue and unwillingness to expend further resources and human lives in the pursuit of victory. Such a win is not a military victory *per se* for the weaker party, but rather owes to the stronger actor's exhaustion. As Clausewitz put it, this is a strategy aimed at achieving a "negative political action."

3. A weak adversary wins through the predominant use of asymmetric strategies and tactics, such as guerrilla warfare, acts of terrorism, or protracted war.

4. A weak adversary wins because of nonmilitary factors, especially public opinion coalescing against the war in the stronger state and negative coverage of the war by the mass media.

5. A strong party is defeated as a consequence of sharp disagreement among the political elite as to the efficacy and morality of the war and strong opposition from antiwar groups, both of which factors play especially prominent roles during political election seasons.

6. The course of the struggle is largely determined by the actions of external forces rather than by the participants themselves. Such actions may include interference from other countries or the provision of military, technical, or economic assistance to one or the other of the belligerents.

7. A strong party is defeated as a result of the pressure exerted by and the condemnation of the international community.

The dependent variable in all of these hypotheses can assume two opposite forms: the defeat of the stronger party or the political victory of the weaker party in an armed conflict. The fact of defeat or victory is not always easy to ascertain, but in most cases it is possible to do so by comparing the objectives of the parties to the conflict and the situation in which they find themselves when the conflict ends. Thus, it is possible to determine when the weaker party managed to eke out a victory over its stronger adversary and to discern how much influence different factors brought to bear on the outcome.

In this book, the strong party is represented by the great powers, and we will consider the following factors as reasons for its defeat:

1. Absence or loss of the will to continue fighting.

2. A protracted war without clearly defined indicators of success.

3. The difficulty of waging war and achieving obvious victory in anti-guerrilla, counterinsurgency, or counterterrorist campaigns.

4. Support provided by other strong actors to the weak adversary.

5. Negative popular and elite attitudes toward a protracted and unsuccessful war (as evinced in public opinion polls, draft dodging, or shaming in the media).

6. A schism in the political elite, manifested during elections.

7. The economic exhaustion of the strong state.

8. The international community's negative view of the war, disapproval of the strong opponent, and condemnation of the aims and means of the war.

This model is presented graphically in figure 1.1.

The reasons for a weak party's victory can also be considered within the theory of asymmetric conflict. The following factors determine the possibility of victory of the weak over the strong:

1. A steady will to win, reflected in the ability to mass mobilize resources for the struggle for a long time.

2. A protracted war, signaling the absence of defeat in the struggle against a superior adversary.

3. The use of guerrilla and terrorist strategies and tactics.

4. Popular support for the war (with active support provided to guerrilla fighters and terrorist groups, participation in the armed struggle).

5. Unity of the political elite and the whole of society, willing to fight and to overcome disagreements.

Figure 1.1. Factors Determining the Defeat of a Strong Party (Great Power) in an Asymmetric Conflict

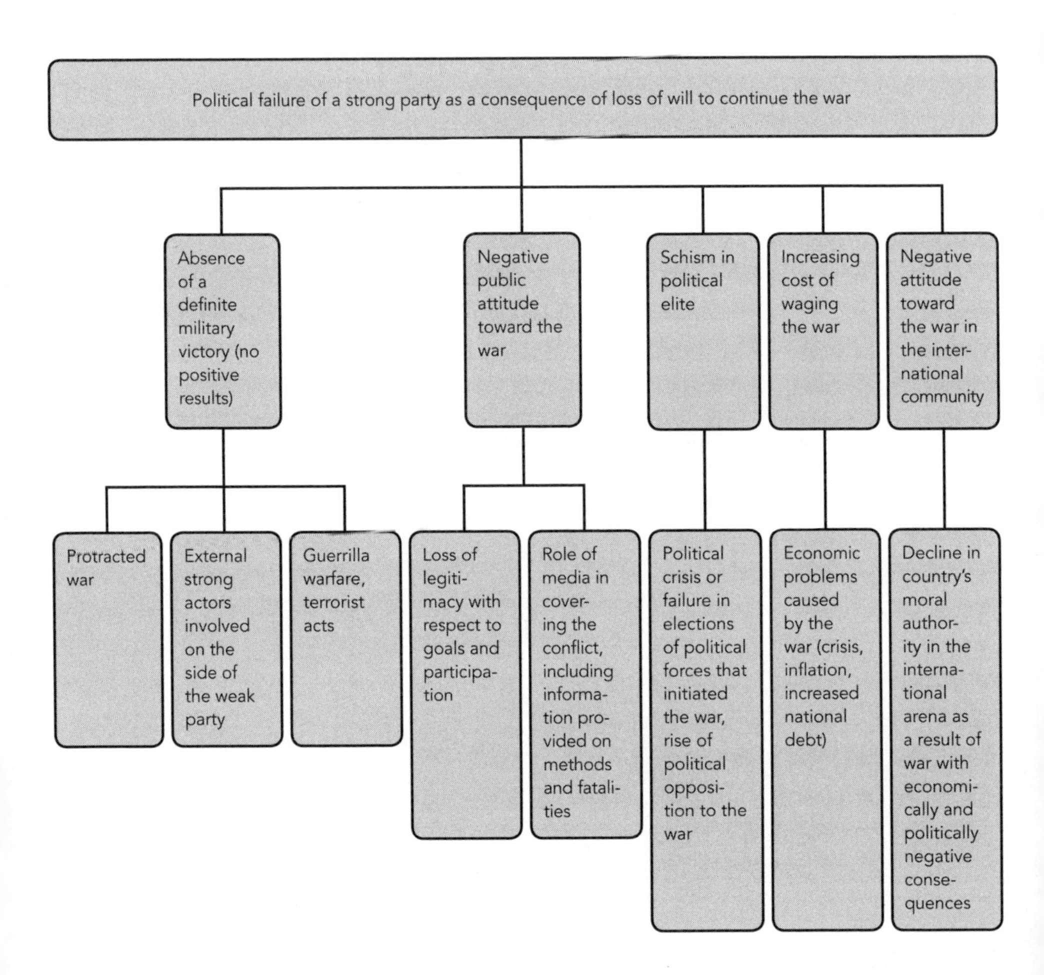

6. Material, military, technical, and other assistance provided by external actors, primarily other great powers.

7. Appeal to the international community for the support of one's just objectives and condemnation of the adversary for its immoral objectives and methods of warfare.

This model is presented graphically in figure 1.2.

The schemes are not identical or simply the reverse of each other. In this book, more attention is paid to the factors that determine the defeat of a great power as opposed to those accounting for the victory of a weak adversary, while keeping in mind that the distinction between weak and strong is not absolute.

The distinctive feature of asymmetric conflicts is that they are conducted on an international stage. This is true even of many domestic conflicts, which become internationalized through the direct or indirect participation of other, external actors. Examining such conflicts makes it possible to identify a set of asymmetric characteristics, structural and dynamic, that together form a particular, stable phenomenon in international relations. As a result, it is important to separate the asymmetric conflict phenomenon from a situation of armed struggle between opponents that are unequal in power and status. Describing as asymmetric any confrontation between unequal adversaries does not have heuristic value; most conflicts contain elements of inequality between the adversaries and on that limited ground could be called asymmetric. For this reason, it is important to differentiate the theory, phenomenon, and model of asymmetric conflict, as well as how the concept of asymmetry is applied in analyzing conflicts.[118]

The *concept of asymmetry* is used to analyze individual elements of a conflict to highlight their incommensurability or incongruence between the belligerent parties.

Asymmetric conflict theory seeks to identify recurrent patterns in confrontations between adversaries unequal in status and power (the basic asymmetries) while grounded in a holistic approach, that is, while taking into account all elements of the conflict that may lead to the defeat of the stronger party. This theory, like any other, reflects only the possibility that

Figure 1.2. Factors Determining the Victory of a Weak Party in an Asymmetric Conflict

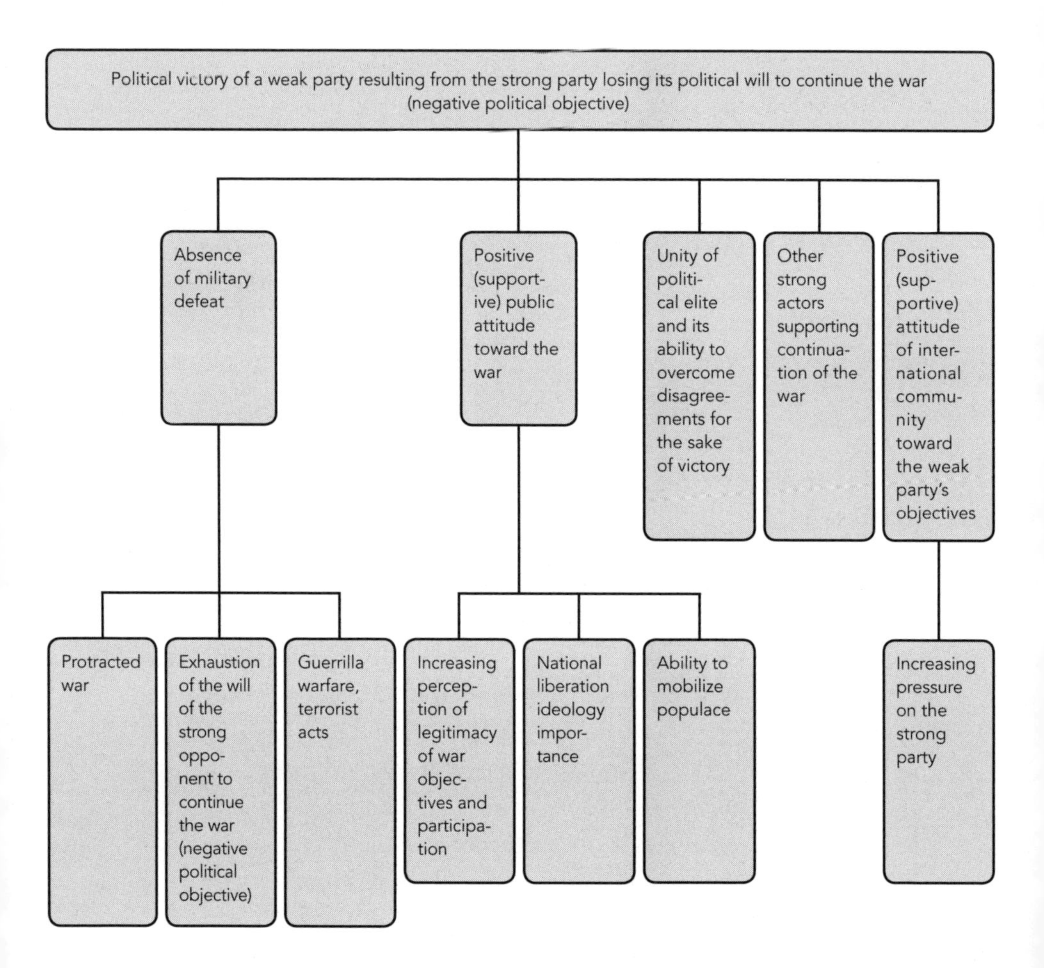

conflict will develop, since asymmetric conflict is not inevitable and does not always end with victory of the weaker party.

The *asymmetric conflict phenomenon* as originally described by Andrew Mack in relation to the great powers has as its leading characteristic the political defeat of the stronger opponent, an outcome that sometimes becomes clear only after the end of the struggle. The indicators of political defeat in an armed conflict that may not coincide with military defeat are cessation of the struggle and relinquishing of the war's objectives, the reasons why the party went to war in the first place.[119] The analysis of conflicts involving great powers, an early step in developing a theory of asymmetric conflict, deals primarily with highly counterintuitive examples of strong actors'—the great powers'—defeat in asymmetric conflicts, though it is also true that statistically, the great powers have been the most frequent participants in such conflicts in the postwar period.

The *model of asymmetric conflict* elaborates a matrix of factors that undermine the will and fighting spirit of the stronger adversary and force it cease fighting, against its own interests. The reasons for the victory of the weaker party can be examined in similar fashion, and there is also an analytical model to aid in the examination.

The present study adumbrates a complex understanding of the asymmetric conflict phenomenon whose most paradoxical feature is the stronger party's defeat in an armed confrontation between asymmetric antagonists. Such an approach includes other interpretations of asymmetry in armed conflicts (traditional, political-legal, structural, mobilizational, tactical and strategic), but the common feature is the great incommensurability between antagonists in power capability and world status. In my view, the asymmetry in resources and power, accompanied by an asymmetry in status, is the main feature of asymmetric conflict. Other asymmetries follow from the basic one and reflect the desire of the weaker party to change the situation of the conflict.

When analyzing the asymmetric conflict phenomenon in specific examples involving the great powers, this book pays special attention to asymmetry in structural characteristics (differences in subjecthood, status, and power capabilities) as well as dynamic ones (differences in strategies and tactics). This is a step toward discovering which asymmetric characteristics or combinations of characteristics proved most significant to the outcome of the conflict. I will use two approaches, one a quantitative statistical assessment of armed conflicts after the end of the World War II and the other a qualitative examination of particular examples of asymmetric conflict, to judge which asymmetric characteristics had a decisive impact on conflict outcome in the view of conflict participants and experts.

Systemic factors that played a crucial role in the emergence of the asymmetric conflict phenomenon should not be overlooked. International relations in the postwar period contributed greatly to the possibility of political defeat of the great powers in conflicts with incommensurately weaker actors. Systemic factors, therefore, help define the specifics of great power functioning in postwar international relations. In my opinion, it was the combination of international relations and domestic politics that led to the emergence, or more precisely the recognition, of the paradoxical constraints placed on great powers in struggles against obviously weaker adversaries.

This analysis draws on historical methods to test certain analytical constructs suggested by political and social scientists: a hypothesis is formed based on the principles and methods of identifying the most general recurring patterns in different societies. It attempts to expand the range of conflicts usually seen as asymmetric and to analyze critically the asymmetric conflict concept. The concept is examined in a number of post–World War II armed conflicts where the great powers suffered political defeat. It should be kept in mind that military victory is not coterminous with political victory, and in the postwar period a disconnect between military victory and the achievement of the stronger party's war objectives (i.e., political victory) became almost commonplace.

The concept of asymmetric conflict does not imply that all wars involving great powers in the post–World War II period ended with their defeat. That would be a simplification. Nevertheless, the concept emerged because the failures of the great powers were not isolated cases but rather represented a recurring phenomenon observable in the postwar order.

The list of armed conflicts involving the great powers after the end of World War II was compiled from several databases created by leading research centers, among them Uppsala University, Sweden; the Peace Research Institute Oslo (PRIO), Norway; and the COSIMO project at the University of Heidelberg, Germany. The list is provided in an appendix to the book, and the conflicts involving great powers that are discussed in the book were drawn from the list.

CHAPTER 2

Identifying the Asymmetry Factor in Armed Conflicts

This chapter attempts to identify the asymmetry factor in post–World War II armed conflicts by analyzing the statistical data compiled in conflict databases. As the term "asymmetric" has penetrated more broadly into the contemporary literature on armed conflicts, researchers often refer to asymmetry in status and power or in strategic characteristics, but rarely undertake the necessary calculations to verify the extent of the presumed asymmetry, its manifestations, or its impact. This chapter seeks to remedy the situation by analyzing data from established conflict databases to determine the degree to which asymmetry is evident in postwar armed conflicts. Focusing then more narrowly on features of asymmetry in armed conflicts involving the great powers provides additional insight into the scale of this phenomenon in international relations and the degree to which it is concentrated in the armed conflicts of great powers. This information may then help establish whether asymmetric conflict is associated exclusively with the great powers, as in Andrew Mack's initial hypothesis,[1] or whether it is a significant factor in conflict between other kinds of entities.

DATA SOURCES AND METHODS OF ANALYSIS

A study of this sort, which seeks to identify recurring patterns associated with armed conflicts, relies on databases on wars and conflicts and proceeds mathematically to derive statistical correlations between the various characteristics of war. A few introductory words on the data sources and analytical methods used in this chapter follow.

Databases are useful because they amass standardized information on numerous characteristics of the phenomena of interest. The main principles followed in compiling databases were largely developed in the first half of the twentieth century by researchers conducting the first extensive research on war. These databases, created by Pitirim Sorokin, Quincy Wright, and Lewis Richardson, are still widely used. Building on this tradition, many researchers and research teams have constructed their own databases, analyzed the data mathematically, and identified patterns of conflict by establishing correlations between variables.[2]

The complex studies of warfare undertaken in the second half of the twentieth century almost always required the development of new data sets. Among the better known projects are US historian Jack Levy's Great Power Wars data set;[3] Zeev Maoz's Dyadic Militarized Interstates Dispute data set, which draws correlations between different variables of armed conflicts;[4] the Correlates of War database, developed by J. David Singer and subsequently by Melvin Small;[5] Ted Robert Gurr's database on ethnic conflicts and national minorities;[6] Ernst Haas' UN collective security database;[7] Kalevi J. Holsti's data set on major armed conflicts of the period 1945–1995;[8] André Miroir, Éric Remacle, and Olivier Paye's database on third-party interventions in conflicts and the degree to which such interventions accord with international law;[9] and Patrick M. Regan's database on third-party interventions.[10] One of the most comprehensive and regularly updated databases on armed conflicts is maintained by researchers at Uppsala University, Sweden, in conjunction with the Peace Research Institute, Oslo (PRIO), under the leadership of Peter Wallensteen.[11] This is the Uppsala Conflict Data Program (UCDP) database and will feature prominently in the statistical analyses in this chapter.

Kristine Eck, who worked for the Department of Peace and Conflict Research, Uppsala University, home to the UCDP, has prepared *A Beginner's Guide to Conflict Data: Finding and Using the Right Dataset.* It offers an overview of available data sets and examines several methods for dealing with large blocks of information. Eck notes that historical research as a qualitative investigation relies on in-depth histories and analyses of conflicts, and suggests that historians therefore find lists of conflicts to be of little use. Nonetheless, she adds, such lists afford researchers a quick overview of the field, and speed the process of selecting conflicts to study in more detail.[12]

A Beginner's Guide also contains descriptions of about 60 of the most prominent conflict data sets. Of these, 44 data sets are curated by researchers and the rest are known as events data sets generated by a computerized pattern-matching program that screens the world's leading news services. According to Eck, studies have shown that "computer coding is as valid as human coding [and] is clearly reliable since the results are transparent and easily reproducible."[13] Databases also allow matching existing statistical information about the events under study. Though collecting this information is a labor-intensive process, the amassed data make it easier to perform factor analysis, identify variables, and check for interconnections between them. The credibility of data sets is ensured by the high professional level of the researchers involved in these projects and the "second look" that the data receive as the information is shared across platforms and made available to public scrutiny. A conceptual framework for acquiring conflict data has been elaborated in the course of these research projects, and verifiable indicators and ways to measure conflict characteristics also have been developed.

Databases on armed conflicts usually cover the following information for each conflict: the direct participants and their characteristics, the beginning and end dates, the conflict's intensity (commonly measured by the number of casualties), the indirect participants, the region, and the subject of conflict. Other important characteristics often included in conflict databases are the behavior of the great powers (whether and in what form they participate), the outcome of the conflict (e.g., victory or defeat, a change in political organization, loss or gain of territory), and specific factors of the conflict's resolution—for instance, whether international organizations were involved, whether peace treaties were signed, or whether there was no formal resolution.

Warfare and armed conflict data sets often are supplemented by the use of databases that register information on conflict participants, and especially on the national strength (power) of different countries. National strength or power is a complex statistic derived from a number of measurable variables, among them total military expenditure, military personnel, energy consumption, iron and steel production, urban population, and total population, to yield what is known as the Composite Indicator of National Capability (CINC) score. The CINC was developed as part of the Correlates of War project, which was launched in 1963 at the University of Michigan under the leadership of political scientist J. David Singer and is currently supported by Pennsylvania State University, Pittsburgh. The six variables that determine the CINC score represent country assets that can be drawn on to wage war.

The Correlates of War project is a chronologically extensive and useful source of comparative data. Initial efforts in building the database were directed toward compiling data on temporal and spatial variation in interstate and extrasystemic war in the post-Napoleonic era. As the background page on the project's website notes, "the fundamental goal of the project was not just to measure the temporal and spatial variation in war but rather to identify factors that would systematically explain this variation. Accordingly, early efforts were undertaken to measure many of those factors that purportedly accounted for war such as national capability, alliances, geography, polarity, and status in the post-Napoleonic period." The project has compiled several data sets, including data on non-state wars, alliances (e.g., formal treaties, pacts, ententes), peaceful and violent changes of territory, national and bilateral trade flows, and membership in intergovernmental organizations. This wealth of data allows additional comparisons of armed conflict participants to be performed along multiple dimensions.[14]

In sum, databases on armed conflicts are a unique resource that helps historians trace and compare the essential characteristics of armed conflicts before they undertake a more detailed investigation of events. The initial data acquisition should be considered both quantitative and qualitative research, insofar as it aims to identify and record the most significant characteristics of armed conflicts in a standardized format that facilitates subsequent statistical analysis. Standardized data acquisition also makes it possible to identify the unique characteristics of specific events. This is a necessary step in isolating specific historical contexts for further study.

Problem Statement and Research Methods

The goal of this chapter is to identify the occurrence and impact of asymmetry as a determinable factor in contemporary armed conflicts, as well as any possible correlations between characteristics inherent to asymmetric armed conflicts. Asymmetric characteristics reflect marked disparities in the fundamental characteristics of the conflict participants or their means of interacting. Researchers classify such characteristics as either structural (i.e., the status, interests, and resources of the participants; the causes and outcomes of the conflict; the social and political situation in which the conflict takes place) or dynamic (i.e., strategies and tactics used, the intensity of interaction, the motivation of the parties involved, the duration of the conflict).

The analysis in this chapter draws on existing databases on armed conflicts to determine the degree to which asymmetry is manifested in post–World War II conflicts. The problem can be broken down into the following strands for the purpose of statistical calculation:

1. The share of armed conflicts that involved parties of markedly different status (e.g., a sovereign state versus a dependent territory; a state versus a political group);

2. The share of armed conflicts in which the adversaries' power and resources were significantly incommensurate;

3. The share of armed conflicts in which asymmetric strategies and tactics were used (contemporary analysts and military experts are unanimous in viewing guerrilla and terrorist strategies as asymmetric);

4. The share of armed conflicts in which developed countries were defeated by their weaker adversaries (a state or a non-state actor) or failed to achieve unequivocal victory, as well as situations in which a state was defeated by a political group or was forced to change the existing relationship (e.g., by expanding the rights of a weaker party, by opting for more symmetric relations).

Then, again with the aid of conflict databases, an attempt is made to uncover the connections between asymmetric conflict characteristics and lay the groundwork for advancing a theory of asymmetric conflict. Thus, the first part of the problem is to identify those armed conflicts characterized by asymmetry (in the status of participants, resources, strategies, tactics, ability to mobilize resources for a prolonged period, the meaning of war for society), and the second part entails establishing the existence of interconnections or correlations between these different characteristics of asymmetry.

Databases Used to Calculate the Asymmetry Factor

The following statistical work draws mostly on two sets of data: the database of armed conflicts compiled by the UCDP,[15] in cooperation with PRIO,[16] and the database on national and international conflicts compiled by the COSIMO project at the University of Heidelberg, Germany, under the leadership of Professor Frank R. Pfetsch.[17] These two databases have some features in common and some unique features. The UCDP database records data on armed conflicts only; the COSIMO database also includes conflicts in which violence was not used, as well as data on separate phases of a conflict. The time frames do not completely coincide: the UCDP database covers conflicts from 1946 to 2006, while the COSIMO database covers the years 1945–1999. Therefore, the conflict statistics would be expected to vary somewhat, but this difference does not prove an impediment to analysis. Each project developed definitions of armed conflict and war that took into account qualitative and quantitative variables. The qualitative variables are the characteristics of the parties to the conflict, the nature of their interaction, and various dynamic parameters. The number of casualties sustained and the duration of the conflict are traditionally seen as key quantitative variables.

The UCDP project defines armed conflict as "a contested incompatibility that concerns government and/or territory where the use of armed force between two parties, of which at least one is the government of a state, results in at least 25 battle-related deaths."[18] The project participants argued in a 2002 analysis of the data that the threshold they chose "is high enough for the violence to represent a politically significant event, although the precise local and international impact may vary."[19] In an earlier exposition of the data, published in 2001, the authors' argument was more straightforward: "We feel that the lower threshold adopted here—25 deaths in a single year—is high enough to represent a politically significant event, although the evaluation of the value of human life at this scale may vary across nations."[20]

The UCDP details the separate elements of the definition as follows:

Use of armed force: The use of arms to promote the parties' general position in the conflict, resulting in deaths.

Arms: Any material means, including both manufactured weapons and nontechnical materials such as sticks, stones, fire, and water.

25 deaths: A minimum of 25 battle-related deaths per year and per dyad (see below) in an incompatibility.

Party: A government of a state or any opposition organization or alliance of organizations. The UCDP distinguishes between primary and secondary parties. Primary parties are those that form an incompatibility by stating incompatible positions. At least one of the primary parties is the government of a state. Secondary parties are states that enter a conflict with troops to actively support one of the primary parties. The secondary party must share the position of the primary party it is supporting in the incompatibility.

Government: The party controlling the capital of a state.

Opposition organization: Any nongovernmental group that has announced a name for itself and uses arms to influence the outcome of the stated incompatibility.

Dyad: A configuration comprising two conflicting primary parties, at least one of which must be the government of a state. In interstate conflicts, both primary parties are state governments. In intrastate and extrasystemic conflicts, the nongovernmental primary party includes one or more opposition organizations.

State: An internationally recognized sovereign government controlling a specific territory or an internationally unrecognized government controlling a specific territory whose sovereignty is not disputed by another internationally recognized sovereign government previously controlling the same territory.

Incompatibility: The stated general incompatible positions.

Incompatibility concerning government: An incompatibility concerning the type of political system, replacement of the central government, or a change of its composition.

Incompatibility concerning territory: An incompatibility concerning the status of a territory, such as the change of the state in control of a certain territory (interstate conflict), secession, or autonomy (internal conflict).[21]

The COSIMO project gives the following definition of conflict:

the clashing of overlapping interests (positional differences) around national values and issues (independence, self-determination, borders and territory, access to distribution of domestic and international power); the conflict has to be of some duration and magnitude of at least two parties (states, group of states, organizations or organized groups) that are determined to pursue their interests and to win the case. At least one party is the organized state. Possible instruments used in the course of a conflict are negotiations, authoritative decisions, threat, pressure, passive or active withdrawals, or the use of physical violence and war.[22]

The COSIMO researchers then added more parameters to differentiate a war from a crisis or a different type of conflict:

1. The fighting occurs between at least two opponents with organized, regular military forces.

2. The fighting is not sporadic but continuous and lasts for a considerable period of time.

3. The fighting is intense, that is, it results in victims and destruction. The number of victims and the amount of destruction are both high.[23]

Thus, the qualitative characteristics of armed conflicts are generally consistent across both the UCDP and the COSIMO databases: (1) the armed conflict or war has at least two parties, one of which is a state; (2) the conflict is considered to be an incompatibility of the parties' interests with respect to governance or territory; (3) the interaction involves the use of military force by the state or regular troops; and (4) the fighting has resulted in casualties.

With respect to the quantitative indicators, the number of casualties was introduced as a comparison parameter in the earliest studies of war. For instance, Lewis Richardson introduced a logarithmic scale from magnitude 1 to 7 to record all cases of violent death, including homicide and wars of different kinds, in which major wars with more than 3,000 deaths were registered as a magnitude of 3 or greater.[24] For almost 30 years, the threshold of war was identified based on a different quantitative parameter: 1,000 battle deaths in the course of one year of fighting or the entire duration of a military struggle. This parameter was introduced by J. David Singer and the Correlates of War project; it was long used, and is still used in some studies, to calculate casualties only among combatants, whereas contemporary wars are characterized by dramatic increases in civilian deaths. The use of this parameter undoubtedly has affected the armed conflict statistics compiled by various researchers, for as a rule, most conflict databases treat the number of casualties sustained during the course of fighting as a measure of the intensity or magnitude of a conflict.

The two main databases that are the source of the data analyzed in this chapter have specific characteristics that must be noted. The UCDP database codifies information on 25 characteristics of 251 armed conflicts that occurred from 1946 to 2006. The variables (conflict data) were coded as follows:

(1) a unique identifier, applicable to all conflicts;

(2) the location of the conflict;

(3) side A—the primary party to the conflict;

(4) side A 2nd—the name(s) of state(s) supporting side A with troops;

(5) side B—the country or opposition actors;

(6) side B 2nd—the name(s) of state(s) supporting side B with troops;

(7) conflict incompatibility—a general coding of the conflict issue;

(8) the name of the territory over which the conflict was fought, if the incompatibility had to do with territory;

(9) the year of observation;

(10) the intensity level of the conflict per calendar year: two different intensity levels are coded, minor armed conflicts and wars;

(11) a cumulative intensity variable, which indicates whether the conflict has exceeded 1,000 battle-related deaths since its onset;

(12) the type of conflict: extrasystemic armed conflict, interstate armed conflict, internal armed conflict, or internationalized internal armed conflict;

(13) the start date, recorded as the date of the first known battle-related death;

(14) start precision, coded to highlight the level of certainty for the start date variable;

(15) start date2: the first time the conflict reaches 25 battle-related deaths in one calendar year;

(16) start precision2: follows the same logic as the start precision variable;

(17–20) variables that refer to episodes within conflicts (start, continuation, termination, end, and precision date);

(21) country code for side A;

(22) country code for side A2;

(23) country code for side B;

(24) country code for side B2;

(25) region of location.

This database records information on *armed conflicts*, and has a low threshold for accepting a conflict as an armed conflict: 25 casualties sustained in the course of fighting. Armed conflict intensity is coded in two categories:

minor: between 25 and 999 battle-related deaths in a given year;

war: at least 1,000 battle-related deaths in a given year.[25]

The UCDP database also records the dynamic in every year of conflict existence. This record allows changes in some of the variables—level of intensity, third-party intervention, and so on—to be tracked for every year of the conflict. However, as a result of this approach some conflicts fall into both categories, minor conflict and war, according to their level of intensity. Thus, 110 of the 251 armed conflicts in the UCDP database reached the level of war, and 201 were coded as minor armed conflicts. Seventy-seven minor conflicts reached the threshold of war in certain years based on the number of casualties and fell into both categories, while 33 armed conflicts were indicated in only one category, war, and 124 minor armed conflicts never reached the threshold of war. The UCDP database is regularly updated, which allows testing hypotheses for the whole post–World War II period of 1946–2006.

The COSIMO database has compiled information on 692 conflicts, structured according to 25 variables for each conflict, for the period 1945–1999. Conflict variables are coded as follows:

(1) code number of conflict;

(2) basic conflict identification number;

(3) name of conflict (named after the most important parties);

(4) start of conflict;

(5) end of conflict;

(6) duration of conflict (in years);

(7) region of conflict;

(8) political system of the initiating party;

(9) political system of the affected party;

(10) disputed issues in conflicts;

(11) intensity of conflict (four levels);

(12) economic and political type of state-participants;

(13) location in Cold War system;

(14) behavior of neighboring states;

(15) behavior of great powers;

(16) direct participants;

(17) external participants;

(18) mediator;

(19) instruments of the conflict initiator (e.g., diplomatic, military, political, economic, information policy, secret agencies and services);

(20) minimum estimates of casualties;

(21) maximum estimates of casualties;

(22) sparking events;

(23) resolution/forms of political outcomes;

(24) modalities of conflict resolution (territorial, military, political);

(25) treaties.

Each category is further subdivided, with numbers and letters assigned to specific items. To take an example used later in the chapter, under "Instruments of the conflict initiator," "10" is the category number for internal instruments. Under 10 are several further specific indicators with a plus or a minus sign attached. The first few are as follows:

10A+ Ready to talk with opposition

10A– Demonstrations or strike for or against government

10B+ Relief of curfew, lifting of a state of emergency

10B– Acts of terror, mass demonstrations, etc.

In addition, the following characteristics were identified without a code number assigned: instrument of the affected party, initiator, aggressor, and number of participants.[26] The appendix to this book provides the sample from the COSIMO database that was used for the research reported here, including the indicators.

One of the tasks of the COSIMO project is to determine the ratio of violent to nonviolent conflicts, as well as factors that influence conflict escalation. The COSIMO database distinguishes the intensity of conflict as follows: latent conflict, crisis, severe crisis, and war. As well, the concepts of basic conflict and follow-up conflict are introduced.[27] This offers a way to record the beginning of a conflict and the resumption of a conflict after a break in the fighting. However, this approach inevitably increases the number of conflicts included in the database. The COSIMO database partitions protracted conflicts in which phases of military struggle alternated with absence of interaction, a methodological procedure that is reflected in the numbers assigned to resumed conflicts. Thus, eight follow-up conflicts were indicated in Lebanon, five in Afghanistan, and six in the war between Iraq and Kuwait. A comparison of the two databases shows that the COSIMO database lists four episodes related to Cyprus, whereas the UCDP database lists two. However, the UCDP database includes some conflicts that are not taken into account in the COSIMO database, such as anti-Soviet movements in the Baltic republics and in Ukraine in the 1940s.

The COSIMO database identifies four levels of conflict intensity: (1) latent conflict, completely nonviolent; (2) crisis, mostly nonviolent; (3) severe crisis, involving the sporadic, irregular use of force: a "war-in-sight" crisis; and (4) war, the systematic, collective use of force by regular troops.[28] On this scale, armed conflicts are conflicts of the third and fourth intensity levels, and according to the data presented, 276 out of 692 conflicts were characterized by conflict of the third intensity level and 113 by the fourth intensity level; that is, they reached the threshold of war. However, using the intensity scale suggested by the COSIMO project introduces the possibility of error. In checking the quantitative parameter of third-level conflicts, I found that 90 cases had no casualties at all, and 13 cases had fewer than 25 casualties. This means that 103 of the 276 third intensity level conflicts do not meet the quantitative criterion of armed conflict. When I checked the quantitative determinant for fourth-level conflicts (war), 26 cases had zero casualties and one case had fewer than 1,000 casualties (200–300). Unfortunately, the auxiliary documents of the COSIMO database and publications related to this database do not explain this discrepancy. If we take into account only the cases meeting both qualitative and quantitative criteria for armed conflict, then the total is 259. This figure was reached as follows: the 103 conflicts with fewer than 25 casualties were subtracted from the overall 276 conflicts assigned to the third level of intensity, leaving 173. Then, the 27 conflicts with fewer than 25 casualties were subtracted from the overall 113 conflicts assigned to the fourth level of intensity, leaving 86. After this sieving process, summing yields the number of armed conflicts: 173 + 86 = 259. This number represents one list on which I base subsequent calculations in this chapter.

To avoid calculation errors, I retained the common quantitative parameter—at least 25 casualties in the course of armed struggle—as a starting point for determining armed conflict, based on the data in both projects and taking into account that the qualitative indicators do not differ. According to this parameter, the COSIMO database overall, across all four intensity levels, has 304 cases that meet the basic quantitative criterion of armed conflict, and in 192 cases the casualties exceeded 1,000. The 304 cases form a second COSIMO list, on which I perform calculations in this chapter.

Therefore, despite certain discrepancies in methodology, the numbers of armed conflicts included in the two databases are close and comparable: 251 for UCDP and 259 or 304 for COSIMO. The characteristics (variables) of armed conflicts in the two databases also partly concur and partly supplement each other.

Therefore, in identifying the asymmetry factor, I will use as the determinant of armed conflict the low threshold of 25 casualties a year, including civilian deaths, or 25 casualties in the course of the whole conflict if the conflict lasted less than a year. This condition allows us to consider the low-intensity armed conflicts that became widespread after World War II. With this basic screening of the database figures to establish comparable study sets, we can proceed to test some of the assumptions mentioned earlier concerning asymmetry in postwar conflicts.

IDENTIFYING ASYMMETRY IN THE STRUCTURAL AND DYNAMIC CHARACTERISTICS OF ARMED CONFLICTS

1. Determining the share of armed conflicts that involved parties of markedly different status

The first step in determining whether the asymmetry factor was a crucial feature of post–World War II armed conflicts is to construct the study sample. Based on the UCDP database, of the 251 armed conflicts (conflicts that met the quantitative screening criterion of at least 25 battle-related deaths, and in which at least one party was the government of a state), only 30 (12%) could be labeled symmetric; that is, they were conflicts between sovereign states, even if the forces were not fully commensurate (as in the cases of Argentina–United Kingdom, United States–Panama, and United States–Grenada). The other 221 (88%) armed conflicts were anticolonial and civil wars involving external actors. Twenty-one of these conflicts (8.4%) are categorized in the UCDP database as extrasystemic, as the struggle occurred between a state and a nonstate participant outside the state's territory. In the UCDP database, these engagements are characterized as colonial and imperial wars. Of the 221 asymmetric armed conflicts, 159 (63.34%) are labeled intrastate, as one party was a state and another was a nongovernmental group or movement, and 41 (16.33%) are labeled intrastate internationalized conflicts owing to the participation of external forces that supported one or both conflict parties (table 2.1).[29]

Table 2.1. Share of Asymmetric Conflicts in the UCDP Database, 1946–2006

Type of conflict	Number	Share (%)
Intrastate	159	63.34
Extrasystemic	21	8.4
Intrastate internationalized	41	16.33
Total asymmetric conflicts	221	88.07
Total armed conflicts	251	100

Source: Calculations based on UCDP data for 1946–2006 (http://www.pcr.uu.se/publications/UCDP_pub/Conflict_List_1946-2006.pdf).

Similar calculations can be done using the COSIMO database to determine the number of armed conflicts in which one of the parties was a nonstate actor, whether a national liberation movement, a political opposition party, or another kind of armed group striving to achieve certain political objectives. However, because of the project's methodology, when this approach is used to identify status asymmetry, the same conflict may fall into different categories and be counted more than once, so that the sum of the shares exceeds 100 percent. To avoid such overlap, instances of multiple categorizations were specifically checked.

Derivations of the share of asymmetric conflicts according to the COSIMO database show that nonstate actors were direct participants in 232 cases (76.3%) out of 304 armed conflicts resulting in at least 25 casualties. In 7 of these cases (2.3%) nonstate actors rendered various forms of support to one of the parties to the conflict. Thus, it is impossible to sum up the data. In 187 cases (61.3%) nonstate actors initiated the conflict, and in 15 cases (~5%) they were the first to use force to achieve their objectives—in other words, they were the aggressors. The number of anticolonial and national liberation armed conflicts is 60 out of 304, or 19.7 percent (table 2.2).

The same calculations can again be performed for the 259 armed conflicts in the COSIMO database that reached the third or fourth intensity levels and met the quantitative criterion of at least 25 battle-related deaths. In 213 of these cases (82.24%) nonstate actors were direct conflict participants, and in 8 cases (3.1%) they participated indirectly. In 172 cases (66.41%) the opposition initiated the conflict and in 15 cases (5.79%) it was the first to use force. Fifty-five of these conflicts (21.24%) are regarded as anticolonial and national liberation conflicts.

Therefore, according to calculations based on the UCDP database, up to 88 percent of armed conflicts in 1946–2006 could be labeled asymmetric because of the participants' status. According to calculations based on the COSIMO database, from 76.3 percent (of 304) to 82.24 percent (of 259) of armed conflicts were asymmetric because of the status of the parties. The shares of intrastate conflicts in both databases are also close: UCDP, 79.67 percent, and COSIMO, 76.3 percent. It is difficult to compare the proportion of anticolonial wars in both databases, for in the UCDP database these armed conflicts are singled out as "extrasystemic"

Table 2.2. Share of Asymmetric Conflicts in the COSIMO Database, 1945–1999

Type of conflict	Number of armed conflicts at all intensity levels with at least 25 casualties	Share (%)	Number of 3rd and 4th intensity level conflicts with at least 25 casualties	Share (%)
One of the conflict participants a nonstate actor	232	76.3	213	82.24
Indirect participation of a nonstate actor	7	2.3	8	3.1
Anticolonial and national liberation conflicts	60	19.7	55	21.24
Total	304	100	259	100

Note: "Total" reflects the number of conflicts that were considered in order to identify status asymmetry. However, this total is not equal to the sum of the conflicts enumerated in the table because in some cases the same conflict falls into multiple categories. For example, the conflict in Algeria, 1945–1946, was recorded in COSIMO data as an anticolonial and national liberation conflict, with direct and indirect participation of nonstate actors. Thus, the conflict counted for all three lines of a table. The purpose of this table to show how each asymmetric feature of a conflict is present on total in data in comparison.

Source: Calculations based on COSIMO data for 1945–1999 (http://www.hiik.de/en/kosimo/data/codemanual_kosimo1b.pdf).

conflicts (8.4%), whereas in the COSIMO database a similar category unites "anticolonial" and "national liberation" as the subjects of conflict (21.24%). Nevertheless, even though certain minor discrepancies exist between the databases, the empirically observed predominance of armed conflicts between state and nonstate actors is fully confirmed statistically.

2. Determining the share of armed conflicts in which the adversaries' power and resources were significantly incommensurate

To identify asymmetry in power and resources, we can proceed from the calculations of status asymmetry made in the previous section, for it is logical to assume that a state has a greater capacity to mobilize material and nonmaterial resources than a nonstate actor. Hence, conflicts characterized by asymmetry in status are also likely to exhibit asymmetry in the present power and resources of the belligerent parties. In addition to the sample of conflicts between symmetric actors in status (conflict between states), we should subsample within this category to exclude from further analysis such cases in which there was obvious inequality of power between parties equal in terms of status. Six instances of warfare can be categorized as such: China–Myanmar, 1969; Cyprus–Turkey, 1974; Kuwait–Vietnam, 1978–1988; Argentina–United Kingdom, 1982; Grenada–United States, 1983; and Iraq–Kuwait and the international military coalition, 1991. These conflicts were not counted as asymmetric because they represent more conventional military disputes between states, with similar statuses of and strategies used by the participants.

The logic that symmetry in power and resources generally aligns with symmetry in status could be wrong in specific cases, for a number of reasons. For example, in the cases of armed struggle in Afghanistan, Colombia, and Lebanon, nongovernmental paramilitary forces had resources for fighting that were at least compatible with those of the central authorities. Moreover, great powers may augment the resources of a nonstate participant by providing military, technical, and economic assistance. For this reason, when specific historical contexts are analyzed (qualitative research), it also makes sense to examine cases in which developed countries were indirectly involved, for the great powers have the capability of mitigating resource inequality.

Assessing the resources of belligerent parties is a complex task, as it requires considering both material and intangible resources, or countable and uncountable indicators. Leading international relations experts often have had recourse to a composite, somewhat vague variable labeled national strength. The American giant of international relations Hans Morgenthau has singled out several indicators that could be used for assessing national strength: "geography; industrial capacity and military preparedness; natural resources and population, national character, national morale, the quality of diplomacy and government."[30]

French political scientist Raymond Aron took a different tack in the 1960s, drawing a distinction between the concepts of strength and power. Strength for him was a potential, a "complex of material, human and moral resources." The power concept embodied the real strength: the ability to activate these forces under specific circumstances to achieve specific goals. Power with respect to states denoted the application of available resources to the conduct of foreign policies in times of war and peace.[31] At the same time, Aron recognized the uncertainty surrounding any assessment of power. In recent years, researchers have begun to actively use the concepts of hard and soft power; this gives additional nuances to the interpretation of strength in international relations and the ways in which it might be evaluated.[32]

The Correlates of War project, led for a long time by J. David Singer, has collected information on the material potential of many countries, beginning in the nineteenth century. For quantifiable indicators, Singer and his colleagues use the six variables listed earlier: military size and expenditure, total and urban population, energy consumption, and iron and steel production. The project participants believe that these indicators make it possible to correctly compare the material capabilities that constitute the national power of different countries. The project's website also notes that the notion of national power is linked secondarily to "the question of effective political institutions, citizen competence, regime legitimacy, and the professional competence of the national security elites. While these are far from negligible, they contribute to national power and the efficiency with which the basic material capabilities are utilized, but they are not a component of such capabilities."[33] These indicators are difficult to represent in quantifiable form, but they can be taken into account in specific cases, which is especially important when analyzing asymmetric conflicts.

Both the UCDP and the COSIMO databases make reference to the political system of conflict participants, as it is seen as a source of national power. However, this indicator is poorly applicable for assessing the power of parties in intrastate conflicts, at least for the research presented here, because of its methodological complexity, internal contradictions, and controversial manifestations. If the most important indications of nonmaterial strength are the will to win and the ability to mobilize the populace to fight and continue the struggle, how can we unambiguously determine which political system offers the best way of doing that? Which system is preferable: democracy, tyranny, autocracy, transformation of the political system, dissolution of the state, or nationalism? This indicator clearly requires qualitative research methods and cannot be used convincingly for quantitative research. However, when a comparative analysis of individual cases of intrastate or other armed conflicts is performed using qualitative methods, the parties' mobilization resources should be taken into account, for these may qualify or influence other measures of power and resources.

We are left with the following indicators of participants' power that can be measured (making it possible to use quantitative methods for comparative analyses): human resources, the size of the armed forces, the quantity and quality of combat equipment and the means to conduct war, and the funds available to spend on war. Let us consider the possibility of using some of these indicators to evaluate whether asymmetry in any of them could have an impact on the course of a conflict.

Human resources. This indicator might be used to determine the size of the population of individual countries, but it seems irrelevant for evaluating the strength of a nonstate participant in armed conflict, first because there is usually no statistically verifiable information about the size of the opposition, and second because overlap or miscounting of countable human resources is possible if these resources are counted in favor of the state (based on the formal criterion of living in a given state's territory) even though the populace may in fact stand against the state. Hence, this indicator is correctly used only to compare forces in an interstate conflict, as was done in T. V. Paul's research,[34] or when conducting an in-depth investigation of individual cases that includes a detailed assessment of the human resources of belligerent parties and possible external assistance.

Size of the armed forces and other armed groups. A comparison of the number of war participants, including both official troop size and the size of other armed groups, in intrastate conflicts also looks questionable from the standpoint of accuracy. It is well known that opposition groups tend to exaggerate the number of their supporters while governments tend to do the opposite. It is also difficult to identify the size of armed forces taking a direct part in warfare in intrastate conflicts. Another factor complicating such an evaluation is the possible contribution of external assistance, which should also be quantified but may be unknown.

Armaments and funding. Military science has always used the quantitative and qualitative characteristics of armaments to compare the strength of opponents. However, a precise measurement and matching of arms is generally possible only for conventional arms. In the case of state conflict participants, this parameter is verifiable because of the existing armament accounting system; moreover, most countries provide official statistics on the size of their armed forces. In the case of nonstate actors, this parameter is poorly verifiable, for official statistics are lacking. Researchers can get only a general idea of the availability and use of different weapon types and about arms supplies by a third party. The same can be said about the amount of military expenditure: for the state actor, these data are more or less verifiable, while for the nonstate actor, such information seems to be almost irrelevant.

In sum, assessing the strength of the sides in intrastate conflicts seems an exercise in approximation, though this does not refute the basic assumption about the power and resource dominance of the state actor. Hence, a comparison of the material and nonmaterial resources of adversaries in an intrastate conflict or in a conflict between a state and a nonstate actor relies to a large extent on the use of qualitative research methods and diligent evaluation of information from different sources as to the quantity and quality of material and human resources.

3. Determining the share of armed conflicts in which asymmetric strategies and tactics were used

At present, most analysts refer to two types of warfare as asymmetric: guerrilla warfare and terrorism. Several principles of armed struggle that are typical of guerrilla wars and political terrorism run counter to conventional warfare. These include (1) avoiding direct clashes with a superior adversary; (2) striking against vulnerable, often nonmilitary targets (infrastructure and, for terrorists, civilians); (3) dragging out the fighting in order to avoid military defeat and weaken the will of the stronger adversary to continue the struggle; and (4) waging war using irregular militants. Such an understanding of asymmetric strategies has dominated US and Israeli military doctrinal documents and political research in recent years. A significant literature on the problem of asymmetric armed conflicts considers such unconventional strategies to be decisive for the outcome; that is, asymmetric conflict is exposed as conflict characterized by the implementation of different forms of interaction between participants that are defined by an initial incommensurability in status and power. As discussed in chapter 1, this idea has been explored in monographs by Gil Merom, Ivan Arreguín-Toft, and Ekaterina Stepanova.[35]

Using the databases at our disposal, we can attempt to identify the share of armed conflicts in which terrorist attacks or guerrilla strategies were used. The UCDP database does not include information that would allow us to identify conflicts of this sort. Information in the COSIMO database does allow the extent of asymmetric strategies to be determined. The variable that takes into account different "instruments" of the conflict initiator and aggressor includes "terror attacks" (indicator 5K–) among military instruments. Among "internal instruments" in the COSIMO database, the following can be considered asymmetric strategies: "act of terror, mass demonstrations" (10B–), "putsch, revolt, private wars" (10K–), and "organized resistance, rebellion" (10N–). It is difficult to identify methodological differences between indicator 5K–, "terror attacks," and indicator 10B–, "acts of terror, mass demonstrations." The only difference recorded in the auxiliary documents of the COSIMO database is the description of the 5K– indicator as "military instruments" and the 10B– indicator as "internal instruments." Moreover, this variable includes the broad notion of "mass demonstrations." This discrepancy creates questions rather than providing answers. However, without other comparably detailed databases containing the necessary information on armed conflict characteristics, we can only say that this database has such a discrepancy.

According to the COSIMO database, terrorist actions were used as a military instrument ("terror attacks," parameter 5K–) in nine (3.47%) out of 259 armed conflicts. In eight of these cases this strategy was used by the conflict initiator and in one case by both parties, the Indonesia–East Timor conflict of 1975–1976.

Indicator 10B–, "internal instruments," is recorded in 73 cases (28.19%); in 68 of them this strategy was used by the conflict initiator and in six cases by both parties: Palestine, 1946–1948;[36] Gabon–Congo, 1962; the Islamic revolution in Iran, 1979–1981; Sri Lanka, 1983–1987 and again in 1987–1995; and Lebanon, 1988–1990.

Indicator 10K–, "putsch, revolt, private wars," is recorded in 32 cases (12.36%); in 3 of them, both parties to the conflict used this strategy. All three cases are attributed to Lebanon: in 1975, 1975–1976, and 1988–1990.

The use of "organized resistance, rebellion" (indicator 10N–) is recorded in 84 cases (32.43%). In three of those cases both parties resorted to this strategy: the civil wars in Zimbabwe, 1983; Laos, 1963–1975; and Somalia, 1991–1999.

Calculation of these indicators for the 304 conflicts in which at least 25 casualties were recorded does not significantly differ from the previous calculations. Indicator 5K– is recorded in 10 cases (3.29%), in one of which both parties used this strategy (Indonesia–East Timor, 1975–1976). Indicator 10B– ("acts of terror, mass demonstrations") is recorded in 80 cases (26.32%), in 74 of which this strategy was used by the conflict initiator and in 6 of which it was used by both parties. Indicator 10K– ("putsch, revolt, private wars") was recorded as a combat strategy in 33 cases (10.86%), in 25 of which this strategy was used by the conflict initiator and in 3 of which it was used by both parties (Lebanon, 1975, 1975–1976, 1988–1990). Indicator 10N– ("organized resistance, rebellion") is recorded in 90 cases (29.61%), in 73 of which the initiator resorted to this strategy and in 3 of which it was used by both parties (Zimbabwe, 1983; Laos, 1963–1975; and Somalia, 1991–1999). According to these calculations, the most widespread asymmetric strategies in both COSIMO lists (304 armed conflicts and 259 armed conflicts) are "acts of terror, mass demonstrations" (10B–), respectively 26.32 percent and 28.19 percent, and "organized resistance, rebellion" (10N–), respectively 29.61 percent and 32.4 percent. Calculations for both lists show a more frequent use of asymmetric strategies by the conflict initiator (table 2.3).

The co-occurrence of indicators was also checked, that is, cases in which multiple asymmetric strategies were used in the course of a single conflict (table 2.4).

The calculations were based on the two lists of armed conflicts (sample sizes 304 and 259), and two co-occurrences of indicators 5K– and 10N– were noted. Both instances were intrastate conflicts in Indonesia, in 1975–1976 and again in 1990–1999. For conflicts across all four intensity levels

Table 2.3. Use of Asymmetric Strategies in Armed Conflicts, COSIMO Database, 1945–1999

Indicators	Number (%) of conflicts at all intensity levels with at least 25 fatalities in which this strategy was used (N =304)	Number (%) of conflicts of 3rd and 4th intensity levels in which this strategy was used (N = 259)
5K–: "terror attacks"	10 (3.29)	9 (3.74)
10B–: "acts of terror, mass demonstrations"	80 (26.32)	73 (28.19)
10K–: "putsch, revolt, private wars"	33 (10.86)	32 (12.74)
10N–: "organized resistance, rebellion"	90 (29.61)	84 (32.43)

Source: COSIMO data for 1945–1999 (see table 2.2).

in the COSIMO database with casualties of at least 25 people (304 cases), indicators 10B– and 10N– co-occurred in 37 cases, indicators 10B– and 10K– in 7 cases, indicators 10K– and 10N– in 8 cases, and indicators 10B–, 10K–, and 10N– in 2 cases. For armed conflicts at the third and fourth intensity levels (259 cases), indicators 10B– and 10N– co-occurred in 34 cases, indicators 10B– and 10K– in 7 cases, indicators 10K– and 10N– in 5 cases, and indicators 10B–, 10K–, and 10N– in 1 case. The calculations are presented in table 2.4.

Indicators 10B– and 10N– co-occurred most frequently in both lists, in 37 of 304 conflicts and in 34 of 259, which can be explained by these two strategies being the most widespread of all violent strategies. This statistic also confirms the common belief that various opposition movements rely heavily on terrorist strategies during uprisings and revolutions. According to these calculations, the share of co-occurring appearances in the list of 304 armed conflicts is 37 of 80 (46%) for indicator 10B– and 37 of 90 (41.11%) for indicator 10N–. For the list of 259 conflicts the share of co-occurring strategies is 34 of 73 (46.48%) for indicator 10B– and 34 of 84 (40.48%) for indicator 10N–.

Table 2.4. Use of Multiple Asymmetric Strategies in Armed Conflicts, COSIMO Database, 1945–1999

Co-occurring indicators	Number of asymmetric conflicts at all intensity levels with at least 25 casualties in which this combat strategy was used (N = 304)	Number of conflicts of 3rd and 4th intensity levels with at least 25 casualties in which this combat strategy was used (N = 259)
5K–, 10N–: "terror attacks"; "organized resistance, rebellion"	2	2
10B–, 10K–: "acts of terror, mass demonstrations"; "putsch, revolt, private wars"	7	7
10B–, 10N–: "acts of terror, mass demonstrations"; "organized resistance, rebellion"	37	34
10K–, 10N–: "putsch, revolt, private wars"; "organized resistance, rebellion"	8	5
10B–, 10K–, 10N–: "acts of terror, mass demonstrations"; "putsch, revolt, private wars"; "organized resistance, rebellion"	2	1

Source: COSIMO data for 1945–1999 (see table 2.2).

The total number of conflicts in which asymmetric strategies were used, taking into account the co-occurrence of some indicators, is 149 (57.53%) for the set of third and fourth intensity level conflicts in the COSIMO database and 189 for the set of conflicts with fewer than 25 casualties (62.17%). Minor conflicts, with fewer than 25 causalities, present a situation of sporadic violence that often takes the form of terrorist acts. Such conflicts are often understudied because of the small level of violence; however, they have the potential to escalate to higher intensities of violence.

The COSIMO data thus allow the conclusion that asymmetric strategies were a widespread phenomenon in postwar conflicts. However, from the quantitative data alone it is difficult to assess to what extent these strategies influenced the outcome. It is also important to keep in mind that in many cases, armed opposition groups acquired the necessary attributes of regular troops, aspired to legalize their status, and abandoned openly unlawful means of struggle such as terrorist attacks. It seems that the influence of asymmetric strategies on the dynamic and outcome of struggle between unequal antagonists should be assessed by studying individual cases, as it is difficult to use quantitative methods to measure the impact of this variable with sufficient precision and accuracy.

4. Determining the share of armed conflicts in which developed countries were defeated by their weaker adversaries

The COSIMO database allows us to determine the share of defeats and victories in armed conflicts, where "outcome" is divided into three categories:

1. *Territorial outcome* (T1: separation of territory; T2: territorial loss; T3: annexation, unification, incorporation of territory; T4: denouncement of territorial claims; T5: status quo, initiator upholds territorial claims).

2. *Military outcome* (M1: stalemate, ceasefire, indecisive outcome; M2: victory of initiator; M3: defeat of initiator; M4: continuation of fighting; M5: withdrawal of troops).

3. *Political outcome* (P1: no agreement reached, status quo ante; P2: some issues still in dispute; P3: partial success; P4: conclusion of a consensual agreement; P5: change of regime; P6: emergence of two different independent regimes; P7: fall of regime; P8: government position weakened; P9: government position strengthened; P10: opposition movement strengthened; P11: suppression of opposition; P12: admission or inclusion of opposition into the government; P14: denouncement of claims; P15: increased influence of external power; P16: decreased influence of an external power; P17: compromise).[37]

Some of the indicators listed above can be seen as signs of victory for the weaker party. These are T1 and T2 among the territorial indicators; M2, M3, and M5 among the military indicators; and P4, P5, P6, P7, P8, P10, P12, and P17 among the political indicators. Other indicators cannot be unambiguously categorized as signaling the victory of one or the other of the parties to the conflict.

I turn now to calculating the share of armed conflicts in which a weaker party was victorious, as manifested in increased authority or elevated status. These conflicts can be grouped as follows: (1) intrastate conflicts, regardless of whether they involved external forces; (2) intrastate conflicts and anticolonial wars involving developed countries, primarily the great powers; and (3) intrastate conflicts in the territory of the great powers (e.g., Northern Ireland–United Kingdom or the Chechen Republic–Russia).[38] In general, I will consider asymmetric those conflicts in which one of the parties was a nonstate actor and another was a state (thus exhibiting asymmetry in status and resources). This classification amounts to identifying those conflicts in which the developed countries were direct participants and those in which the developed countries fought the opposition movements in their own territories.

The calculations focus on two separate lists: one of asymmetric conflicts and one of asymmetric conflicts involving the great powers. Of the armed conflicts in the COSIMO database, 210 can be considered asymmetric in terms of the status of the participants, and the great powers participated directly or indirectly in 116 of them. The results of the struggle in these conflicts are calculated for each list separately. Conflicts in which territorial indicators were noted in the database can be summed up. However, the database often records several political indicators for the same conflict, so that it is impossible to sum up the manifestations of these indicators. In total, 39 situations can be identified in which the initiator won, and in 23 of the 39 situations the weaker party was the initiator. It is noteworthy that the Vietnam War was recorded as a military victory for the United States in the COSIMO database, even though the outcome of this war is usually seen as a defeat for the United States (it is often forgotten that it was a political rather than a military defeat).

Twenty-three asymmetric conflicts involving the great powers were recorded, and in 15 cases the weaker party was the initiator and won. Signs of military victory of the weaker party in the category "victory of initiator" were recorded in approximately 11 percent and 13 percent of asymmetric conflicts, respectively. It is important to note that even though nonstate actors and opposition groups were the most frequent initiators of armed conflict in the post–World War II period, they seldom achieved a military victory. A military victory for the weaker party occurred more frequently in armed conflicts involving the great powers, which is indirect confirmation of hypotheses about existing domestic and international restrictions that forced strong actors to terminate a conflict without achieving a military victory. However, this observation should be tested using qualitative research methods. The most significant manifestations of the weaker party's political victory were associated with indicators "change of regime" (11.43% and 10.34%) and "compromise" (16.66% and 9.48%). The indicators "opposition movement strengthened" (7.14% and 5.17%) and "admission or inclusion of opposition into the government" (3.80% and 3.44%) were recorded in fewer cases than "fall of regime" (6.19% and 6.89%), a negative sign of a weaker party's victory. The results are presented in table 2.5.

The following conclusions can be drawn, based on the use of UCDP and COSIMO data for identifying the asymmetry factor in armed conflicts in the post–World War II period:

1. There is a clear predominance of asymmetric conflicts based on status asymmetry. The calculations indicate that up to 88 percent of all armed conflicts in the UCDP database are characterized by a pronounced asymmetry in the participants' status, as are about 82 percent of those in the COSIMO database.

2. In the case of intrastate conflicts, the power and resources of the conflict participants can be compared on the basis of the formal assumption that asymmetry in resources and power is a corollary to status asymmetry. For anticolonial wars and intrastate internationalized conflicts the power ratio can be established only with a high degree of conditionality, which supports using qualitative research methods for individual cases.

3. Asymmetric strategies are a widespread phenomenon and were used in 57.53 percent to 62.17 percent of conflicts. However, a formal analysis

alone cannot establish whether asymmetric strategies had an impact on the outcome of the struggle or were an inevitable consequence of the initial asymmetry. Adding to the problem, some indicators in the COSIMO database are not properly defined, as they merge different strategies (such as indicator 10B–). The influence of asymmetric strategies should be identified based on a detailed study of those specific historic contexts in which a weaker party was victorious.

4. The outcomes of armed conflicts indicate a low number of victories for parties weaker in military and political terms. The military victory of a weaker party is seen in about 11 percent of all armed conflicts and in about 13 percent of conflicts in which the great powers became involved. However, military victories were not significantly associated with political indicia of victory (compromise and regime change). As noted earlier, weaker parties *did not* achieve a military victory more often if the great powers became involved in the conflict. This circumstance requires a more thorough study of manifestations of the asymmetry factor in armed conflicts involving the great powers.

5. Nonstate actors initiated the largest number of conflicts and were direct participants in more armed conflicts than were the great powers or other countries. This parameter is another indisputable piece of evidence that status asymmetry dominated the outcome of post–World War II armed conflicts.

In conclusion, analyzing conflict databases has helped identify a pronounced status asymmetry, typical for post–World War II armed conflicts, and the widespread use of asymmetric struggle strategies. However, attempts to identify the significance of asymmetry in other conflict characteristics underscore the need to use qualitative, historical methods of analysis in individual cases.

In the next sections we will try to determine the degree to which asymmetry was operative in conflicts involving the great powers.

Table 2.5. Results of Struggle in Asymmetric Armed Conflicts, COSIMO database, 1945–1999

Indicator	Number out of total number of asymmetric conflicts (N = 210)	Share of all asymmetric conflicts (%)	Number out of total number of asymmetric conflicts with great power involvement (N = 116)	Share of asymmetric conflicts involving great powers (%)
Territorial outcome	11	5.24	7	6.03
T1: Separation of territory	7	3,33	4	3.45
T2: Territorial loss	4	1.9	3	2.59
Military outcome				
M2: Victory of initiator	39/23[a]	18.57/10.95	23/15	19.82/12.93
M3: Defeat of initiator	47/6[b]	22.38/2.86	40/7[c]	34.48/6.03
M5: Withdrawal of troops	4	1.9	4	3.44
Political outcome				
P5: Change of regime	24	11.43	12	10.34
P6: Emergence of two different independent regimes	5	2.38	3	2.58
P7: Fall of regime	13	6.19	8	6.89

Indicator	Number out of total number of asymmetric conflicts (N = 210)	Share of all asymmetric conflicts (%)	Number out of total number of asymmetric conflicts with great power involvement (N = 116)	Share of asymmetric conflicts involving great powers (%)
P8: Government position weakened	8	3.8	2	1.72
P10: Opposition movement strengthened	15	7.14	6	5.17
P12: Admission or inclusion of opposition into the government	8	3.8	4	3.44
P17: Compromise	35	16.66	11	9.48

Notes: a. These figures show the total number of victories recorded for the initiator (the first number) and the number of conflicts in which the weaker party or a non-governmental movement won. The Hungarian Uprising of 1956 was recorded as a "victory of initiator," where the opposition was the initiator.

b. A defeat of the initiator as the stronger party is recorded in the following cases: the Cuban Revolution of 1961; the riots in Brunei in 1962; the French defeat in Indochina, 1945–1954; the Arab–Israeli War of 1948–1949; the war in Indochina, 1977–1978 (DRV–Khmer); and the Soviet war in Afghanistan, 1979–1988.

c. Here the first number denotes the number of all cases in which "defeat of initiator" was recorded, and the second number denotes cases in which the state was initiator, and thus formally the stronger party lost. It is of interest that the war in Afghanistan, 1978–1988, was recorded as a "defeat of initiator" for the Soviet Union.

Source: COSIMO data for 1945–1999 (see table 2.2).

IMPACT OF THE ASYMMETRY FACTOR ON CONFLICTS INVOLVING THE GREAT POWERS

Using the COSIMO database, I attempted to evaluate the involvement of the great powers, as well as other countries and nonstate actors, in armed conflicts of the post–World War II period. As described earlier in the chapter, the list of 304 armed conflicts with at least 25 casualties, culled from all four intensity levels in the COSIMO data, was used for calculations. The results are presented in table 2.6.

The United Kingdom is the absolute leader in terms of direct participation in post–World War II armed conflicts, a position that can be explained by the rise of national liberation movements in its vast colonial and mandate territories after World War II. The two other clear leaders in terms of direct participation in armed conflicts are France and the USSR/Russia, with 17 instances each, followed by China (16 instances) and the United States (11 instances).

However, the data on indirect participation in conflicts offer a different picture, confirming the widely accepted view that the superpowers actively battled for influence in the postwar world through participating indirectly in armed conflicts in third world countries: the United States in 78 instances and the USSR/Russia in 55. The other great powers were less involved in such actions, though the indirect participation of France and China exceeded the degree of each country's direct participation in armed conflicts: France had 20 instances of indirect involvement versus 17 instances of direct participation, while for China the respective figures are 19 and 16. The United Kingdom shows a pronounced domination of direct participation over indirect participation, 27 instances versus 16.

The United Kingdom (14), China (11), and the USSR/Russia (8) led in initiating conflicts. The United States was the most active participant in conflict resolution (33 cases), followed by the United Kingdom (10), France (8), and Russia (7).

As for the entities not among the great powers, the most active direct participants in armed conflicts were Syria, Iraq, and India, with 17 instances each; Israel and Indonesia, with 11 instances each; Iran and Pakistan, with 9 instances each; South Africa, with 8; the Democratic Republic of Vietnam (North Vietnam), with 7; Turkey, with 6; and Algeria, with 5. With the notable exception of Iran (and less so of Turkey and Algeria), the indirect participation of these countries does not exceed their level of direct participation. For Iran, 22 instances of indirect participation were recorded versus 9 instances of direct involvement. Turkey and Algeria are also exceptions, but to a lesser extent: for Turkey, 9 instances of indirect participation were recorded versus 6 instances of direct participation, while for Algeria the respective numbers are 6 and 5.

According to the COSIMO database, nonstate actors were direct participants in 232 (76.32%) of 304 armed conflicts and initiated 187 (61.51%) such conflicts. Though the "nonstate actor" category is an aggregative one, this statistic highlights an obvious feature of armed conflicts in the post–World War II period, in which nonstate participants actively engage in armed conflict.

The next question of interest is to what extent the asymmetry factor was manifested in armed conflicts in which the great powers were involved, either directly or indirectly. Again referring to the list of 304 asymmetric armed conflicts culled from the COSIMO database, I have identified 116 instances in which the great powers were involved. Let us determine the following:

1. The share of conflicts in which asymmetric strategies were used: (a) in total, (b) in the case of direct involvement of a great power, and (c) in case of indirect participation of a great power.

2. The results of armed conflicts involving the great powers: (a) in total, (b) in the case of direct involvement of a great power, and (c) in case of indirect participation of a great power.

3. Distribution trends of the great powers' direct and indirect involvement in armed conflicts in the post–World War II period. I will try to identify any such trend by examining several hypotheses:
   - The degree of direct great power involvement in armed conflicts depends on possessing colonies.
   - Competition among the superpowers and the ideological struggle between the West and the East were the reasons for great power participation and their support of belligerent parties.

Table 2.6. Participation of Different Countries and Nonstate Actors in Armed Conflicts, COSIMO Database, 1945–1999

Participant in armed conflicts	Direct participant	Indirect participant	Initiator	Aggressor	Intermediary in resolution
Great powers					
United Kingdom	27	16	14	1	10
USSR/ Russia	17	55	8	3	7
France	17	20	1	3	8
China	16	19	11	1	0
USA	11	78	4	3	33
Other states and nonstate actors					
Syria	17	11	7	1	3
India	17	5	4	1	4
Iraq	17	7	11	2	1
Israel	11	5	3	2	0
Indonesia	11	1	4	2	0
Iran	9	22	1	0	4
Pakistan	9	6	4	1	3
South Africa	8	5	4	2	1
North Vietnam (DRV)	7	4	4	0	0
Turkey	6	9	1	2	1
Algeria	5	6	3	2	2
Italy	0	4	0	0	1
Angola	2	3	0	0	0
Belgium	1	3	0	0	3
Taiwan	2	1	0	0	0

Participant in armed conflicts	Direct participant	Indirect participant	Initiator	Aggressor	Intermediary in resolution
Brazil	3	0	1	1	1
The Netherlands	2	0	0	0	0
German Democratic Republic	2	0	0	1	0
Federal Republic of Germany	0	1	0	0	0
Japan	0	0	0	0	0
Nonstate actors	232	7	187	15	3

Note: Countries are listed in order of their direct participation in armed conflicts, from most (United Kingdom) to least (Japan). "Nonstate actors" include national liberation movements, groups in opposition to the central government, and similar participants.

Source: COSIMO data for 1945–1999 (see table 2.2).

- The degree of direct involvement by the great powers in asymmetric armed conflicts has constantly declined in the postwar period, while their indirect participation has increased.

4. The most common forms of great power behavior in armed conflicts.

1. The share of conflicts in which asymmetric strategies were used

Table 2.7 presents the share of armed conflicts in which asymmetric strategies were used and the great powers were involved.

Asymmetric strategies were used in a total of 74 (63.79%) of 116 conflicts involving the great powers (data not shown). Indicators 10B– and 10N– were both recorded for 13 of these conflicts. Indicator 5K– did not co-occur with any other. The aggregate use of asymmetric strategies for all asymmetric conflicts for the period 1945–1999 is 54.5%, less than that for conflicts involving the great powers (63.79%) (not shown). The greater use of asymmetric strategies in conflicts where the great powers participated indirectly can be explained by the large number of conflicts in which the great powers took an indirect part.

2. The results of armed conflicts involving the great powers

Table 2.8 presents the outcomes of armed conflicts in which the great powers were directly or indirectly involved. In some cases, the database recorded both direct and indirect involvement of great powers in the same conflict, so the total number of conflicts is not always equal to the sum of the conflicts with direct and indirect great power participation. Also, totals may not be provided in all cases because several military and political outcomes might have been recorded for the same conflict.

The data in table 2.8 confirm once again that great power participation in asymmetric armed conflicts was largely indirect. The indicators of military conflict outcome M1, M2, and M3 are the easiest to assess. A defeat for the opposition occurred almost twice as often as a military victory for the opposition (29 vs. 15 cases), and almost the same 2:1 ratio is observed for the direct and indirect participation of the great powers in armed conflicts; that is, they were almost twice as likely to be involved, directly or indirectly, in conflicts in which the opposition sustained a military defeat. This observation does not refute the core of the

Table 2.7. Use of Asymmetric Strategies in Conflicts Involving the Great Powers, COSIMO Database, 1945–1999

	Indicator			
Great power involvement	**Number (%) of records, 5K–: "terror attacks"**	**Number (%) of records, 10B–: "acts of terror, mass demonstrations"**	**Number (%) of records, 10K–: "putsch, revolt, private wars"**	**Number (%) of records, 10N–: "organized resistance, rebellion"**
Direct participation of great powers	1 (0.86)	8 (6.89)	3 (2.58)	11 (9.48)
Indirect participation of great powers	2 (1.72)	27 (23.27)	11[b] (9.48)	33[c] (28.44)
Total	**3 (2.58)**	**30[a] (25.86)**	**13 (11.20)**	**41 (35.34)**

Note: a. The total does not represent a sum, as this strategy was recorded with both the direct and the indirect involvement of the great powers.

b. In one case, Afghanistan II (Soviet intervention), 1979–1988, both direct and indirect involvement of the great powers was recorded.

c. In five cases, both direct and indirect involvement of the great powers was recorded: Morocco, 1944–1956; Indonesia, 1945–1949; Malaya-Indonesia (Sarawak/Sabah), 1963–1966; Indochina II (Vietnam War), 1964–1973; and Yemen PR, 1965–1967.

Source: COSIMO data for 1945–1999 (see table 2.2).

asymmetric conflict model, which sees the military victory of a weaker party as less likely than a political victory.

When the outcomes of all asymmetric conflicts and those involving the great powers are compared, the most obvious feature is the large number of political indicators of the weaker party's victory in conflicts in which the great powers were indirectly involved. Thus, with the exception of indicator P17, "compromise," which was recorded in almost one-third of cases, all other political indicators were recorded most frequently for conflicts in which the great powers participated indirectly.

Table 2.8. Outcomes of Asymmetric Armed Conflicts Involving the Great Powers, COSIMO Database, 1945–1999

Indicator	Number (%) of all asymmetric conflicts involving the great powers (N = 116)	Number (%) of asymmetric conflicts in which the great powers participated directly	Number (%) of asymmetric conflicts in which the great powers participated indirectly
Territorial			
T1: Separation of territory	4 (3.44)	1 (0.86)	3 (2.58)
T2: Territorial loss	3 (2.58)	1 (0.86)	2 (1.72)
Military			
M1: Stalemate, ceasefire, indecisive outcome	22 (18.96)	4 (3.44)	18 (11.51)
M2: Victory of initiator (weaker party)	15 (12.93)	4 (3.44)	13 (11.20)[b]
M3: Defeat of initiator (weaker party)[a]	29 (25.00)	10 (8.62)	22 (18.96)[c]
M5: Withdrawal of troops	4 (3.44)	1 (0.86)	3 (2.58)

Indicator	Number (%) of all asymmetric conflicts involving the great powers (N = 116)	Number (%) of asymmetric conflicts in which the great powers participated directly	Number (%) of asymmetric conflicts in which the great powers participated indirectly
Political			
P5: Change of regime	14 (12.06)	1 (0.86)	13(11.20)
P6: Emergence of two different independent regimes (partition)	3 (2.58)	–	3 (2.58%)
P7: Fall of regime	8 (6.89)	2 (1.72)	6 (5.17)
P8: Government position weakened	2 (1.72)	–	2 (1.72)
P10: Opposition movement strengthened	6 (5.17)	–	6 (5.17)
P12: Admission or inclusion of opposition into the government	4 (3.44)	–	4 (3.44)
P17: Compromise	13 (11.20)	4 (3.44)	10 (8.62)[d]

Notes: a. As noted in table 2.5, in more cases the initiator (in the form of a strong state) inflicted a military defeat on the opposition (29 of 36 cases).

b. Data on direct and indirect participation of great powers concurred in two cases: Indonesia, 1945–1949; and Yemen (Aden), 1965–1967.

c. In this case, the great powers took both direct and indirect part in armed conflict in three instances: Greece, 1946–1949; Burma, 1949–1961; and Malaya 1948–1960.

d. In one case, Malaya's war for independence of 1948–1960, the great powers were involved on both sides of a conflict and were both directly and indirectly involved.

Source: COSIMO data for 1945–1999 (see table 2.2).

3. Distribution trends of the great powers' direct and indirect involvement in armed conflicts in the post–World War II period

To determine any trends in the distribution of direct and indirect participation of the great powers in armed conflicts of the post–World War II period, I examined three hypotheses. The existing research literature suggests that the degree of direct great power involvement in armed conflicts depends on (1) possessing colonies (e.g., the United Kingdom, France), (2) possessing dependent territories, or (3) experiencing internal conflicts in the great power's own territory. Indirect participation, by contrast, stems from the ideological struggle between West and East in the course of the Cold War or is a manifestation of allied commitments. Table 2.9 presents the calculations.

The figures in table 2.9 fully confirm the association between the great powers' possession of colonies and dependent territories and the degree of their involvement in armed conflicts. All conflicts directly involving the United Kingdom and France took place in their colonies. The United States, the Soviet Union, and China took a direct part in an insignificant number of armed conflicts: the United States, 2; the USSR/Russia, 5; and China, 4.

For the Soviet Union, indirect participation in armed conflicts significantly exceeded direct participation (33 vs. 5 instances). Direct participation in conflicts was motivated by the desire to retain control in Central Europe (Poland and Hungary, 1956) and in Iran (1945). The war in Afghanistan looks like an exception rather than part of a pattern of Soviet behavior. The most recent case recorded in the database is the military operation in the Republic of Chechnya.

Data for the United States show a similar pattern, with very little direct involvement in armed conflicts and a much more pronounced level of indirect participation, respectively 2 versus 50 instances. The first cases recorded in the database are the conflict in Cambodia, 1956–1970, and the Vietnam War, 1964–1973.

For China, the gap between indirect and direct participation in armed conflicts is less pronounced than in the case of the superpowers: 15 versus 4 instances. China was directly involved in armed conflict in the civil war of 1944–1949, in Burma in 1949–1961, and in Tibet in 1950–1951 and again in 1954–1959.

Table 2.9. Participation of the Great Powers in Asymmetric Armed Conflicts, COSIMO database, 1945–1999

Parameter	Countries				
	UK	France	USA	USSR/ Russia	China
Total number of conflicts with direct participation	15	12	2[a]	4	4
Conflicts with dependent territories or internal conflicts in one's own territory (direct involvement)	15	12	–	1	1
Total number of conflicts with indirect involvement	12	15	50	39	15
Conflicts with one's own former dependent territories (indirect involvement)	7	8	5[b]	3	5[c]
Conflicts with dependent territories or internal conflicts in the territory of other countries (indirect involvement)	5	7	45	30	10

Notes: a. The database recorded two instances of direct US involvement in armed conflicts: Cambodia (1956–1970) and Indochina (1964–1973).

b. The United States was also involved in 11 conflicts in Latin American countries.

c. All five cases of China's indirect involvement in armed conflict were in Indochina.

Source: COSIMO data for 1945–1999 (see table 2.2).

An analysis of indirect great power involvement in armed conflicts demonstrates that the United Kingdom and France typically became enmeshed in conflicts in the territory of their former colonies. For the United Kingdom this was observed in 7 of 12 cases of indirect involvement, and for France in 8 of 15 cases. For the USSR/Russia and the United States, indirect involvement in conflicts for ideological reasons leaves no shred of doubt. China also supported revolutionary and national liberation movements, but its involvement was far less than that of the superpowers.

Let us now test the assumption that superpower competition and ideological confrontation between the West and the East were the main reasons for great power involvement in armed conflicts and for their support of belligerent parties. The data obtained are presented in table 2.10, where "X" indicates the participation of a country in an armed conflict.

Great powers belonging to different political-economic systems supported opposite-side adversaries in 27 (23.27%) of the total 116 conflicts involving the great powers. The Soviet Union took part in 21 such conflicts, the United States in 19, China in 13, the United Kingdom in 10, and France in 2. Nine armed conflicts saw the involvement of three great powers, either simultaneously or at different times during the conflict. The most intense clash of great power interests took place during the 1945–1954 war in Indochina, which involved four great powers. In total, 15 conflicts were recorded in which the Soviet Union or the United States supported the antagonists. In other cases, the pairing of antagonistic great powers was as follows: the United Kingdom and Soviet Union in 3 instances (Palestine, 1946–1948; Malaysia [Indonesia], 1963–1966; Yemen, 1965–1967), China and the United States in 3 (Burma, 1948–1961; Burma, 1948–1999; Thailand, 1965–1968), and China and the United Kingdom in 2 instances (Malaysia, 1948–1960; Yemen-Oman, 1963–1979).

That the great powers entered an asymmetric armed conflict on opposing sides in 23.27 percent of 116 conflicts underscores the significance of this finding, but does not prove that competition among the great powers was the reason for their decision to enter a struggle in third world countries. Moreover, conflict nodes are evident in which the struggle was episodic, lapsing and resuming over time: Indochina, Southeast Asia, the Middle East, and Africa. The data suggest a more restrained position of the great powers with regard to becoming involved in armed conflicts in which other great

Table 2.10. Participation of the Great Powers on Opposite Sides of Asymmetric Armed Conflicts, COSIMO Database, 1945–1999

Conflict	USSR	USA	China	UK	France
Iran–USSR, 1945–1946	X	–	–	X	–
China, 1945–1949	X	X	X	–	–
Indochina, 1945–1954	–	X	X	X	X
Palestine, 1946–1948	X	–	–	X	–
Greece, 1946–1949	X	X	–	X	–
Malaysia, 1948–1960	–	–	X	X	–
Burma, 1948–1961	–	X	X	–	–
Burma, 1948–1999	–	X	X	–	–
Angola, 1961–1974	X	X	X	–	–
Malaysia-Indonesia, 1963–1966	X	–	–	X	–
Yemen-Oman, 1963–1979	–	–	X	X	–
Zaire, 1964–1965	X	X	–	–	–
Indochina, 1964–1973	X	X	X	–	–
Yemen, 1965–1967	X	–	–	X	–
Thailand, 1965–1968	–	X	X	–	–
Namibia, 1966–1990.	X	X	–	–	–
Nigeria, 1967–1970	X	–	–	X	X
Sri Lanka, 1971	X	X	–	X	–
India, 1971	X	X	X	–	–
Yemen, 1972	X	X	–	–	–
Indochina, 1973–1976	X	X	X	–	–
Rhodesia, 1976	X	X	–	–	–
Angola, 1976–1991	X	X	–	–	–
Indochina, 1977–1978	X	–	X	–	–

Table 2.10. *(continued)*

Conflict	USSR	USA	China	UK	France
Indochina, 1978–1991	X	X	X	–	–
Afghanistan, 1979–1988	X	X	–	–	–
Nicaragua, 1981–1990	X	X	–	–	–
Total	21	19	13	10	2

Note: Conflicts are listed in chronological order by start date. Zaire, so named from 1971 to 1997, is today the Democratic Republic of Congo. Rhodesia, so named from 1965 to 1979, is today Zimbabwe.

Source: COSIMO data for 1945–1999 (see table 2.2).

powers were involved. This revelation refutes the widespread opinion that the great powers sought to fight for spheres of influence, including ideological ones, in third world countries during the Cold War.

To detect trends in great power participation in asymmetric armed conflicts in the postwar period over time, I investigated direct versus indirect participation rates. The results are presented in table 2.11, where the letter "d" beside a number indicates direct participation and the letters "id" indicate indirect participation in the conflict.

As already noted, France and the United Kingdom took the most active indirect part in asymmetric armed conflicts, and in all cases these were conflicts in colonies and former dependent territories. The peak of direct participation for these countries came during the first post–World War II decade, when the United Kingdom was involved in 10 such conflicts and France in 6. The absolute peak of British direct participation in asymmetric armed conflicts was reached during the first postwar decade and was followed by a significant decline in both direct and indirect participation in any form of armed conflict. The share of asymmetric conflicts for the United Kingdom is 62.79 percent of all armed conflicts in which it took part in the post–World War II period. For France the first two postwar decades give practically the same figures of direct involvement in asymmetric conflicts. The period extending from the high point of détente (1975) to the end of the Cold War

Table 2.11. Breakdown of Great Power Participation in Asymmetric Armed Conflicts, COSIMO Database, 1945–1999

Country	1945–1954	1955–1964	1965–1974	1975–1991	1992–1999	Overall participation, 1945–1999
UK	10 d + 4 id =14	3 d + 4 id = 7	2 d + 2 id = 4	1 id	1 id	15 d + 12 id = 27
France	6 d	5 d + 2 id = 7	1 d + 1 id = 2	11 id	1 id	12 d + 15 id = 27
USA	8 id	2 d + 11 id = 13	12 id	18 id	1 id	2 d + 50 id = 52
USSR/ Russia	0 d + 7 id = 7	2 d + 11 id = 13	11 id	1 d + 6 id = 7	1 d + 4 id = 5	4 d + 39 id = 43
China	4 d + 4 id = 8	4 id	5 id	2 id	—	4 d + 15 id = 19

Source: COSIMO data for 1945–1999 (see table 2.2).

(1991) shows a surge in France's indirect participation in asymmetric conflicts, in sharp contrast to the decade of 1965–1974 and the 1990s. In fact, asymmetric conflicts account for almost 73 percent of France's overall participation in armed conflicts in the postwar period.

A pronounced predominance of indirect versus direct participation in armed conflicts is generally typical of the United States, and the share of participation in asymmetric conflicts also reflects this trend—50 instances of indirect involvement versus 2 instances of direct participation. Starting from the first postwar decade, US intervention in asymmetric conflicts went up. The 1990s constitute an exception. The asymmetric conflicts in which the United States participated as a share of the total number of conflicts in which the United States was involved is 58.43 percent.

The degree of Soviet participation in asymmetric conflicts is lower than in the case of the United States, which also reflects the trend of general participation in armed conflicts in the post–World War II period. The USSR/Russia took part in 72 armed conflicts, including 43 asymmetric ones (59.7%). The Soviet Union's participation in armed conflicts peaked in 1955–1974, after which activity declined, whereas for the United States the next 15 years (1975–1991) were marked by increasing participation in asymmetric conflicts. This observation again refutes a widely shared view, namely, that minimum intervention in armed conflicts overseas was typical of the United States after the war in Vietnam, a position attributed to the so-called Vietnam syndrome and the unexpected US political loss.

This observation does, however, reflect the importance to the United States and its foreign policy of counteracting the "communist threat," which became the basis for the provision of military and economic assistance to certain countries after the political defeat of the United States in Vietnam. Some have thought that anticommunism peaked in the United States in the 1950s, during the McCarthy era. However, the desire to prevent new countries' transition into the communist camp was most evident in US foreign policy of the years 1960–1980. The effect of this focus on US foreign policy and the provision of aid is explained by the American diplomat George F. Kennan, reflecting on the time when the US government was discussing rendering military assistance to Greece in 1947. Kennan wrote in his memoirs:

> Throughout the ensuing two decades the conduct of our foreign policy would continue to be bedeviled by people in our own government as well as in other governments who could not free themselves from the belief that all another country had to do, in order to qualify for American aid, was to demonstrate the existence of a Communist threat. Since almost no country was without a Communist minority, this assumption carried very far. And as time went on, the firmness of understanding for these distinctions on the part of our own public and governmental establishment appeared to grow weaker rather than stronger. In the 1960s so absolute would be the value attached, even by people within the government, to the mere existence of a Communist threat, that such a threat would be viewed as calling, in the case of Southeast Asia, for an American response on a tremendous scale, without serious regard even to those main criteria that most of us in 1947 would have thought it natural and essential to apply.[39]

Returning to the analysis, China in these statistics demonstrates a more moderate participation in asymmetric armed conflicts, with a maximum level reached in the first postwar decade. In general, China's share of involvement in armed conflicts is lower than that of the other great powers, and the share of its involvement in asymmetric conflicts is 54.29 percent.

The above calculations of the trends in armed conflict participation by the great powers over time are in accord with conclusions drawn by Istvan Kende, one of the first researchers to focus on wars in third world countries. He noted that in the first postwar decades, 91 percent of external forces in intrastate armed conflicts were represented by colonial powers, while in 1967–1976 this figure drastically declined, to 50 percent (10 of 20 wars). According to Kende, external actors took part in 74 of 120 armed conflicts in 1945–1976. In the 1970s the participation of France and the United Kingdom declined (to one and three wars, respectively), while the participation of the US armed forces drastically increased (to 11 of 20 armed conflicts), and the participation of other developed countries also increased (to 14 of 20 armed conflicts).[40]

4. The most common forms of great power behavior in armed conflicts

To determine the most common forms of great power behavior in armed conflicts, we can use the corresponding variable in the COSIMO database. In total, 20 types of behavior were coded and registered. To clearly compare the behavior of the great powers, I reviewed all armed conflicts in the database (692), conflicts with at least 25 casualties (304), conflicts of the third and fourth intensity levels with at least 25 casualties (259), asymmetric armed conflicts (210), and asymmetric armed conflicts involving the great powers (116). The results are presented in table 2.12.

In general, a proportional decline in indicators can be observed as the list of conflicts is reduced to only asymmetric armed conflicts involving the great powers. The dynamic of two indicators, A and I, significantly differs from that of others. Indicator A, "Two great powers remain neutral and inactive toward each other and in their relations with other conflict parties," gradually declines during the transition from all conflicts, including nonviolent ones (48.99%), and reaches its minimum though still significant value in asymmetric armed conflicts involving the great powers (15.51%). This characteristic indicates that in almost half of all conflicts, the great powers occupied a neutral position. In armed conflicts, this parameter does not vary much: it is 41.12 percent in conflicts with at least 25 casualties,

Table 2.12. Behavior of Great Powers in Armed Conflicts, COSIMO database, 1945–1999

Great power behavior (COSIMO database coding)	All conflicts in COSIMO database (N = 692)	Conflicts with at least 25 casualties (N = 304)	Armed conflicts of the third and fourth intensity levels (N = 259)	Asymmetric conflicts (N = 210)	Asymmetric conflicts involving the great powers (N = 116)
A: Two great powers (A, B) remain neutral and inactive toward each other and in their relations with other conflict parties	339 (48.99%)	125 (41.12%)	106 (42.40%)	87 (41.43%)	18 (15.51%)
B: Two great powers (A, B) mediate together between two states/ governments (K, L) or their internal, nongovernmental groups (x, y, z, w)	28 (4.05%)	14 (4.61%)	14 (5.41%)	14 (6.67%)	6 (5.17%)
C: Two great powers (A, B) together call on the conflict parties (K, L, x, y, w, or z) to resolve their conflict by peaceful means	24 (3.47%)	15 (4.93%)	15 (5.79%)	12 (5.71%)	4 (3.44%)
D: One great power (A or B) calls on the parties (K, L, x, y, w, or z) to resolve their conflict by peaceful means	18 (2.6%)	10 (3.29%)	7 (2.7%)	4 (1.9%)	4 (3.44%)
E: Two great powers (A, B) together dictate a settlement at the expense of two states (K, L)	—	—	—	—	—

Great power behavior (COSIMO database coding)	All conflicts in COSIMO database (N = 692)	Conflicts with at least 25 casualties (N = 304)	Armed conflicts of the third and fourth intensity levels (N = 259)	Asymmetric conflicts (N = 210)	Asymmetric conflicts involving the great powers (N = 116)
F: A great power uses a smaller power (K) to gain influence in another smaller state (L) (e.g., substitute wars at the height of the Cold War as a conflict by proxy)	3 (0.43%)	2 (0.66%)	2 (0.77%)	2 (0.95%)	1 (0.86%)
G: A great power (A) supports one state (K) that is in conflict with another state (L)	31 (4.48%)	16 (5.26%)	13 (5.02%)	8 (3.81%)	6 (5.17%)
H: A great power (A) supports an internal, nongovernmental group (x) against another group (y)	7 (1%)	6 (1.97%)	5 (1.93%)	5 (2.38%)	4 (3.44%)
I: A great power (A) supports a smaller state (K) against an internal, nongovernmental group (x)	74 (23.34%)	49 (16.12%)	44 (16.99%)	43 (20.48%)	39 (33.62%)
J: A great power (A) supports an internal, nongovernmental group (x) against state/government (K)	—	—	—	—	—

Table 2.12. *(continued)*

Great power behavior (COSIMO database coding)	All conflicts in COSIMO database (N = 692)	Conflicts with at least 25 casualties (N = 304)	Armed conflicts of the third and fourth intensity levels (N = 259)	Asymmetric conflicts (N = 210)	Asymmetric conflicts involving the great powers (N = 116)
K: A smaller power supports a great power against an internal, nongovernmental group	37 (5.35%)	12 (3.95%)	10 (3.86%)	8 (3.81%)	9 (7.75%)
M: Two great powers (A, B) together support different internal, nongovernmental groups (x, y, l, e, or m) in another state (K)	3 (0.43%)	2 (0.66%)	2 (0.77%)	2 (0.95%)	2 (1.72%)
N: Two great powers (A//B) are confronted in a nonviolent conflict (Cold War constellation)	16 (2.31%)	4 (1.3%)	2 (0.77%)	—	—
O: Two great powers (A//B) are ideologically confronted in a nonviolent conflict within or via a smaller state (K or L)	3 (0.43%)	1 (0.33%)	1 (0.38%)	1 (0.47%)	—
P: Two great powers (A//B) are economically confronted in a nonviolent conflict within or via a smaller state	2 (0.28%)	1 (0.33%)	-	-	—

Great power behavior (COSIMO database coding)	All conflicts in COSIMO database (N = 692)	Conflicts with at least 25 casualties (N = 304)	Armed conflicts of the third and fourth intensity levels (N = 259)	Asymmetric conflicts (N = 210)	Asymmetric conflicts involving the great powers (N = 116)
Q: A great power (A) enters into a political conflict with a smaller power (K)	47 (6.8%)	11 (3.6%)	7 (2.7%)	4 (1.9%)	2 (1.72%)
R: A great power (A) enters into a military conflict with a smaller power (K), e.g., a colonial war	18 (2.6%)	14 (5.41%)	13 (5.02%)	8 (3.81%)	7 (6.03%)
S: Two great powers (A//B) enter into a military conflict with a smaller state (K), e.g., imperial wars	—	—	—	—	—
T: Two great powers (A//B) in a military conflict (e.g., "classic" European wars before 1945)	—	—	—	—	—
U: Other	29 (4.2%)	20 (6.58%)	18 (6.95%)	10 (4.76%)	6 (5.17%)

Source: COSIMO data for 1945–1999 (see table 2.2).

42.40 percent in armed conflicts of the third and fourth intensity levels, and 41.43 percent in asymmetric armed conflicts. A significant decline in "neutrality of great powers" was recorded only for asymmetric armed conflicts involving those powers.

Indicator I, "A great power supports a smaller state against an internal, nongovernmental group," is distinguished by a significant increase in its share. For extreme groups of conflicts, such difference does not seem very impressive—23.34 percent of all conflicts, including nonviolent ones, and 33.62 percent for asymmetric armed conflicts involving the great powers. However, the indicators for armed conflicts show this indicator first declining to 16.12 percent, while for asymmetric conflicts the indicator increases, to 20.48 percent, and then to 33.62 percent when great powers are involved. Such changes can be explained by the low level of great power participation in such conflicts in general, as well as by the fact that great powers more often supported official governments in their struggle against opposition movements rather than the nongovernmental actors (here we should note the across-the-board absence of indicator J, "A great power supports an internal, nongovernmental group against state/government").

Indicator R, "colonial wars," is also manifested, but insignificantly: in 2.6 percent of all conflicts, 5.41 percent of all armed conflicts, and 6.03 percent of asymmetric armed conflicts involving the great powers.

Several indicators and their values raise doubts and encourage testing of empirical material. Indicator F ("conflict by proxy") turns out to be surprisingly poorly manifested, though the literature often mentions a received opinion that such a strategy was widely used indirectly by the superpowers in their struggle for global influence. If we assume that the data coding in the COSIMO database is accurate to impeccable, such an observation would contradict the Cold War analytics.

The meaning of some coded indicators is not entirely clear—for instance, indicator K, "A smaller power supports a great power against an internal, nongovernmental group." Indicators E, J, S, and T are not recorded in the database, and indicator U, "other," was recorded in a significant number of cases. The absence of recorded values is understandable in the case of indicators S, "imperial wars," and T, "war between great powers," while with other indicators the reason for the absence is not clear. Letter "L" is not present and coded in the data set at all.

CONCLUSIONS

Work with the UCDP and COSIMO databases as described in this chapter to identify the impact of the asymmetry factor in postwar conflicts involving the great powers leads to the following conclusions:

1. The share of asymmetric conflicts involving the great powers correlates with their overall participation in armed conflicts. The values are especially close for the USSR/Russia, the United States, and China, from 58.43 percent (United States) to 54.29 percent (China) to 51.39 percent (USSR/Russia). The highest level of involvement in asymmetric conflicts, 73 percent, was noted for France. This finding is no doubt connected to the wars that France waged in its former colonies. In this respect, it should be noted that the experience of the United Kingdom, which had larger colonies, suggests a different solution to the problem. After a decade of heavy involvement in small wars in 1945–1955, the United Kingdom rather quickly acknowledged the formal independence of its former colonies and created a new structure of legal, political, and economic relations in the form of the Commonwealth of Nations, an intergovernmental organization that today comprises 53 states, almost all former territories of the British Empire.

2. Quantitative data on great power participation in armed conflicts in the post–World War II period do not exhibit a pattern of linear trends valid for all countries in this category. The lack of a common pattern likely results from the fact that, despite the existence of "great power" status with common country features, the notion of "superpower" also emerged during the same period, and with the superpowers there came a decline in the role of the traditional great powers in world politics. Furthermore, a new great power emerged, China, and the characteristics of its "greatness" did not necessarily correspond to traditional ones. Some trends typical for the world of the great powers (colonial empires) from the late nineteenth century to the first half of the twentieth century were still around in the postwar period. However, new manifestations of grandness emerged or became stronger, and they were not associated with colonial possessions or the direct use of force. The legacy of the

colonial world and of imperial forms of interaction between the center and periphery of the world system became a source of problems for the great powers rather than a way to preserve their high status. In the postwar period, the so-called new liberal order developed with the strengthened role of nonmilitary factors—trade, finance, international conflict resolution mechanisms, and mediation services. In this new world order, great powers and superpowers implemented their interests in different ways and found new means of exercising their influence.

3. The frequent participation of the great powers in asymmetric conflicts in the post–World War II period points strongly to the conclusion that such military actions were not something new, as is often claimed. Keeping this observation in mind, one can assert that the military doctrines of the great powers and the composition of their armed forces had to take into account kinds of warfare that were different from conventional war between sovereign powers, or what is usually referred to as "big war."

One final note concerning the databases: their compilation is the result of diligent labor on the part of many people, who collect, process, and code the data. Any activity of such complexity is not free from errors and inaccuracies. Thus, misprints in country coding were uncovered in the COSIMO database, as well as irregular updating of indicator codes, and these inconsistencies inevitably raise questions and could lead to mistakes in using the data. I managed to obtain answers to my inquiries from Professor Frank R. Pfetsch, one of the founders of the COSIMO project. No doubt, databases enable scholars to form a more globally comprehensive picture based on a long time scale and a large amount of data, but this work is only a preliminary to conducting further research using qualitative and historical methods. Employing a combination of quantitative and qualitative approaches helps researchers avoid mistakes, inaccuracies, and unsubstantiated generalizations while developing a more robust understanding of their research topic.

CHAPTER 3

The Dissolution of the British Empire and Asymmetric Conflicts in Dependencies

THE DISSOLUTION OF THE BRITISH EMPIRE

The British Empire was one of the largest and most powerful empires in modern history. It managed to create a whole "British world," the so-called Pax Britannica. The dissolution of the empire occurred relatively recently, and an imperial legacy can be traced in the lives of many nations that were part of or associated with it. The end of the British Empire is often claimed to have been relatively peaceful; however, decolonization was accompanied by protracted wars involving the British armed forces and those of some of its dominions. These wars were costly and strenuous for the metropolis. Moreover, it had to justify them morally before the court of domestic and international public opinion.

After the end of World War II, the United Kingdom was militarily the third most powerful country in the world, after the United States and the Soviet Union. It retained the military and material capabilities to exercise control over its colonial possessions. A readiness to transform the British Empire was evident even before the beginning of World War II, but that did not mean complete surrender to national liberation movements in the colonies. Moreover, British society supported the idea of empire and its presumed civilizing influence on the people of its colonies. Right up to the final dissolution of the empire, there were debates in the United Kingdom about what "empire" meant, and about the need to transform the empire into a commonwealth of states.

Using the asymmetric conflict model to explore the reasons for the dissolution of the British Empire affords a new perspective on this issue. It questions some established opinions, and weighs the influence of the various factors that affected the process. To carry out this analysis, it will be helpful to identify the circumstances that set events on track for a fast if not always peaceful outcome. Such circumstances align with three groups of factors: military and power factors, domestic and economic factors, and international factors.

Military and power factors. These factors refer primarily to the material capability of the colonial power to maintain order in its dependent territories and subdue the resistance of local forces striving for independence. Specifically, these factors are the military power of the United Kingdom and its military strategies, and the distinctive ways in which it waged war in its colonies and dependent territories. This category should also include the military capabilities of the belligerent parties that managed to inflict political if not military defeat on the United Kingdom.

Domestic and economic factors. These factors are related to the desire to preserve the empire and to pursue certain foreign, military, and economic policies to achieve this goal. Domestic factors include the position of the major political parties and the population with regard to preserving the empire, and the associated costs of doing so. Parliament, the ruling parties, political forces, and the general public all had conflicting opinions on whether to shore up or dissolve the empire. Economic factors include the state of the economy, which conditioned the nation's ability to sustain its foreign policy course.

International factors. These factors have to do with the policies of other powers and international organizations that affected the capability of the United Kingdom to preserve the empire. Here it is important to emphasize the position of the United States as the new leader of the Western world, as well as the position of the Soviet Union as the other new postwar superpower, and the position of the newly independent countries. Because the asymmetric conflict model implies political rather than military defeat of the stronger party in an armed struggle, the ideological contest between West and East for influence in newly independent countries should be listed among the international factors, as both sides often saw a new country's choice of ideological patron as a political victory or defeat.

From Empire to Commonwealth

British imperial dominance peaked in the early twentieth century, when the British Empire comprised vast territories in Africa, the Middle East, Asia, North America, and Australia. In 1913, 427,467,000 people were imperial subjects.[1] After the end of World War I, the United Kingdom was given a mandate to govern the former colonies of the Ottoman Empire and Germany, and the British population reached its maximum. In 1939, the population of the empire was 500 million, or roughly one-fourth the world's population, while the population of the metropolis (the United Kingdom of Great Britain and Northern Ireland) was just under 47 million.[2]

The transformation of the empire started with the formal definition of Dominion status, as declared at the 1926 Imperial Conference, held in London. The nature of British relations with its Dominions was laid out in the Balfour Declaration: "There are autonomous Communities within the British Empire, equal in status, in no way subordinate one to another in any aspect of their domestic or external affairs, though united by a common allegiance to the Crown, and freely associated as members of the British Commonwealth of Nations."[3] A formal transformation of the British Empire into the British Commonwealth was stipulated by the 1931 Statute of Westminster, though its ratification lingered on into the post–World War II years. The Dominions of Canada, the Commonwealth of Australia, New Zealand, Ireland, and the Union of South Africa were in fact individual states. Their troops took part in World War I, representatives of these states participated in the Paris Peace Conference and signed the Treaty of Versailles, and the states joined the League of Nations.

In 1939, the Royal Institute of International Affairs published *The British Empire: A Report on Its Structure and Problems*, which dealt separately with the Dominions, as well as with Newfoundland, Southern Rhodesia, India, Burma, and the "colonial empire." The report stated that "the Empire can better be described than defined," and clarified that the notion of a "colonial empire" applied to crown colonies, mandated territories, and protectorates. Regardless of the way they had joined the empire, however, whether through conquest, cession, occupation, treaty, or League of Nations mandate, dependent territories were united by a common system of administration and common legislation. According to the report, the colonial empire was inhabited by some 58,350,000 people. This section of the report closely examined the system of governance in different territories. The degree of self-government depended, according to the document, on the degree of development of the local population.[4] However, as noted by the Russian scholar Galina Ostapenko, "there was no significant difference in degree of dependence."[5]

The Royal Institute's report listed the problems of the empire in order of importance:

1. *The problem of preserving the unity of the Commonwealth and empire on foreign policy issues.* The difficulties of preserving such a unity stemmed from the special position of some Commonwealth countries (e.g., Ireland), differences in country status, differences between the security of certain territories and the security of the Commonwealth and the empire in general, and economic and political disagreements between Commonwealth countries. The report paid special attention to relations with the United States in the context of a common foreign policy. It noted that "the Empire as a whole endorses the Monroe Doctrine" and that "the preservation of status quo is one of the fundamental purposes of the United States, and is also one of the major interests of the British Empire. If the empire were to collapse and sovereignty of any of its important components were to pass into the hand of any other Power, the United States would obviously be exposed to new dangers, in spite of the oceans that separate America from Europe and Asia." Finally, it perspicaciously noted that "in the event of war there would be a serious danger of disintegration of the Empire if its members differed on the fundamental question whether or not they should take part."[6]

2. *The defense problem.* The defense problem was defined by the way this task was fulfilled within the empire. The report underlined that "the geographical, strategic, and political facts that are of the greatest importance for Imperial defense tend to confirm the functional dissimilarity

that derives from historical causes." On the one hand, vast areas did not require any special protection, for they bordered on friendly countries (the US-Canadian border) or natural frontiers (Australia) or faced no serious threats (African possessions and India); on the other, the defense of strategic regions required significant resources. The report emphasized that a system of imperial defense was "to a considerable extent centralized rather than co-operative," and that "in proportion to wealth or population an exceptionally high fraction of the cost and responsibility is borne by the United Kingdom, while most of the Dominions, according to their several strategic and geographical positions, more or less openly rely on it for help in defence." The report considered whether imperial defense could be federalized, noting that "the nations of the Commonwealth independently reserve the right to decide, not only what defensive arrangements they shall make, but also when and how those defensive arrangements shall be put into operation by active participation in war." The United Kingdom provided technical assistance, supplied standardized equipment, trained officers, and developed aviation and navy capabilities. Its armed forces were a "small but highly trained expeditionary force" that could be used in crisis situations. In other respects, the armed forces of the empire were represented by the local forces of the dominions or other territories. India possessed a large army, with "Indian troops trained and officered in the higher ranks by British officers (subject to progressive Indianization)"; the report discussed whether the Indian army could be "potentially available for Imperial purpose outside India."[7]

3. *The colonial question*. The colonial question was considered in two aspects: (1) as "the methods of administration and the relations between native and immigrant peoples," and (2) as "the relations between the Empire and foreign countries." Relations with other countries were considered exclusively from the standpoint of claims laid by "dissatisfied powers"—that is, Germany, Italy, and Japan—to certain territories or access to supplies of raw materials and overseas markets. The report noted that the British Empire considered its colonies and dependent territories a crucial source of economic prosperity and intended to defend its interests in the world of power politics, rejecting the simplified Wilsonian approach to

> the colonial problem. It specifically stated that "the colonial problem is extraordinary difficult, not only because of its own complexity, but also because it is being considered in the atmosphere of the conflict between democracy and totalitarianism, between the League of Nations and power politics. The principal Imperial Powers are democratic states and members of the League of Nations. The principal dissatisfied Powers, on the other hand, are totalitarian states whose ideology elevates the sovereignty of the State at the expense of the ideal of international community and interdependence embodied in the League, and regards expansion in the colonial and in other spheres as essential to national prestige and to the fulfillment of the 'national destiny.'"[8]

In 1946, after the end of World War II, the Royal Institute of International Affairs published another report. This report, titled *British Security*, reviewed the traditional foreign policy of the United Kingdom before the war and analyzed the changes that had taken place during the war. It articulated what it called the three great British interests of modern times: "protection of these islands from attack by invading forces; the maintenance of the all-important British trade; and the development and security of oversea[s] possessions." In its foreign policy the United Kingdom traditionally had relied on "the policy of the free hand" and had refrained from entering into binding obligations to any of the great powers. However, the experience of two world wars, according to the report, "has made it clear that the policy of restricting commitments with the object either of maintaining neutrality or of preserving freedom of decision no longer offers the best hope of security. The affairs of all parts of the world have become so interdependent and events in any one have such wide repercussions elsewhere that the achievement of world peace and stability calls for the co-operative efforts of all, and particularly of the major, peacefully disposed States."[9]

The British position was defined by the fact that "Britain stands in the world at the head of an Empire and the centre of a Commonwealth with interests stretching right across the surface of the globe." The report named three sources of British strength—its economic, strategic, and political positions—but noted the war's negative impact on the United Kingdom: "As compared with the Powers with the greatest natural resources, such as the United States and Russia, she is probably weaker than before the war.... She has become, on balance, a debtor instead of a creditor country, since a large part of the foreign investments upon which she drew heavily for financing her own effort and that of her allies in the last war, as well as in the early part of this war, have gone."[10] Nevertheless, the recovery of industrial and agricultural production and world trade would strengthen the position of the empire and the Commonwealth countries. The United Kingdom still enjoyed a strategic position globally, grounded in the triple foundation of land, sea, and air forces; however, in the future the burdens of ocean security would likely have to be shared with other countries, through international alliances.

The Royal Institute report identified new factors that had an impact on security policy. The first was "[an] ideological cleavage in public opinion in most countries arising from the impact of fascist and communist theories upon the established order of society throughout the world. The precise effect of this disturbance is not easy to measure; but its main result, in Britain as elsewhere, has been to introduce into the discussion of foreign affairs a factor of partisan conflict, of which any British Government must take account in estimating how far they can rely on united support at home for action abroad." Moreover, the need for collective security measures had become much more important because of the relative weakening of the developed countries through war. Nuclear weapons were seen as a new factor in international relations and one whose impact was difficult to predict. Evaluating the sources of British political strength, the authors of the report believed that two of the most important conditions for stability in foreign relations were a shared outlook on foreign affairs by the different political parties and continuity in foreign policy, which should be maintained by successive governments. The report held that British society should be kept well informed in a timely manner about foreign policy issues and that secrecy was inadvisable, for transparency in the matters of government policy and foreign affairs would be expected to contribute to national unity: "Foreign affairs are not an obscure, mystical subject, beyond the comprehension of the ordinary man," the authors wrote. "They are as much a matter for common sense, for free and open discussion, and for public judgment as other affairs of state."[11]

The report marked a sea change in British thinking about national security, which was no longer parsed as relations between its overseas possessions and the imperial head but as embedded in international relationships with other powerful nations. The countries and regions identified as having an impact on British security were the Commonwealth, the United States, the Soviet Union, Europe, the Middle East, and the Pacific. Further, the conditions and mechanisms for creating international security would have to be pursued, and here the role of the United Nations was noted. The report also unambiguously stated that the United States had an absolute military resource superiority in the world and therefore had a major role to play in ensuring international security, as the United Kingdom once had.[12] As for the Soviet Union, the report's authors emphasized the need for close cooperation to ensure security in Europe and stability in the Middle East, and to formulate common approaches to India and China. At the same time, they noted that mistrust and ignorance about each other's history created serious psychological obstacles to Anglo-Soviet cooperation. The three main regions in which the interests of the two countries were closely intertwined were Europe, the Middle East, and the Far East. The report's assessment of the Soviet Union was reserved, merely pointing to the need to find common interests on which they could cooperate (the possibility of cooperation was confirmed by specific examples drawn from recent history).[13] As for the Pacific Region, the United Kingdom needed a "permanent accord with the United States and the USSR" to pursue the following interests and goals: "1) the maintenance of the political association with the Dominions; 2) a general responsibility towards all the Pacific territories which are associated with her; 3) good relations with China; 4) commercial interests, including her investments in Pacific countries, trade and exchange of products; 5) communications, in which oceanic shipping and air routes are of prime importance; 6) the defence of individual territories; and 7) on a different plane, general support in the Pacific as elsewhere for 'liberal' tendencies and regimes."[14]

Thus, it appears that after the end of World War II, the United Kingdom intended to restore colonial control while taking into account changes in the international situation and the new alignment of forces. Bernard Porter has noted that the rapid Japanese occupation of the British colonies in the Far East became possible as a result of the negative attitude of the empire's subjects: "In the early 1940s her colonies in the east toppled one by one before the Japanese wind: Hong Kong in December 1941; then

Malaya—with her people apparently not lifting a finger to stop it, which put a damper on Britain's euphoria; then, in February 1942, that great new 'invulnerable' imperial bastion in the east, Singapore (because its guns were all pointing out to sea and the Japanese came in from the land); and then Burma."[15] According to Keith Jeffery, "the failure by Britain to protect Imperial subjects had a long-term effect.... The Empire was gravely, if not fatally, injured."[16] British historian Michael Howard wrote that after the fall of Singapore, "the charisma on which British rule in the East had rested for a hundred of years and which British defence planners had been so anxious to preserve was destroyed forever." American journalist Walter Lippmann, in his column in the *Washington Post*, argued that "the loss of Britain's Far Eastern Empire transformed overnight an imperialist's war into a war of liberation." This editorial caused a debate in British political circles over the lessons of this defeat. One lesson was that colonial nations did not resist the Japanese troops owing to the unpopularity of the British rule. Another lesson was that many dependent territories represented "plural societies," multinational and multiconfessional, a circumstance that had to be taken into account in the postwar reorganization of the system of governance.[17]

The United Kingdom's main military purpose in the remaining colonies was to prevent them from falling under communist influence, rather than to preserve the status quo. Even before the United States did, the United Kingdom began viewing the struggle in Asia and Africa as part of a global struggle against communism. The official documents of the British government provide clear evidence of this fact.[18] The Russian scholar I. I. Zhigalov quotes the words of Winston Churchill, who said that by spring 1945, the Soviet threat had in his eyes "replaced the Nazi enemy."[19] Classified British government documents confirm that in early 1946, considerable attention was devoted to the problem of Soviet influence in Europe, the Balkans, the Middle East, India, and China.[20]

In late October 1948, the Labour government agreed to drop "British" from the name of the Commonwealth as a result of India's decision to introduce a republican form of political system in the country. Prime Minister Clement Attlee's memorandum of December 30, 1948, stated that the term "British" should be avoided so as not to evoke associations with the British Empire, "[which] tendency has increased in recent times." At the same time he noted that the "Commonwealth includes the Colonies and other dependent territories, some of which are under the administration of Australia, New Zealand and South Africa." However, "the phrase 'Commonwealth and Empire', which conveniently implies a distinction between the self-governing and dependent parts of the whole, has no constitutional authority and is permissible only in colloquial use."[21] This decision generated a wave of indignation on the part of Conservatives, headed by Winston Churchill, in the British Parliament.[22] In 1949, the name "British Commonwealth" was officially changed to "Commonwealth of Nations," and India joined it. According to the British historian W. Roger Louis, "India's decision strengthened the Whiggish view of the Empire's progress and purpose including the belief that British rule had been designed originally to allow dependent peoples to advance towards self-government and to reach fulfillment in the Commonwealth."[23] Taking a slightly different approach, David French argued that despite the clear inability of the United Kingdom to maintain the status of great power alongside the United States and the Soviet Union after World War II, "British defence policy, like her foreign policy, was designed to preserve as much as possible of Britain's world power in increasingly adverse circumstances."[24]

W. Roger Louis, who edited *The Oxford History of the British Empire*, noted that the 125 historians who contributed to the five-volume project often had as much difficulty agreeing on when the empire began as on when it ended. According to him, the independence of Malaysia, Singapore, Aden, and Rhodesia was referred to as "the death rattle of British imperialism." The project participants "reaffirmed that historical judgment changes dramatically from one generation to the next" under the impact of ideological "engagement in relation to the times" and because of the complexity of the phenomenon itself.[25]

In 1964, the Commonwealth consisted of 18 independent states and the United Kingdom, as well as nine colonies in Africa, two in in Asia, six in the Americas, thirteen in Oceania, and two in Europe. The population of the empire, including that of the United Kingdom, was around 68 million people, having shrunk to almost one-seventh of its earlier size. The population of the colonial power was around 70 percent of the overall number of imperial subjects.[26] In 1966, the position of colonial secretary was dissolved and responsibility for relations with dependent territories was transferred to the secretary of state for Commonwealth affairs. In 1971, the Singapore Declaration of Commonwealth Principles was adopted; it stated that the Commonwealth was "a voluntary association of independent sovereign states." Along with a commitment to international peace, mutual understanding between nations, human rights protection, democracy, and elimination of the gap between poor and rich countries, the Singapore Declaration held that "we oppose all forms of colonial domination and racial oppression" and "will therefore use all our efforts to foster human equality and dignity everywhere, and to further the principles of self-determination and non-racialism."[27]

Charles Edmund Carrington, who held the Abe Bailey Chair of Commonwealth Relations at the Royal Institute of International Affairs, London, from 1954 to 1962, claimed that the British Empire ceased to be the world's leading power after the fall of Singapore in 1942,[28] and that from that moment on, "the whole world moved into a phase of social development to which the French have recently given a name, the useful word 'decolonization.'"[29] Carrington identified three phases of the British Empire's collapse and referred to decolonization as the "transfer of power" to those nations that were mature enough to administer their own affairs. In the first stage some "old" colonies received the status of Dominion; in the second stage South Asia was liberated, between 1947 and 1957; and in the third phase Africa was decolonized, under the catchphrase "Independence now," between 1958 and 1961. The fourth phase, according to Carrington, entailed completing the transfer of power to small states. He argued that the United Kingdom should retain military bases in strategic locations (Malta, Gibraltar, Cyprus, Aden [now part of Yemen], Kenya, Singapore, Hong Kong) by internationalizing the bases through the system of the United Nations, the Commonwealth of Nations, the North Atlantic Treaty Organization (NATO), and the South East Asia Treaty Organization (SEATO).[30]

By the mid-1990s, all four of Carrington's phases had been completed, though often independence was granted after a period of direct confrontation and even war:

1. In the first phase, which occurred mainly before World War I, Dominion status was granted to Canada (1867), the Commonwealth of Australia (1901), New Zealand (1907), the Union of South Africa (1910), and Ireland (1921).

2. In the second phase, which lasted from the end of World War II to the Suez Crisis of 1956, independence was given to Transjordan in 1946; Bhutan, India, and Pakistan in 1947; Brunei, Burma, Palestine, and Sri Lanka in 1948; and Sudan in 1956.

3. In the third phase, which ran from 1957 to the end of the 1960s, when long-lasting military campaigns on the periphery of the empire came to an end, independence was granted to the larger states in Africa, Asia, and Latin America: Ghana and Malaya in 1957; Singapore in 1959; Somalia, Nigeria, and Cyprus in 1960; Sierra Leone, Kuwait, and Tanganyika in 1961; Western Samoa, Jamaica, Trinidad and Tobago, and Uganda in 1962; Kenya, Singapore, Sarawak, North Borneo, and Zanzibar (including Pemba Island) in 1963; Malta, Malawi (Nyasaland), and Zambia in 1964; Gambia and the Maldives in 1965; Barbados, Botswana, Guyana, and Lesotho in 1966; South Yemen in 1967; and Nauru and Swaziland in 1968.

4. In the fourth phase, which lasted from the 1970s until 1997, independence was gained mainly by small, often island states in the final dissolution of the empire. During this time the following territories also acquired sovereignty: Bahrain, Qatar, and the United Arab Emirates in 1971; the Bahamas in 1973; Grenada and Tuvalu in 1974; Seychelles in 1976; Dominica and the Solomon Islands in 1978; Kiribati, Saint Lucia, Saint Vincent, and the Grenadines in 1979; Vanuatu and Zimbabwe in 1980; Antigua and Barbuda and Belize in 1981; and Saint Kitts and Nevis in 1983. Sovereignty over Hong Kong was transferred to China in 1997, and the ceremony is regarded as symbolizing the end of this period.[31]

As relayed by the editors in the introductory note to one of the volumes of *British Documents on the End of Empire*, "in 1945 Britain had over fifty formal dependencies; by the end of 1965 the total had been almost halved and by 1985 only a handful remained."[32] Trevor Owen Lloyd in *The British Empire, 1558–1995* lists the following stages of empire dissolution: (1) 1899–1922: fighting and reorganizing; (2) 1922–1945: the defeat of the idea of an imperium; (3) 1945–1960: independence by degrees; (4) 1960–1983: independence at once.[33]

Thus, in the first two postwar decades the British Empire underwent a radical change and practically ceased to exist. To what extent this process was voluntary is an open question. Official British government documents of the time refer to the process of the empire's dissolution as a "transformation," "power transfer," or "evolution," which suggests that it was perceived as inevitable and even desirable. Evidence in favor of the voluntary nature of empire dissolution is at hand in British efforts to create political, economic, administrative, and legal institutions in the colonies according to the British model, as well as in the constant discussions by officials, politicians, and members of Parliament on reorganizing the empire into a commonwealth. That most former colonies joined the Commonwealth can be considered confirmation of the voluntary transformation of the empire. At present, 53 countries are members of the Commonwealth of Nations, and almost all are former British colonies or dependent territories. Burma, Ireland, Sudan, Somalia, and some other countries did not join the Commonwealth. Pakistan left the Commonwealth in 1972, then rejoined in 1989. South Africa left the Commonwealth in 1961 and rejoined in 1994.

The authors of *The Oxford History of the British Empire* note that some of the contributors to that work challenge the assumption of a voluntary and peaceful "self-dissolution" of the empire and consider the Commonwealth to be a continuation of empire by other means.[34] Ronald Hyam, editor of the 17-volume project *British Documents on the End of Empire*, indicates that this selection of documents shows that the main British objective was to contribute to the development of dependent territories and prepare them for independence, rather than to fulfill British imperial ambitions.[35] Moreover, Niall Ferguson writes, "What is very striking about the history of the Empire is that whenever the British were behaving despotically, there was almost always a liberal critique of that behavior from within British society. Indeed, so powerful and consistent was this tendency to judge Britain's imperial conduct by the yardstick of liberty that it gave the British Empire something of a self-liquidating character."[36]

Indeed, multiple lines of evidence support a critical attitude of British politicians and British public figures toward imperial order. *British Rule in India Condemned by the British Themselves*, published in London in 1915 by the Indian National Party, is one of this kind. The book is a collection of quotations from British public figures speaking about extreme poverty in India, which was a direct consequence of the British rule. It is noteworthy that the preface was written by William Jennings Bryan, the US secretary of state during President Wilson's administration. Bryan talks about justice and law, about the British refusal to give Indians an opportunity to control their treasures, and about British rule being worse than "Russian despotism": "How long will it be before the quickened conscience of England's Christian people will heed the petition that swells up from fettered India and apply to Britain's greatest colony the doctrines of human brotherhood that have given to the Anglo-Saxon race the prestige that it enjoys?"[37] However, the domestic critique of the British Empire, which appears in abundance in this publication, usually had in mind improving the system of governance rather than destroying it.

Even Soviet historiography, despite its fierce condemnation of the cruelties associated with British imperialism, expressed the view that "dissolution of the British Empire is distinguished by the relatively peaceful course of this process and the fact that most colonies making up the Empire did not completely sever their relations with the metropolis but became formally equal participants of the Commonwealth."[38] Soviet historians interpreted the Commonwealth as a transformation of imperial control into neocolonialism[39] rather than as a mechanism for the free association of independent states. The term "neocolonialism" reflected the Marxist-Leninist understanding of the imperial order as a system of exploitation and oppression of indigenous peoples. Even today this assessment is to a large extent present in the Russian research literature, though interest in empires, including the history and legacy of the Russian Empire, has prompted Russian researchers to develop a new perspective on this phenomenon.[40] Today both Russian and Western authors often consider empires, including the British Empire, as predecessors to globalization, economic and political integration, and efforts to create a single global space.

Despite the changed form of interaction with its dependent territories, the United Kingdom nevertheless aspired to retain its interests and levers in its former colonies. It formulated several key political, economic, and security interests during the transformation of its colonial empire:

- preserving the system of economic preferences and access to the most crucial sources of raw materials and natural resources, which ensured favorable conditions for the British economy;

- preserving close political ties with dependent countries and acting against competitors from the Western and communist countries; and

- preserving its military presence in strategic regions of the world to protect its interests and those of its allies and dependent territories.

Empire had enabled the United Kingdom to retain its great power status. The Russian historian Galina Ostapenko interpreted the formula of British Prime Minister Winston Churchill as follows: "The empire provides the United Kingdom with an independent position among the strong powers of the world." Moreover, according to Ostapenko, Churchill did not see any contradiction between the notions of "empire" and "freedom."[41] Indeed, in his letter to US President Franklin D. Roosevelt in December 1943, Churchill wrote that "the principles of imperialism already have succumbed to the principles of democracy," but also observed that "if imperialism is dead, it seems very reluctant to die down." According to Churchill, "the imperialism of Germany, Japan, Italy, France, Belgium, Portugal and the Netherlands will, we hope, end or be radically revised by this war. British imperialism seems to have acquired a new life [as a result] of the infusion, into its emaciated form, of the blood of productivity and liberty from a free nation through lend lease. British imperialism is also being defended today by the blood of the soldiers of the most democratic nation on earth."[42] As David Kaiser notes, this is not what the American president wanted to hear about the future of the British Empire.[43]

A discussion of the system of international trusteeship at a conference in San Francisco between April and June 1945 resulted in a heated debate over the inclusion of such concepts as independence or self-government in the UN Charter. Lord Cranborne, British representative and leader of the Conservatives in the British Parliament, argued that many parts of the British Empire had greater freedom than some independent states in Europe, and that an international system of trusteeship was not a way to preserve imperial control but rather would prepare colonial nations for independence. China, the Soviet Union, and several smaller countries sought to write "independence" into the document, while the representatives of France and the Netherlands were against this wording. The Soviet delegates insisted on including provisions pertaining to the "complete national independence and self-determination of all colonial territories." There were serious disagreements on that issue within the US delegation, though later the American delegation spoke against including the word "independence."The discussion of the colonial issue turned mainly on the problem of international security. W. Roger Louis would later note that "the history of the colonial question at San Francisco can be viewed as an attempt to resolve the two issues of security and colonial accountability, on the one hand, and the larger question of the future of dependent peoples on the other."[44]

Those who study the dissolution of the British Empire often argue that the empire was primarily an economic phenomenon and emerged as a result of trade and consumerism, accompanied by some insignificant state intervention. Until the mid-1950s, half the global trade was provided by the British pound sterling, which only gradually was driven out by the US dollar. Military force was used to conquer territories, fight against other European empires, and suppress riots in colonized territories. However, according to Niall Ferguson, military and financial power alone would have been insufficient to create a world empire. The colonization process—which included the mass resettlement of Britons in conquered territories and the creation of the "British order," with special rules for public space organization—played a significant role in developing and maintaining the British Empire. Ferguson writes:

> When the British governed a country—even if they only influenced its government by flexing their military and financial muscles—there were certain distinctive features of their own society that they tended to disseminate. A list of the more important of these would run as follows: 1) The English language; 2) English forms of land tenure; 3) Scottish and English banking; 4) The Common Law; 5) Protestantism; 6) Team sports; 7) The limited or "night watchman" state; 8) Representative assemblies; 9) The idea of liberty.[45]

Alan Burns also echoed Ferguson's words in his 1957 book *In Defence of Colonies*.[46]

Transforming the empire under new circumstances, the British authorities strived to implement their idea of appropriate organization of the local community in economy, politics, culture, education, health care, and many other spheres. The concept of nation-building, popular now and frequently used in the process of conflict resolution and postcolonial recovery, is present in official British documents on the colonial question. In analytical material prepared by the British Colonial Office in May 1950, prior to the Anglo-American talks on the American position in the United Nations with regard to the colonial issue, officials stated that dependent territories had to meet five conditions before they could be considered ready for self-determination: "1) the people must be healthy and vigorous; 2) they must have education, and technical knowledge and skills; 3) they must be able to produce all they possibly can for their own needs; 4) they must have something to sell to the outside world in exchange for the things they need but cannot produce themselves; 5) they must be able to govern and administer their affairs with reasonable honesty and efficiency."[47] The policy of the British Empire, according to the Colonial Office, was aimed at preparing and creating these conditions.

The publication of the British *Documents on the End of Empire* was a unique attempt to provide researchers with primary sources on the logic, drivers, and problems of the empire's dissolution. The preface stated that "the central themes" of the publication were "the political constraints, both domestic and international, to which British governments were subject, the economic requirements of the sterling area, the geopolitical and strategic questions associated with priorities in foreign policy and in defence planning, and the interaction between these various constraints and concerns and the imperatives imposed by developments in colonial territories." According to the editors, two central topics of colonial policy dominated between 1945 and 1951: economic recovery and the Russian expansion.[48]

In the early 1960s, the swift collapse of the empire spread to Africa. In his famous speech in February 1960 in Cape Town, South Africa, British Prime Minister Harold Macmillan emphasized the role played by nationalism in Britain's granting independence to the countries of Africa: "The growth of national consciousness in Africa is a political fact, and we must accept it as such.... I sincerely believe that if we cannot do so we may imperil the precarious balance between the East and West on which the peace of the world depends."[49] As the British historian Alistair Horne has noted, this speech elicited indignation from British Conservative politicians, who claimed that Macmillan had gone too far, that his speech was premature and provoked radical political activity in Africa.[50] However, other British politicians saw nationalism as the main antidote to communist influence in the colonies, along with good relations with the former colonial power, and this was the idea voiced by Macmillan. Commissioner General for South East Asia and the British representative to SEATO Lord Selkirk (George Douglas-Hamilton) expressed the same idea, writing to the prime minister in August 1961, "Whether we like it or not we have to recognize that China, both militarily and ideologically, is becoming increasingly the dominant force throughout South East Asia. The only long-term effective answer to Communist China is nationalism, coupled with recognition by each State that it has an obligation to defend its own territory. I was glad to note recently that this idea seems to be more readily recognized in Washington than it was. We must clearly do everything we can to promote nationalism as a counter to communism and avoid policies (especially with an imperialistic flavor) which may lead nationalists and communists to join forces against us."[51]

Thus, both military and nonmilitary factors played crucial roles in the dissolution of the British Empire. Any persistent countermeasures undertaken by the British authorities and the forceful suppression of national liberation movements in dependencies could have resulted in a breach of economic and political ties and, more dangerous, pushed the former colonies into a search for third-party support.

ASYMMETRIC ARMED CONFLICTS ACCOMPANYING THE COLONIES' STRUGGLE FOR INDEPENDENCE

The loss of colonial possessions was of economic, military, and strategic importance for the British Empire and was accompanied by forms of warfare often called small wars or emergencies in British historiography, though the level of casualties and the durations of the conflicts are consistent with the criteria for war accepted in the research literature.[52] According to my calculations based on the COSIMO database, over the period of 1945–1999 the United Kingdom was engaged in 27 asymmetric armed conflicts with dependencies; in 15 cases participation was direct and in 12 cases it was indirect. Seven of the 12 cases of indirect participation were conflicts with the empire's former dependencies, and 5 cases were conflicts with the dependencies of other countries. Twenty-one armed conflicts involving the United Kingdom occurred in the first two decades after World War II. In all these conflicts, regular British Army troops fought against various irregular troops and national liberation movements that used guerrilla strategies. The most bloody and protracted were the armed conflicts that preceded the independence of India, Malaya, Kenya, Yemen, and Nigeria. The independence of Sudan, Cyprus, Malawi, and Israel was also achieved by local forces fighting against British rule.

Around one-third of the asymmetric armed conflicts in which the United Kingdom was involved in the first two decades after World War II took place in countries that were not British possessions: in Greece, Indochina, Indonesia, Lebanon, and Angola. In Greece, it aimed to prevent the victory of communist forces during the Greek Civil War. In Indochina, it participated in disarmament of the Japanese troops south of the 16th degree north latitude, following the decision of the Potsdam Conference. After the Democratic Republic of Vietnam (DRV) was established in September 1945, its troops occupied the southern part of the peninsula but were driven out by British troops by the spring of 1946.

British policy in Indochina aimed at helping restore French control, as well as fighting against local pro-communist and nationalist forces, which were seen as a threat to the British possessions in South and Southeast Asia. However, as contemporary British scholars note, an attempt to restore the empire through military action by the British Army, which had a significant share of Indian and African soldiers, took place under conditions that greatly differed from those of the prewar period. According to British historians Christopher Bayly and Tim Harper, British troops poured into Burma, "reversing the humiliating defeat which they had suffered at Japanese hands three years earlier. The British went on to occupy Thailand, much of former French Indo-China and Dutch Indonesia," but this "revivified British Empire" faced "a variety of powerful, armed and embittered nationalist leaderships determined to claim immediate independence."[53] In the Dutch colony of Indonesia, British and Australian troops were present to accept the surrender of Japanese troops, and they did not fight against establishment of the independent government of

Sukarno. In April 1946, British troops left Indonesia and transferred power to the Dutch colonial administration.

Sir Julian Paget, lieutenant colonel in the British Army, provides data on 34 small wars in which the United Kingdom participated in 1945–1966. He labeled three of these as "limited wars": Korea in 1950, the Suez Crisis of 1956, and Kuwait in 1961. He defined 11 military operations as counterinsurgency campaigns: Greece, 1945; Palestine, 1946–1948; Aden, 1947; Malaya, 1948–1960; Kenya, 1953–1955; Cyprus, 1954–1958; Togo from 1957; Masqat and Oman from 1957; Brunei, 1962–1963; Malaysia, 1963–1966; and Aden from 1963. And finally, he defined 20 conflicts as "policing operations": British Honduras, 1948, Singapore, 1950; Aqaba, 1951; British Guiana, 1953; Buraimi (Oman), 1955; Hong Kong, 1956; British Honduras, 1957; Aden, Jordan, and Nassau, 1958; Cameroon, 1960–1962; Jamaica 1960; Zanzibar, 1961; British Guiana, 1962; British Honduras from 1962; Cyprus from 1964; Swaziland from 1964, Zanzibar, 1964; East Africa, 1965; and Mauritius, 1966.[54]

Although there may be objections to classifying certain cases as wars, from 1945 to 1966 there was no single year in which the United Kingdom was not involved in hostilities, though formally it was not in a state of war. Paget argued that the British troops would continue to be involved in such wars for the next 20 years, and his expectation came to pass. Information on 28 hostilities in the British dependencies is presented in table 3.1.

Paget attempted to identify the outcomes of 20 insurgencies worldwide that occurred between 1945 and 1966. According to his calculations, only five ended in unambiguous victory for the insurgents: the conflicts in China, Palestine, French Indochina, Algeria, and Cuba. The results of five unfinished campaigns were ambiguous or incomplete: those in Vietnam and Laos (though these events are now seen without question as victories for the insurgents), Angola, Yemen (from 1960), and South Arabia. The United Kingdom and other major powers, in Paget's opinion, defeated insurgencies in Greece, the Philippines, Laos (1945–1954), Malaya, Kenya, Cyprus, Oman, Brunei, and Malaysia.[55] Ivan Arreguín-Toft makes a similar attempt to identify victors in asymmetric conflicts of great powers; however, he believes that insurgents achieved victory in Malaya and Cyprus.[56]

One of the main problems of the Asian colonial empire after the war was the issue of Indian and Burmese independence. Political factors played

Table 3.1. British Army Operations in Dependencies, 1945–1982

Location	Years of military operation	Year of independence
India-Pakistan	1945–1948	1947
Palestine	1945–1948	1948
Malaya	1948–1960	1957
Gold Coast (Ghana)	1948	1957
Nigeria	1949	1960
Kenya	1952–1960	1963
Sudan	1953–1955	1956
Cyprus	1955–1959	1960
Oman	1957–1959	1971
Togo	1957–	1960
Jamaica	1960	1962
Cameroon	1960–1962	1961
Kuwait	1961	1961
Zanzibar	1961	1963
Aden	1947, 1958, 1962–1967	1967
British Honduras (Belize)	1948, 1957	1964 (self-government)
British Guiana (Guyana)	1954, 1962–1966	1966
Swaziland	1963	1968
Jamaica	1960	1962
Mozambique Channel	1965–1975	1975 (Mozambique)
Uganda	1964	1962
Bermuda	1968–1969, 1973–1977	1968 (self-government)
Northern Ireland	1969–	
Belize	1970–	1981
Cayman Islands	1970	British overseas territory

Table 3.1. *(continued)*

Rhodesia (Zimbabwe)	1979–1980	1980
New Hebrides (Vanuatu)	1980	1980
Falkland Islands	1982	British overseas territory

Note: Conflicts listed in chronological order from starting date of conflict.

Sources: Data compiled from Mileikovsky, *Raspad Britanskoi imperii* [The dissolution of the British Empire]; Paget, *Counter-insurgency Campaigning; The Oxford History of the British Empire*, vol. 4: *The Twentieth Century*; and Bayly and Harper, *Forgotten Wars.*

a decisive role in the United Kingdom's acknowledgment of their independence, though Soviet researchers also noted that the creation of a large Indian army contributed to the collapse of British rule, especially in light of the growing anti-British sentiments among the Indian military.[57] To the annoyance of its British partners, during the war the United States actively lobbied for India's independence. W. Roger Louis in *Imperialism at Bay*, which examines the role of the United States in the decolonization of the British Empire, mentions Roosevelt's attempt to discuss the possibility of Indian independence with Churchill during the latter's visit to Washington, D.C., in the winter of 1942. Churchill's response is recorded in his memoirs: "I reacted so strongly and at such length that he never raised it verbally again." Louis writes that Roosevelt advised Josef Stalin not to mention the word "India" in conversation with Churchill.[58] The British position with regard to its colonial possessions was then formulated as "We hold what we have."[59] Many British scholars have noted the serious disagreements between the United States and the United Kingdom over colonial issues during the war and after its end, in particular the independence of India and Burma.[60]

In light of the powerful national liberation movement in India and India's active participation in World War II, the issue of independence seemed decided. Nevertheless, the British cabinet continued discussing the transfer of power over the course of several months in 1946. The actual wording "transfer of power" came up during a discussion about granting independence to India, Burma, and Ceylon. In a secret letter of July 13, 1946, addressed by the governor-general of India Lord Wavell to Secretary of State, India Office, Lord Pethick-Lawrence, Lord Wavell wrote, "The transfer of political power in India to Indians will affect Great Britain and the British Commonwealth in the principal issues: Strategy, Economics and Prestige." He noted that "the principal advantage that Britain and the Commonwealth derive from control of India is Strategic.... The war of 1939–1945 could hardly have been won without India's contribution of two million soldiers, which strengthened the British Empire at its weakest point." Moreover, the strategic location of India and Ceylon was useful for naval bases, air transport, and securing oil supplies in the area of Iran (Persia) and the Persian Gulf. The importance of having trained military personnel in India was emphasized, as it protected British interests in this vast region. On the economic side, India was a valuable trading partner, an important market for British capital and for the employment of numerous professional personnel from the colonial power. Lord Wavell also noted that "in international Prestige, Great Britain should on the whole gain by her transfer of power, provided that this results in an orderly and friendly India." His general conclusion was that "on the whole Great Britain should not lose, but, on the contrary, may gain in prestige and even in power, by handing over to Indians, provided that the following main conditions are fulfilled: A. Power can be transferred in an orderly manner to a friendly and united India. B. A satisfactory defensive alliance can be secured." The greatest danger, according to the secret letter, was that an independent India might "come under the domination of Russia" or under communist influence as a result of revolutionary movements supported by Russia.[61]

The transfer of power to Pakistan took place on August 14 and to India on August 15, 1947. Burma became independent on January 4, 1948, and immediately left the Commonwealth. Ceylon became a Dominion on February 4, 1948. Having granted independence to Burma, the United Kingdom retained a military presence there, as well as the right to use naval ports and airspace, according to the British-Burma agreement of October 17, 1947.[62] The largest "pearls in the crown of the British Empire" were lost in the first postwar years. Prime Minister Clement Attlee stated in 1948, "No doubt, we could retain India and Burma for two or three more years. But we could do that at the cost of colossal spending of money and manpower, and if we did that, then they, having won the independence, would break away from England completely. We turned the nations, that could become our enemies, into friends. This was worth the risk."[63] During the meeting of the Cabinet Committee on Commonwealth Relations on January 7, 1949, it was noted that India's departure from the Commonwealth "would be a disastrous blow to the prestige and influence of the Commonwealth and would gravely affect the economic position of the United Kingdom and the sterling area generally."[64]

Immediately after the end of the war in Europe, the United Kingdom was drawn into hostilities in Palestine, its mandate territory since 1920 following the League of Nations' decision. The struggle of Israeli and Arab armed groups was well under way in Palestine even before the start of World War II, and British officials were among the casualties. After putting down the Arab uprising in 1939, it was important for the United Kingdom to preserve the loyalty of Arab leaders in the region so as to secure its interests in the Middle East on the threshold of war. One step in this direction was the introduction of quotas on Jewish migration into Palestine, along with tightened control over migrant flows. According to the Soviet historians Stanislav Desiatskov and Alexander Sudeikin, as a result of these measures, "Arab states ceased support to the guerrilla movement of Palestinian Arabs, and soon it started to decline for a while."[65] However, during the war in Europe, owing to the increased number of refugees, the British authorities were forced to modify their position with regard to immigrants from Europe, though the results were somewhat ambiguous and inconsistent. British attempts to maintain the balance between the Jewish and Arab populations by cooperating with the political elites in Palestine had to face the growing contradictions between these communities.[66]

The rise of Jewish extremism and the armed response of Arab groups led to full-scale civil war in 1946–1948, which showed that the British plan to create a joint state in Palestine was not viable. At a cabinet meeting in January 1947, Prime Minister Clement Attlee asked Chief of the Imperial General Staff General Montgomery "whether law and order could be preserved in Palestine in such circumstances." The answer was recorded as follows: "If there were active opposition from either Jews or Arabs alone, the situation could be handled with the military forces now available in Palestine. If there were active opposition from both communities, the situation could not be handled without military reinforcements"—which would have to be provided at the expense of the British occupation forces in Germany—or "this could in the last resort be done without retarding the demobilisation scheme." Foreign Secretary Ernest Bevin indicated that the Palestinian problem was aggravated "when President Truman intervened with his demand for the immediate admission of 100,000 Jews to Palestine."[67] The analytical survey that was prepared by the British Foreign Office on British policy toward Palestine stated that "Palestine was traditionally an area in which United States irresponsibility and vicarious idealism combined with Zionist pressure to produce clear-cut and categorical imperatives."[68] As a contemporary researcher has noted, "massive propaganda, particularly in the United States, whose pressure on the British government to permit Jewish immigration into Palestine greatly added to Britain's difficulties."[69]

The contemporary American expert on terrorism Bruce Hoffman, director of the Center for Security Studies at Georgetown University, Washington, D.C., believes that the reason for the failed British military operation in Palestine in 1939–1947 was a wrong strategy chosen by the British authorities based on the experience with suppressing the Arab disturbances of 1936–1939. According to Hoffman, despite a twentyfold numerical superiority of British troops and police over the Jewish terrorist groups Etzel and Lechi, they failed to oppress the extremist groups because the British authorities did not take into account the important differences between the Arabs and Israelis. Hoffman refers to a document summarizing these differences, prepared by the British representative in Palestine for the secretary of state for colonies. According to the document, "Jewish terrorism" was a city phenomenon with a lower level of intensity, organization, and Jewish community support than the Arab riots.[70] British attempts to use the methods of collective penalty employed by the British Army during the Arab Revolt failed when they were applied to the Jewish community, and even had an opposite effect. Hoffman describes a British police reprisal in July 1947 during a period of martial law in Tel Aviv. Policemen retaliated after the mined bodies of two sergeants were found hanging in an orange grove. The sergeants, who had been kidnapped and murdered, were British military men whom Etzel had offered to exchange for its arrested fighters. Before order was restored, 5 Jews had been killed and 16 injured, and 25 Jewish-owned shops were destroyed.[71]

From the perspective of our analysis, Hoffman's argument narrows the problem down to a military-strategic one and does not take into account several important factors contributing to the British failure that he himself mentions. Without denying the obvious British failure to establish peace and order in Palestine, the British authorities acted under political, international, and psychological pressure rather than merely for strategic reasons. British troops and police could use arms only for self-defense, and high-level permission was needed for the use of arms in other circumstances; this permission was granted by the British high commissioner for Palestine after coordination with the colonial secretary and military secretary. Official British documents also reveal the delicacy of the issue of using force or coercion on the Jewish population of Palestine, in light of international and especially American public opinion. The behavior of Jewish extremist groups was also psychologically difficult for the British soldiers and politicians to understand. Both the military and the civilian population became victims of terrorist attacks. It seemed that British politicians and military troops and the Jewish population of Palestine were united by a common enemy: fascism. Moreover, the British troops, accustomed to direct confrontation with the enemy, were not ready to exhibit restraint after attacks against barracks and military camps and the murder of British soldiers. The struggle intensified after the end of major warfare in Europe, which created a psychologically difficult situation for an exhausted army that had suffered heavy losses. The need to accept the surrender of the Japanese troops in the Far East and restore British order in the occupied territories created an additional burden for the armed forces, politicians, and British society itself. Even if one accepts Hoffman's argument that the British administration in Palestine could not find the right strategy for isolating and eliminating terrorist groups, the hostile attitude of the Jewish population toward the policy of limited migration to Palestine, restrictions on the acquisition of land, and the loss of trust of the Arab population made the situation practically hopeless for the British administration.[72]

According to the notes from the September 20 cabinet meeting, the prime minister said that in his view, "there was a close parallel between the position in Palestine and the recent situation in India.[73] He did not think it reasonable to ask the British administration in Palestine to continue in present conditions," and hoped to terminate the British mandate and withdraw the British administration and British forces from Palestine. It was also stated that the British government should not commit itself "unconditionally to cooperate with other members of the United Nations in implementing any policy requiring the use of force."[74] The December 3, 1947, joint memorandum of Secretary of State Ernest Bevin and Colonial Secretary Creech Jones on the British position before the United Nations concerning Palestine confirmed that "the United Kingdom representatives at the United Nations, while assisting the respective Committees with factual information, have consistently taken the line that we would not comment on the substance of the partition proposal, [and] would not be responsible for enforcing a settlement which was not agreed by both Jews and Arabs or to support a United Nations commission in enforcing it."[75] The inability of the British authorities to ensure security in Palestine, together with a severe domestic economic crisis and the high sensitivity of the world community to the issue of Jewish rights, forced the United Kingdom to announce its withdrawal from Palestine in September 1947. In February 1948 the British mandate in Palestine was terminated, and in May of that same year British troops were completely withdrawn.

One of the most important factors motivating the United Kingdom to cease its attempts at coercion in Palestine was the need to preserve friendly relations with Arab leaders. The United Kingdom considered the Middle East an important strategic region and sought to establish friendly relations with the regional governments in the new context of strengthened nationalist feelings and a striving for complete independence, as in Sudan and Egypt. At the same time, it was unsuccessful in gaining the support of its stronger ally, the United States. A confidential memorandum, "Present State of Jewish Affairs in the United States," prepared by the Earl of Halifax, British ambassador to Washington, and sent on February 25, 1946, stressed that the British policy in Palestine had been met with complete ignorance and misunderstanding in United States, and that "average Americans" and even congressmen knew no more about Arab states and the Arab world "than they do of the moon."[76]

One of the most protracted armed conflicts involving the British armed forces was an emergency in Malaya between 1948 and 1960. Unlike the engagement in Palestine, this campaign of the British forces is usually seen as successful from military and strategic standpoints and as a victory over the communist movement. In *The Malayan Emergency*, a monograph analyzing these events, Donald Mackay calls Malaya "the domino that stood," a reference to the domino theory of successive communist influence in neighboring countries.[77] The British authorities considered Malaya "the only territory in South-East Asia which has been developed as a base for the British forces on which we rely to defend our interest in the area," and Singapore was named "a focal point in sea and air communication in South-East Asia."[78] The United Kingdom went to considerable effort to recover and strengthen its position in Malaya. After the end of the war, reforms were carried out, and existing semiautonomous princedoms were transformed into the Federation of Malaya. When reforms were discussed, it was taken into consideration that greater British control was necessary for security reasons, to preserve control after disarmament of the Japanese troops and because of the region's strategic significance.[79] The reforms included an attempt to normalize relations between the largest ethnic groups, the Chinese (around 45% of the population) and Malays (44%). However, an anti-British movement, headed by the Communist Party of Malaya and mostly comprising ethnic Chinese, was on the rise in the country. A State of Emergency Regulation was introduced in Malaya in the summer of 1948 and not officially lifted until July 31, 1960.

Anthony Clayton has written that at the peak of armed clashes, up to 35,000 military men and 73,000 policemen were involved.[80] A secret CIA memorandum prepared in November 1949, "Current Situation in Malaya," provided the following data on British strength and capabilities in the region: "39,000 British Far East Land Forces, tactical units consisted of seven battalions of British infantry, seven battalions of Gurkha infantry, three battalions of Malaya infantry, one British artillery regiment, one Malay artillery regiment, and one British armored car battalion. Some units were composed mainly by Malays and some Ceylonese, Chinese, and Indians. The Royal Air Force strength in Malaya was 68 aircraft, of which approximately 50 percent were directly involved in the suppression of the terrorists. In addition to the British military forces there were 60,000 to 70,000 police and auxiliaries, mostly Malays of varying levels of training and usefulness." The memorandum stressed that the State of Emergency Regulation that had been introduced in the Federation of Malaya and the Colony of

Singapore had granted "wide power to the two governments: banishment of individuals, a mandatory death sentence for illegal possessions of arms, powers of search persons and premises without warrant, long-term detention of suspected without trial, close control over movements of persons and vehicles, and power to impose curfews and occupy properties. Legislation, making compulsory the registration of Chinese over twelve years of age and carrying of identity cards, was later enacted." During one year of the emergency, "3,000 aliens and 56 British subjects (most, if not all, Chinese and Indians)" were deported under the emergency regulations, and some 7,500 persons were detained, "among them many Chinese squatters, suspected or convicted of aiding the terrorists." The memorandum mentioned the British strategy of complete evacuation and resettlement of some Malayan squatters from "particularly bad areas of a federation."[81] The success of such a campaign was associated with the tactic of General Gerald Templer, who implemented the program of resettlement of squatters. However, the same tactic when implemented by Americans during the Vietnam War did not bring success. According to data provided by Mackay, during the 12 years of the Emergency, 2,473 civilians were killed, 1,385 were wounded, and 810 disappeared. The security forces lost 1,865 killed and 2,560 wounded, a total of 4,425 casualties, and in turn killed just under 6,700 and captured 1,200. In addition, 2,681 terrorists surrendered.[82]

The Malayan Emergency was one of the most protracted and cruel military campaigns of the British in Asia. It required deployment of the largest part of the British troops in Asia, was associated with significant expenses, and ended with the granting of independence to Malaya in 1957. British authorities managed to implement the program of modernization and construct the foundations of a modern economy and political life for the sake of the local population and the British residents. The United Kingdom spent around £86 million on the economic development of Malaya in 1945–1949.[83] After Malaya's independence, the United Kingdom maintained a large military contingent in the region and a military base in Singapore to preserve its strategic position and access to important raw materials, as well as to counter the communist influence. At the same time, it intended to prevent conflicts between the newly independent states in the region—Malaysia, Indonesia, and Borneo—that could draw it into armed clashes.[84]

The military operation in Malaya is usually evaluated as successful, insofar as "the military victory and … the effective mobilization of local political and popular support" were achieved.[85] Mackay also notes that the nature of success was that the campaign "destroyed the presumption that a Communist guerrilla insurgency must always succeed."[86] However, he also writes,

> in terms of the very basic British objective set in 1948, the restoration of effective colonial rule, the campaign can be seen as a tactical success but a strategic failure: the MPC [Malaya Communist Party] was defeated but British rule had to be abandoned. By 1951, however, the overwhelming imperative of British policy had become decolonization, and in that context the campaign was a British triumph. The handover of power in 1957 was not to a victorious guerrilla force, but to a democratically-elected government broadly representative of popular will, and one that could be relied on, at least for some time, to favour British interests. Beyond that, what appears from the history of the Emergency is a confusing picture of overlapping elements, all of which led to success, but none of which can properly be said to have been the single defining factor.[87]

The dissolution of the British colonial empire in Africa was similarly accompanied by military campaigns in Gold Coast, Kenya, and Malawi (Nyasaland). The state of emergency in Kenya was maintained for eight years, from 1952 to 1960. To suppress the Mau Mau uprising, British authorities used both political methods and cruel punitive measures.[88] According to various sources, during the uprising tens of thousands of insurgents were killed. When the state of emergency was lifted in 1960, it was decided to prepare for the transfer of power to the African majority. The independent Kenyan government was headed by Jomo Kenyatta, who had been imprisoned for almost eight years for participating in the uprising.

Fighting against British rule, local forces often applied strategies that today are recognized as asymmetric. David French has argued that "the indigenous populations of India, Egypt, Palestine, Aden, Cyprus, Malaya, Borneo and Kenya refused to meet the British on their own terms and adopted a mixture of guerrilla warfare, urban terrorism and forms of nonviolent opposition.

Although the British did not suffer a series of conventional military defeats, for the insurgent nationalists were never in a position to invade Britain, their resistance was bitter and prolonged. That fact alone, implying as it did rising financial and human costs, was enough eventually to persuade policy-makers in London to negotiate a settlement which usually entailed the withdrawal of British forces and the granting of independence."[89] Insofar as all armed conflicts involving the United Kingdom resulted in the political independence of the former dependencies, this can be considered a political, if not military, defeat for the great power.

In this respect, it is important to present some principles of British counterinsurgency as they emerged over the course of several campaigns. The British military paid special attention to the issue of small wars and antiguerrilla campaigns. A relatively small army and the global possessions of the British Empire defined the principles of governance and maintaining order in dependent territories. The British approach consisted in giving broad authority to officers and relying on local personnel to maintain order. Military operations in the territory of the empire were referred to as "imperial policing," as reflected in the title of a book by Major-General Sir Charles William Gwynn, published in 1939. Gwynn wrote that such operations were not military in the proper sense of the term, even if they were carried out by the army. Unlike in a conventional war, when the army should use maximum force to achieve victory, police operations require the use of minimum force but maximum self-restriction, restraint, and patience. Police operations are aimed not at fighting and destroying the enemy but at maintaining or restoring order among the crown's subjects. Gwynn stressed that the army functions as a reserve force to support the civil administration. He specifically wrote that the suppression of various revolutionary movements, "unless nipped in the bud, is a slow business," and "becomes a battle of wits in which the development of a well-organised intelligence service, great mobility, rapid means of inter-communication and close co-operation between all sections of the Government forces are essential."[90]

During the period of empire dissolution the term "imperial policing" was replaced by "counterinsurgency operation." Today the acronym of this term, COIN (counterinsurgency), is widely used. Julian Paget defined insurgency as "armed rebellion against the Government, in which the rebels have the support or acquiescence of a substantial part of populace." The methods they use "to achieve their aim of overthrowing the Government may include guerrilla warfare, but insurgents may equally well resort to civil disobedience, sabotage or terrorist tactics." Thus, guerrilla warfare is used among other tactics and is "in the sense of the special form of warfare, based on mobile tactics by small, lightly armed groups, who aim to harass their opponents rather than to defeat them in open battle."[91]

Paget, like many other authors of the Cold War period, believed that postwar insurgencies arose for a variety of reasons, chief among which was "Communist cold war campaigning." Not all uprisings were the product of communist influence, but communist countries used insurgencies and anti-Western sentiments to spread their influence in the former colonies, even though they sometimes faced insurgencies themselves in "friendly countries." In analyzing the influence of communist ideology on insurgencies, Paget introduced a term that has become widespread—"warfare by proxy," or "a proxy war." He wrote that the communist powers' support of insurgencies was "the most economical technique of their cold war campaigning; it is traditionally the method of the few fighting the many, and it costs the counter-insurgents far more in terms of money, manpower and effort than it does their enemies."[92]

Paget wrote of the need for a completely new approach to the counterinsurgency campaign and illustrated this claim by rephrasing the famous expression of Mao Zedong about "insurgents living among the populace of a country 'like fish in water.'" "The policy must be to destroy the fish," Paget wrote, "either by removing the oxygen from the water or by draining the pond, rather than by chasing individual fish in the muddy and weedy water with a very small net." To understand the British approach, it is important to identify the needs and tasks of insurgents and those who fight them. The essential requirements of this approach can be summed up as "winning the hearts and minds" of the local population, a phrase attributed to General Templer. According to Paget, "the support of the local population is an important factor for both sides in any counter-insurgency campaign, but particularly for insurgents, who will resort to ruthless intimidation, if necessary, in order to achieve it. Short-term local support is not the same, however, as the long-term objective of winning the hearts and minds of the people permanently. This is the ultimate aim of both sides, and is a predominant factor in all their thinking."[93]

To ensure the support of the local population, civil rather than military actions should be the principal means; however, the armed forces can and should play a decisive role in the implementation of civilian tasks. Paget named five rules for winning the hearts and minds of the population:

1. The first essential is that those in control must understand and respect the feelings and aspirations of the nation. They must also take positive steps to prove this sympathy, so that the people are convinced that victory over the insurgents will bring them a political settlement which will meet their wishes.

2. Secondly, there must be firm and fair Government, so that law and order are maintained and the interests of the populace are safeguarded, not exploited, by those in power. The Government may be either internal or external, but in either case, corruption and dissension must be eliminated, justice must be done (and be seen to be done!) and administration of the country must be as efficient as it can be made in the face of the inevitable insurgent efforts to disrupt it.

3. The third factor is the building up of public confidence by the establishment ultimately of a sound national Government, even if it is at first under auspices of a controlling Power.

4. The fourth and perhaps the best-known measures for winning hearts and minds is what the Americans call "civic action". This is organized aid to a country in variety of forms, and is now an accepted feature of most counter-insurgency campaigns. It covers direct financial aid, technical advice and the provision of amenities such as schools, hospitals, roads, new industries, food and welfare.... A most valuable form of "civic action" is the training of the populace to help themselves when they finally achieve their freedom from the insurgency.

5. Finally, the Public Relations and Information aspects are all-important. Not only must the propaganda, indoctrination and

> falsehoods of the insurgents be countered quickly and firmly, but the Government must also put across their own case forcibly and frequently.... Propaganda is an unpopular word, but it is a powerful weapon for either side in any cold war campaign and any insurgency, and it is one that Britain could wield with more skill and conviction. It can be just as effective as armies in defeating insurgency—and very much cheaper![94]

These principles indicate that insurgency was considered a political problem. The armed forces involved in counterinsurgency operations performed the auxiliary function of maintaining order and eliminating the most dangerous manifestations of insurgencies. However, their activities were subject to the customary law and to civilian government, and military personnel were expected to behave in strict accordance with the law in the difficult context of provocations and tension. Paget mentions the need to strictly investigate and prosecute any cases of violation of laws and discipline by the military, and this was to be done by the courts and be open to the public.

Another important book, *Low-Intensity Operations: Subversion, Insurgency and Peacekeeping*, an account of the British experience with counterinsurgency, was published by General Frank Kitson in 1971. Kitson listed three main factors as responsible for the increased incidence of subversion and insurgency in the post–World War II period: "(1) the changing attitude of people towards authority; (2) the development of techniques by which men can influence thoughts and actions of other men; (3) the limitation imposed on higher forms of conflict by the development of nuclear weapons." Evaluating the spread of insurgency in modern conflicts, Kitson pointed out that "most countries now regard subversion and insurgency as an integral part of one total war and not as a separate subject.... Whether in Europe or overseas, the pattern of conflict is such that it is virtually impossible to imagine an orthodox war taking place without an accompanying campaign of subversion and insurgency, although the reverse is by no means true."[95]

Kitson hewed to the British national tradition of understanding the role and tasks of the military in counterinsurgency operations, especially against the background of the anti-British riots in Northern Ireland and the growing number of international peacekeeping operations the United Kingdom was taking on. He argued that in a democratic country, rulers cannot afford the same brutality and disregard of law applied by dictators to restore order. Relations between the civilian population and the military receive special attention. Kitson also referenced Mao Zedong's analogy likening insurgents to fish in the sea; however, his approach is expressed in tougher terms than Paget's:

> In attempting to counter subversion it is necessary to take account of three separate elements. The first two constitute the target proper, that is to say the Party or Front and its cells and committees on the one hand, and the armed groups who are supporting them and being supported by them on the other. They may be said to constitute the head and body of a fish. The third element is the population and this represents the water in which the fish swims.... If a fish has got to be destroyed it can be attacked directly by rod or net, providing it is in the sort of position which gives these methods a chance of success. But if rod and net cannot succeed by themselves it may be necessary to do something to the water which will force the fish into a position where it can be caught. Conceivably it might be necessary to kill the fish by polluting the water, but this is unlikely to be a desirable course of action.

Recalling the British experience of operations in Aden, Kitson pointed to the dangers attendant on the early withdrawal of armed forces that have been maintaining order in the country and rendering assistance to the local government. Any announcement of the complete withdrawal of external armed forces from the country could lead to the complete loss of support from local forces fighting the insurgents.[96] These conclusions to a large extent correlate with the reasons proffered for the US failure in Vietnam, and now in Iraq.

On peacekeeping, Kitson wrote that although it is a "fundamentally different occupation to the countering of subversions, there is a surprising similarity in the outward forms of many of the techniques involved."[97] He wrote about the difference of peacekeeping operations because external "force acts on behalf, and at the invitation of, both sides to a dispute, and it is supposed to prevent violence without having recourse to warlike actions against either of them."[98] By the time his book was published, the British military had accumulated a vast experience participating in UN peace-keeping operations in Cyprus and Congo. Kitson indicated that "political considerations do govern the efforts of those involved in peace-keeping to a much greater extent than they govern the efforts of those involved in most forms of warfare and the lesson has to be taught and learned."[99]

Kitson devoted an entire chapter to the training of the military for participation in counterinsurgency and peacekeeping operations. According to him, such training should include the following: (1) explaining the fundamental nature of subversion and insurgency in the course of conflict; (2) training military personnel in how to cooperate with representatives of the local police force and civil government; (3) teaching officers how to direct the activities of their own soldiers, including any policemen or locally raised forces; (4) training military personnel in and teaching all ranks about the actual techniques of counterinsurgency; (5) how to collect and make the best possible use of overt information; and (6) providing troops with some understanding of the fundamental nature of the peace-keeping operation.[100]

In conclusion, Kitson noted that in order to make the army ready to counter subversion and insurgency and to take part in peacekeeping operations, it is important to understand how these activities differ from conventional wars and to apply this knowledge in training and organizing the army units for these purposes. He specifically stressed that "the qualities, required for fighting conventional war are different from those required for dealing with subversion and insurgency; or for taking part in peace-keeping operations for that matter. Traditionally a soldier is trained and conditioned to be strong, courageous, direct and aggressive, but when men endowed with use of these qualities become involved in fighting subversion they often find that their good points are exploited by the enemy. For example, firm reaction in the face of provocation may be twisted by clever propaganda in such a way that soldiers find the civilian population regarding their strength as brutality, and their direct and honest efforts at helping to restore order as the ridiculous blunderings of a herd of elephants."[101] Kitson's book demonstrates the change in the nature of the British military's participation in small wars, which were already considered to be limited enterprises with a temporary military presence, an understanding reflected in the characterization of relations with the local population and government.

Later works by British authors on the principles of counterinsurgency and peacekeeping operations largely repeated and developed the main provisions of the works cited above,[102] even as the principles of counterinsurgency operations applicable in 1960–1970 found their way into the management of today's peacekeeping missions. For instance, a 2005 book by British general Rupert Smith, *The Utility of Force: The Art of War in the Modern World*, demonstrates a profound understanding of the nature and tasks of such operations when performed by regular fighting troops.[103] Smith's book summarizes his 40 years of military service, during which he took part in peacekeeping operations in Africa, the Middle East, Asia, and Europe.[104] His main idea is that peacekeeping operations are distinguished from conventional wars in being "wars amongst the people," that is, among the civilian population. This nonmilitary participation in turn determines the tasks, means, and principles of the participation, behavior, and legitimacy of use of armed forces.

Thus, the military campaigns of the United Kingdom in its dependent territories were from the very beginning in service to political, economic, and strategic goals. The amount of force used to quell uprisings no doubt varied with the situation; however, it is evident that there was a conscious striving for minimum use of force, accompanied by a strict self-restraint on the part of political and civil institutions.

FACTORS IN THE BRITISH POLITICAL DEFEAT IN ASYMMETRIC CONFLICTS IN ITS COLONIES

Military Power and the Postwar Economy

Military power and the limitations of the postwar economy were the most important reasons for the British political defeat in colonial conflicts. By the end of World War II, British armed forces had reached their maximum size in British history, at more than 5 million troops. The United Kingdom maintained large troop numbers on its military bases and in its dependent territories in Asia (Singapore, Hong Kong, Malaya), in the Middle East and Africa (Suez Canal, Aden, Oman, Bahrain, Kenya), and in Europe (Malta,

Cyprus). The military presence of the land-based British Army in Europe was significant and included occupation forces in Germany and Austria (1.1 million out of almost 2 million people in the armed forces in late 1946) and in Greece.[105] The armed forces were gradually demobilized in the first postwar years: in 1946 the size of the armed forces was 2.053 million troops; this figure fell to 1.302 million in 1947 and to 689,000 in 1950. By 1952, the British armed forces had increased to 872,000 troops as a result of the war in Korea.

Weakened by war, the United Kingdom was forced to ask the United States for financial assistance, specifying the conditions of its participation in military operations. In February 1947, British Foreign Secretary Ernest Bevin warned the US leadership that if the United States did not provide assistance to the United Kingdom, the latter would have to cease providing military assistance to Greece and Turkey in the fight against leftist and communist forces. The start of the war in Korea in 1950 demanded a considerable increase in British defense spending and in the size of its armed forces; moreover, the country was drawn into the armed struggle in Malaya. The cabinet noted that the United Kingdom needed to increase its defense budget by £100 million in the current year, and even that would not be sufficient. Chancellor of the Exchequer Sir Richard Stafford Cripps prepared the memorandum titled "Defence Requirements and United States Assistance" in July 1950, which stated, "Clearly it would be impossible for us to meet a rise of expenditure of this magnitude without assistance," and a specific expenditure plan was suggested. The defense expenditure was set at £820 million for 1950–1951 (an increase of £40 million over the 1949–1950 level) and at £900 million for 1951–1952. It was noted that "on general economic grounds we consider £950 millions as the maximum defence expenditure we can afford in 1951–1952 and in the following years." It was stated that Britain "should ask for assistance to meet the difference between the total cost of defence in these three years of £3,400 millions and £2,859 millions, i.e. £550 millions." Moreover, this aid "should be in free dollars which we either hold or use to make purchases in any part of the world, and not aid which can be used for purchases made in dollars. This is essential because this additional programme will inevitably restrict our engineering exports, and is, therefore, likely to increase our sterling liabilities. We need to receive aid in a form which enables us to raise our dollar reserves in order to offset these liabilities."[106]

Cabinet documents also noted that the deployment of British troops to Korea could weaken the British position in those areas of Asia where it counteracted the communist threat (Malaya and Hong Kong). Moreover, an increase in defense expenditure "would force the Government to choose between a lower standard of living or longer dependence on United States aid."[107] In discussing British military involvement in Korea, British politicians kept in mind the relations with continental China, officially recognized by Britain but not by the United States. It was feared that a British military presence could threaten the security of Hong Kong and result in a denial of oil shipments to Britain and strategic exports to China. The British policy toward China, the documents noted, had been founded on the principle that Britain "should at all cost avoid action which would force the Chinese Communists into the arms of Moscow."[108]

According to official statistics, the United States allocated US$3.835 billion to the United Kingdom under the Mutual Security Program in 1949–1955, with most of these funds allocated in 1949 ($1.62 billion) and 1950 ($917 million).[109] In total, the United States allocated the following sums under the Mutual Security Program for military objectives in 1949–1955: $14.603 billion to Western European countries and $2.33 billion to countries in the Middle East, Africa, and South Asia, including Turkey (which received $453.026 million), and Greece ($843.894 million).[110] The comments to the statistical data indicate that these sums aggregate payments under various government programs for emergency relief. These were funds allocated under aid programs to France, Italy, Austria, the United Kingdom (British loan), special programs for Turkey and Greece, the European Recovery Program (the Marshall Plan, 1948–1951), and aid programs to Korea and China (Taiwan).[111] Figure 3.1 shows the ratio of military to nonmilitary spending (in billions of dollars) by US assistance programs in the first postwar decade, according to the 1956 *Statistical Abstract of the United States*.

In 1957, after the disastrous British operation during the Suez Crisis, a secret document was adopted, titled "Statement of Defence." This white paper indicated a "fundamental revolution in defence policy of the country."[112] It is symptomatic that the document starts with a statement of principles: a radical revision of defense policy was necessary

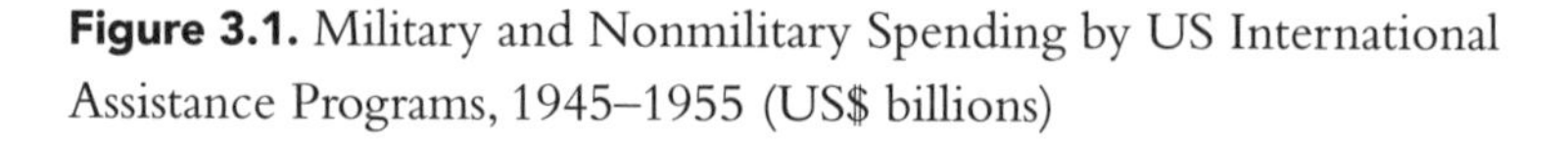

Figure 3.1. Military and Nonmilitary Spending by US International Assistance Programs, 1945–1955 (US$ billions)

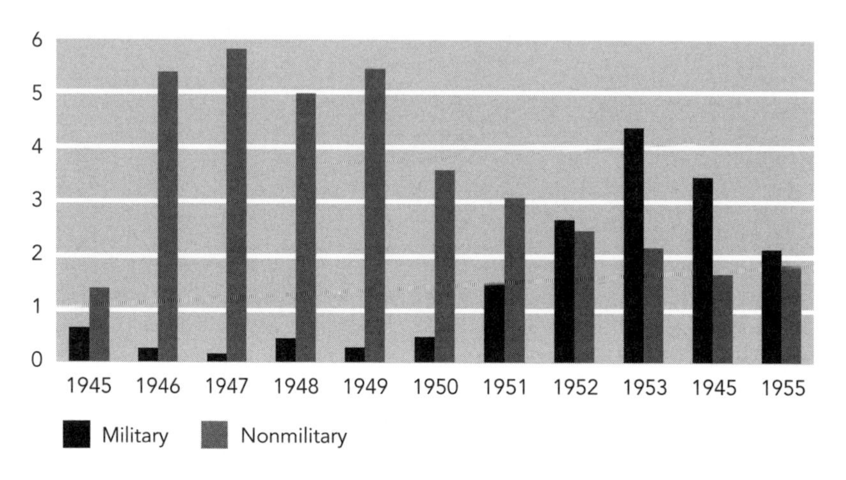

Source: *Statistical Abstract of the United States: 1956,* 77th annual ed. (Washington, DC: U.S. Department of Commerce, Bureau of the Census, 1956), 882, figure 46, "U.S. Government Foreign Grants and Credits, by Program: 1945 to 1955 (billions of dollars)."

for economic, international, and military reasons: "Britain's position and influence in the world depend first and foremost upon the health of her internal economy and the success of her export trade. Without these, military power is of no avail and in any case cannot in the long run be supported."[113] According to the document, conscription was to be cancelled,[114] and the size of the armed forces was to be maintained at the level of 375,000, compared to about 700,000 in 1957. The size of the Royal Navy was to be reduced by about 20 percent, the army by 45 percent, and the Royal Air Force by 35 percent.[115] Professor Keith Hartley, director of the Centre for Defence Economics, University of York, in explaining the reasons for a transition to contract service, referred to the debate in Parliament and among cabinet members. The main argument was that conscription was "extremely wasteful in its use of manpower," "an inefficient method of acquiring military personnel" with "increasing cost, complexity and skilled labour requirements of modern weapons," and was simply not needed with the threat of nuclear deterrence. As

Hartley noted, "Overall, the defence economics problem has compelled the UK to make defence choices in world of uncertainty."[116]

According to the "Statement of Defence," in the then current situation "the overriding consideration in all military planning must be to prevent war rather than to prepare for it," and to develop a nuclear retaliation capability. The main focus in defense should be placed on strategic aviation, the development of a nuclear force, obtaining US assistance in nuclear deterrence, and participation in collective security systems (Western European Union, NATO, the Baghdad Pact, SEATO). It was noted that "Britain must provide her fair share of the armed forces needed"; however, "she cannot any longer continue to make a disproportionately large contribution" to the defense of "the free world, particularly in Europe." That is why it was planned to reduce the presence of the British Army in Germany (from 77,000 troops to 64,000) and to reduce the number of tactical bombers in Germany by half, but equip them with nuclear weapons to compensate for the reduction in force size. The Middle East, the colony in Aden, Singapore, and the Persian Gulf area were identified as strategic regions, which Britain confirmed it would defend with "land, air and sea forces," which therefore had to be maintained in that area and were intended to "preserve stability and resist extension of Communist power in that area." The document confirmed Britain's willingness to continue providing assistance to Malaya and to maintain a military presence in Singapore and a significant military garrison in Hong Kong. In the colonies, the document called for reducing the troop size to the minimum necessary level and using local forces where possible, to be assisted in case of emergency by the "Central Reserve," which could be quickly relocated from the British Isles. It also proposed developing a Commonwealth Strategic Reserve, especially comprising Australian and New Zealand forces.[117] However, the Commonwealth partners, it seemed, were patently "look[ing] to the USA rather than to Britain for defence co-operation," and British attempts to share with them the responsibility for maintaining regional security and counteracting communism were not fully successful.[118]

The changes in the British defense strategy as detailed in a meeting of the cabinet on April 2, 1957, implied the active involvement of local forces in the colonies and former dependent territories to meet Britain's overseas defense commitments.[119] In accordance with the established hierarchy of objectives, the British military strategy in the colonies was aimed at maintaining order and keeping forces loyal to the crown in power. Military responsibility was to a large extent shifted to the local armed forces, which meant a continuation of traditional imperial policy under new conditions. The tradition of raising local armed forces in dependent territories contributed to the colonies' and dominions' gaining independence sooner than the colonial power wanted them to.[120]

Anthony Clayton, analyzing the defense and security policy of the empire from 1900 to 1968, spoke to the issue of "deceptive might" and the growing mismatch between British needs and resources. The main problem was, no doubt, exhaustion of society and resources during World War II.[121] The Soviet Union, its expanding presence in Europe and growing influence on national liberation movements in Asia and Africa, and its nuclear weapon capability were seen as the main security threats of the post–World War II period. Earlier than the United States did, the United Kingdom identified the Soviet threat and world communism as the major problem of the postwar world order. In this respect, it had to be pragmatic in identifying priorities for its military deployments and deciding the degree to which it would participate in military campaigns. According to David French, "in the late 1940s and 1950s many of the factors which had once facilitated British military superiority beyond Europe were fast disappearing." These factors included NATO commitments that forced the United Kingdom to keep a significant military force in Europe, and "international opinion that was increasingly hostile to metropolitan governments who used armed force to support their colonial pretensions."[122]

New strategic documents were later adopted that further reduced defense programs and the armed forces, brought back British troops from overseas military bases, shifted the defense focus to Europe, and developed the rapid response forces. Both Soviet and Western scholars noted that in the postwar United Kingdom, Parliament and the cabinet agreed on the need to balance the United Kingdom's desire to retain its authority in the world with the economic and political limitations on its ability to maintain high defense spending.

In 1967, when "British trade deficits plunged to their worst level in history," according to Roger Louis,[123] another white paper on defense was adopted.[124] According to the white paper, the economic situation in the

United Kingdom had forced the government to reduce military spending and reconsider its defense commitments to its allies and dependent territories. The main source of military threat was still the Soviet Union; hence the focus moved to military deterrence in Europe in alliance with NATO countries. The white paper noted that with the growing military power of the Commonwealth countries, the United Kingdom could decrease its military presence, close most military bases and withdraw personnel, transform its aid into military-technical and expert assistance, and provide troops only in case of urgent need. Moreover, the United Kingdom would participate in large military operations only with its allies, rather than alone. A specific plan for British troop withdrawal from Aden, South Arabia, by 1968 was outlined, as was a more general plan to reduce the British military presence in Singapore and Malaysia by about half by 1970–1971[125] and completely close its bases in both nations by the mid-1970s.[126] Richard Crossman, Lord President of the Council, wrote a memorandum on defense withdrawals that stated that Hong Kong would be the only remaining British base overseas, but he recommended cutting British military commitments as soon as possible in the Middle East, stressing that "in the Arab world a British military presence is an embarrassment to our friends and a provocation to our enemies." As for a British presence in the Far East, he noted with obvious bitterness and resignation, "it is difficult to suppose that after nearly ten years of steadily declining military strength in the area a British military presence would be credible either to our allies or to our enemies. In fact it would be a residual delusion of grandeur with which we would delude only ourselves."[127]

David French believed that "it was an economic crisis, not public opinion or US pressure, which compelled the Wilson government to reshape British defence policy and forsake almost all of Britain's remaining imperial commitments." This decision to withdraw British military personnel from "East of Suez," or about 100,000 troops, in the early 1960s "marked the end of Britain's role as a great imperial military power." If in 1965–1966 one member in four of the British armed forces had been deployed outside Europe, by 1973–1974 the figure was only one in ten, and by 1981 the forces raised outside the United Kingdom "had shrunk to only 3 per cent of total military power."[128]

The nuclear arms race in Europe and the intensification of the Northern Ireland problem greatly affected the increase in defense spending in the

1970s. The Conservative cabinet of Edward Heath (1970–1974) made a commitment to modernize armaments, while Richard Nixon's administration promised to maintain a US military presence in Europe. Budget priorities were considerably revised by the cabinet of Prime Minister Margaret Thatcher (1979–1990), who demanded that welfare objectives had to be reconsidered in favor of a strengthened defense capacity of the country and the modernization of armaments based on new security threats. In 1974–1979 the British defense budget was equal to around 4.75 percent of GNP, or £8.558 billion per year; in 1979–1980 it was planned to increase defense spending by 3 percent. Annual expenses for the counterterrorist operation in Northern Ireland were estimated at £400 million, or more than 20 percent of defense spending.[129] Figures 3.2 and 3.3 show the changes in armed forces and defense spending in the postwar period. Today, the United Kingdom ranks 28th in the world in terms of army size and second in terms of defense spending.

Domestic Factors

Domestic economic and political factors played an important role in decision making on defense policy and strategy. In addition to maintaining a certain level of defense spending, postwar British governments had to ensure economic recovery and achieve a certain standard of living. Churchill wrote to Roosevelt in 1943 that if economic recovery took too much time, "we may again have people shouting that 'We can't eat the Constitution.' They may even add to the non-edibles the Atlantic Charter and the Four Freedoms. This might lead to panic, bankruptcy and revolution."[130] Thus, the problem of economic recovery played a crucial role in defining the budget priorities, and the task of retaining imperial control by any means was not the primary one. In addition, limitations on the military budget were imposed by the postwar promises of Clement Attlee's Labour government to create a welfare state in the United Kingdom, since this required redirecting state spending to social needs.

The United Kingdom was in a precarious economic state after the war and, like many other European countries, was forced to ask the United States for financial aid. Ritchie Ovendale has written that "at the end of the war Britain was in a devastated economic state, in effect bankrupt," and that the victory in the war "was probably, at best, Pyrrhic." Ovendale argues

Figure 3.2. Size of the British Armed Forces, 1945–2001 (thousands)

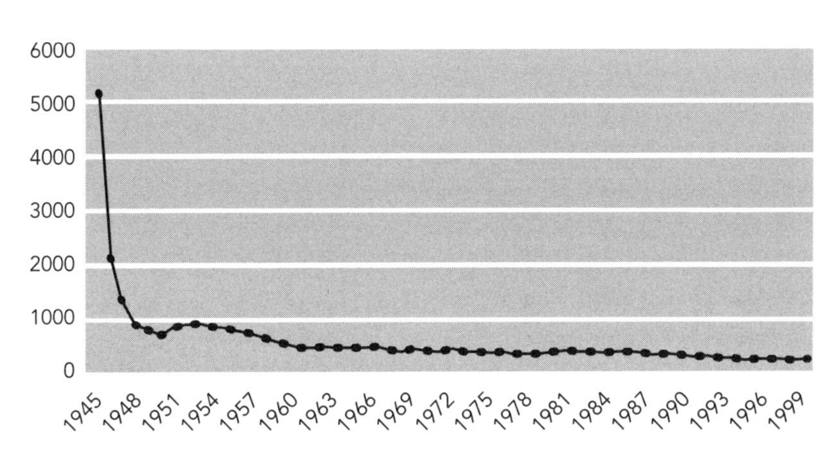

Source: Correlates of War project, National Material Capabilities, 1816–2001 (v3.02) database (http://www.correlatesofwar.org/).

Figure 3.3. Military Spending in Britain, 1945–2001 (US$ thousands)

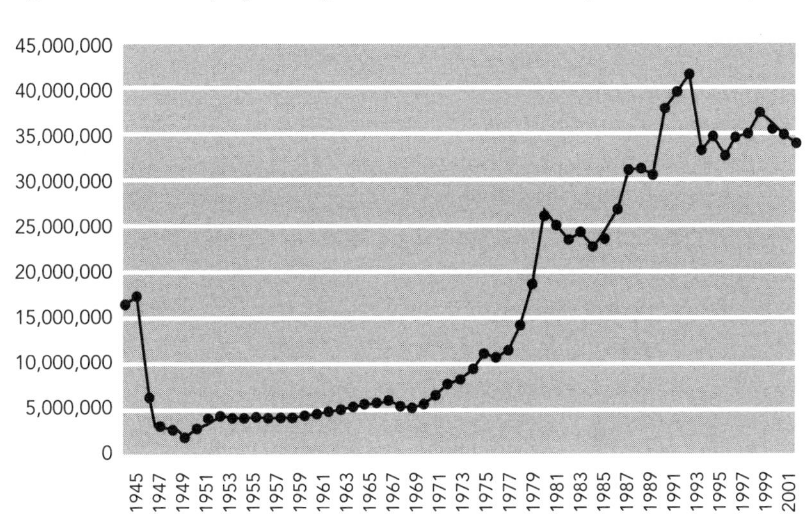

Source: Correlates of War project, National Material Capabilities, 1816–2001 (v3.02) database (http://www.correlatesofwar.org/).

that "British forces occupied parts of Europe and the Far East: Britain had to see that peoples in these areas did not die of starvation even if this meant greater austerity at home." Under these conditions, in Ovendale's opinion, a weakened United Kingdom had no other choice but to become an ally of a stronger power. He also says that "the British people might have wanted to turn inwards but the newly-elected Labour government fought to sustain Britain's role as a world power, and despite the economic difficulties, to ensure that it was the *pax Americana* that replaced the *pax Britannica* and not a world dominated by Russian communism."[131]

Between 1939 and 1945, British foreign debts had increased from £500 million to £3.25 million,[132] including £1.321 billion of Indian debt[133] and that of other dependent territories and indebted countries (e.g., Egypt, Sudan, Iraq, Argentina, Portugal). But the United Kingdom did not want the United States to take part in solving the problem. American historian and economist Charles P. Kindleberger, who participated in the development of the Marshall Plan, has said that the British position reflected its fear of a drop in the pound sterling exchange rate and the negative influence that such a drop would have on the domestic economic situation.[134] In August 1945 the United States suspended Lend-Lease supplies, and Anglo-American talks on a loan continued until December 1945. John Maynard Keynes, head of the British delegation at these talks and chairman of the World Bank Commission, noted that Britain turned to the United States for a loan "primarily to meet the political and military expenditure overseas."[135] The British dependence on the United States that had formed during the war was retained and even strengthened during the first postwar decade. US aid was conditional: the preferential acquisition of American goods and products was to be offered in exchange. In addition to these prerequisites, the United States used financial levers to exert pressure on the United Kingdom concerning its colonies and security in Europe, Asia, and the Far East.

In December 1945, an agreement was signed for a loan of US$3.75 billion, due in 50 years, at 2 percent interest.[136] According to the agreement, the United Kingdom undertook an obligation to abolish the sterling zone a year after the agreement entered into force, to restore the free exchange of the pound to the dollar, to unfreeze the funds in British banks belonging to Britain's dominions, colonies, and other countries of the sterling zone, and to pay them in dollars if the owners wanted that. The United Kingdom undertook an obligation to reduce and then eliminate preferential customs duties protecting the market of the British Empire.[137] In July 1946 the US Congress ratified this agreement, and on July 16, 1947, the United Kingdom introduced the free conversion of the pound but failed to support it longer than seven weeks. In July 1946 the United States cancelled control over wholesale prices, which led to their drastic rise (from 111 to 141 units from May to December 1946). This rise is seen as the reason why all funds under the "British loan" were spent by late 1947. There was a severe crisis in the United Kingdom, and new loans were needed. On December 1, 1948, an agreement was concluded on humanitarian aid to the United Kingdom—the Economic Cooperation Agreement, which remained in force until June 1951. Under the Marshall Plan, the United Kingdom received US$337 million in the form of loans and US$2.351 billion worth of manufactured goods between 1948 and 1951.[138]

The United Kingdom's economic and financial dependence on US assistance enabled the latter to pursue an economic policy that in the long run contributed to the destruction of the sterling zone and the establishment of the US dollar as an alternative world currency by the end of the 1950s. Mint parity existed between the two countries until 1978, with the United Kingdom devaluing the pound three times after the start of World War II. In 1939 the United Kingdom set the pound-dollar exchange rate at £1 = $4.03, reducing by 17 percent the rate that had existed for more than 100 years: £1 = $4.86 (1837–1939). In September 1949 the pound was devalued by 30 percent (to £1 = $2.8) under US pressure.[139] In November 1967 the pound was again devalued, by 14 percent (to £1 = $2.4). Figure 3.4 shows the ratio between the pound and the dollar with mint parity.

The ratio of the market value of the pound sterling to the US dollar shows greater flexibility and a more radical pound devaluation (figure 3.5).

Statistics on export-import flows of manufactured goods between the United States and the United Kingdom for the period 1936–1970 reflect the change of balance in favor of the United States, primarily with Lend-Lease supplies and aid programs from 1941 to 1948 (figure 3.6). Despite the regular statements issued by US officials on the country's commitment to free trade principles, protectionist measures over the import of manufactured goods helped to insulate the US domestic market from overseas competition.

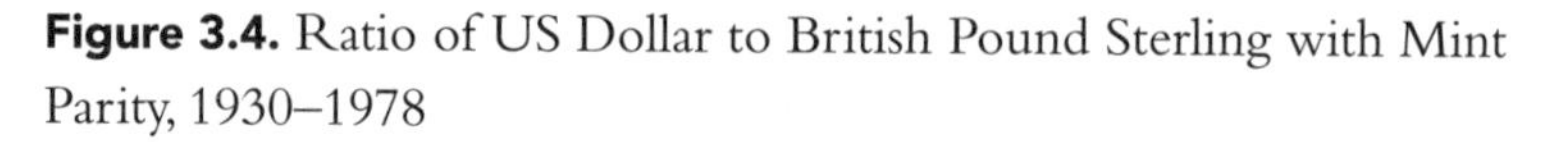

Figure 3.4. Ratio of US Dollar to British Pound Sterling with Mint Parity, 1930–1978

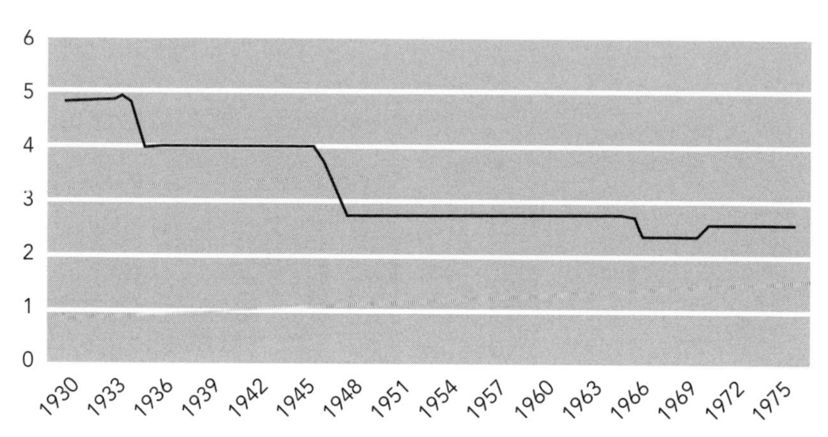

Source: *Historical Statistics of the United States: Earliest Times to the Present: Millennial Edition*, vol. 5, pt. E: *Governance and International Relations*, edited by Susan B. Carter (New York: Cambridge University Press, 2006), 5-561–5-562, table Ee612–614, "Dollar-Sterling Parity: 1789–1978."

Figure 3.5. Ratio of US Dollar to British Pound Sterling in Free Circulation, 1917–1999

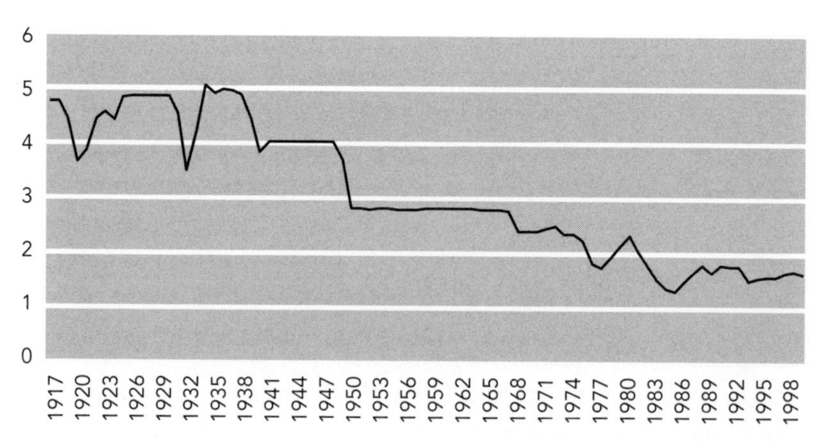

Source: *Historical Statistics of the United States: Earliest Times to the Present: Millennial Edition*, vol. 5, pt. E: *Governance and International Relations*, edited by Susan B. Carter (New York: Cambridge University Press, 2006), 5-569-5-570, table Ee621-636, "Bilateral Exchange Rates—Europe: 1913–1999."

Figure 3.6. Volume of Export-Import Flows between the United States and the United Kingdom, 1936–1970 (US$ millions)

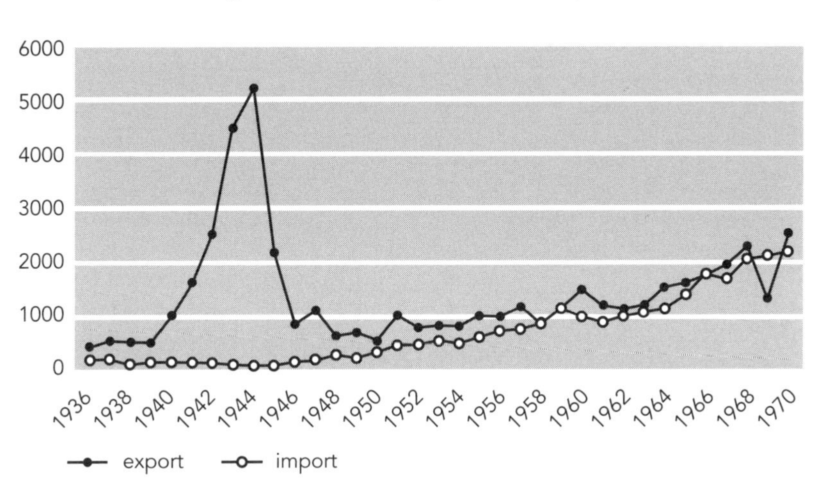

Source: Series U 317–334: Value of Exports (Including Reexports) of U.S. Merchandise, by Country of Destination: 1790–1970 (in millions of dollars), Series U 335–352: Value of General Imports, by Country of Origin: 1790–1970 (in millions of dollars), in *Historical Statistics of the United States: Colonial Times to 1970,* pt. 2 (Washington, DC: U.S. Department of Commerce, Bureau of the Census, 1975), 903, 905.

Contemporary scholars, relying on available memoirs, archives, and statistics for the wartime and early postwar period, note the presence of conflicting and sometimes opposing US and British positions. For instance, according to British historian Alan S. Milward, the United States was insufficiently generous with regard to the United Kingdom when dealing with Lend-Lease funds and the repayment of loans, unlike Canada, which cancelled the United Kingdom's debt under the mutual aid agreement.[140] Meanwhile, British researchers and politicians expected the United States to be interested in rendering unconditional economic and military assistance to the United Kingdom during the war and after its end. From the British perspective, the United Kingdom was the only European country that had successfully fought fascism and defended the interests of the free world. The restoration of British military strength after the war, primarily seen as the recovery of imperial power, was supposed to create an efficient protection from the spread of communist ideology and Soviet influence. Churchill wrote to Roosevelt that "sustaining Britain as a first class power has for many years been the cornerstone of America's foreign policy."[141] David Kaiser notes that the correspondence of Churchill and Roosevelt "illustrates the emerging new balance of power in the western world, entirely dominated by the United States despite Churchill's unceasing efforts to reserve a distinct military, political, and economic role for the British Empire," and also demonstrates Roosevelt's resistance to the idea that the United States would "police the world, and therefore developed his concept of 'four policemen'—the United States, the Soviet Union, the British Empire, and China—who would jointly keep the world under control," which contradicted Churchill's views.[142] Some secret messages of the British Foreign Office indicate that the United States regarded Britain as "a second-rate Power, a 'junior partner' of the United States." At the same time, there was "no real reason to fear that in fact Britain and the Commonwealth will be forced to play the role of a second-class Power, even though their physical strength is exceeded by the strength of other Powers. There are other things in leadership besides dollars and guns.... Moreover, there seems to be little doubt that Britain will continue to exercise enormous influence upon the ways in which Americans look at the world. In our view, consciously or unconsciously, willingly or unwillingly, they will for long continue to see many crucial things through the British window."[143]

According to a report on an April 1944 London visit from Under Secretary of State Edward Reilly Stettinius to Secretary of State Cordell Hull of May 22, 1944: "As the result of our conversations in London, we are more deeply convinced than ever that the United States must play an aggressive role in the creation of the international machinery necessary to ensure world security and economic stability. It is clear that the British attach great importance to the active participation of the United States in the world problems of the post-war era. We feel that in order to ensure our participation they will go far toward meeting our wishes on the form and character of the machinery for international cooperation." The British position on the need for a wide network of military bases is described in the detailed report: "Mr. Churchill repeatedly emphasized the need for international funds to support international bases, even under a trusteeship arrangement. He believes that in this way the United Nations will learn how expensive it is to maintain a security system such as the British have maintained in the past through national bases under Empire organization."[144] Indeed, the United States started allocating funds after the war to support British military bases. In 1953 the United Kingdom received a loan of US$5.6 million for Kenya and more than US$1 million for Tanganyika under the Mutual Security Program, in order to expand and modernize ports and docks.[145]

Charles P. Kindleberger has claimed that the allocation of huge funds under the provisions of the Lend-Lease Act was an issue of concern for the US Department of the Treasury, which was worried that "the American public would be upset by a generous provision of aid without some sort of means test," also taking into account the British desire to preserve the market of the empire from external impact and restore it after the war. Furthermore, he believes that US humanitarian aid to Europe in the post–World War II years in the form of merchandise supplies also failed to accomplish its objective, insofar as the goods were sold and "the aided government received considerable sums of its own money, raising the question of how to handle them." This "contributed measurably to the demoralization of both the US foreign service and Congress," and, "[by] wasting some of the aid-receiving country's resources" caused disputes between aid recipients and the United States and resulted in senseless spending.[146]

An analysis of the funds allocated by the United States in the form of grants and loans in 1945–1955 demonstrates that the United Kingdom received the largest assistance, comparable only with the aid given to France. It is noteworthy that, as researchers have pointed out, US financial aid to France played a negative role in France's relations with its former colonies. British historian Andrew Williams wrote that, ironically, American financial assistance enabled France to wage an unpopular, bloody, and politically disastrous military campaign in Indochina for almost 10 years.[147] Sami Abouzahr calculated that "by the time the French abandoned the effort after the catastrophic defeat at Dien Bien Phu (1954), the US was financing 80 per cent of the French war effort, and had committed itself financially, politically and emotionally to preventing a Communist victory there." Abouzahr argued that "to French prime minister Bidault, Marshall Aid was nevertheless a blessing that would allow him to 'avoid the abandonment of French positions'. European Recovery Program (ERP) appropriations let France finance the war in Indochina at the expense of domestic reconstruction projects favoured by the European Cooperation Administration, the body in charge of administering Marshall Aid."[148]

Only a small part of US financial assistance was in the form of grants. Much more was in the form of loans, and some of the grants were transformed into loans.[149] The last repayments of the British loan were to be made in 2006. Data on the volume of funds paid by the United States in the form of foreign grants and loans, including the recipient country's obligations under previous loans, show that the United Kingdom paid most of its debt in the late 1950s and 1960s. According to the *Statistical Abstract of the United States*, the United Kingdom received US$1.034 billion between 1950 and 1963 under the programs of military assistance and $500,000 between 1964 and 1969.[150] Ritchie Ovendale notes that in 1956, "the period of foreign aid was ending and Britain had to find means of increasing by £400 million the credit side of its balance payment."[151] Figure 3.7 illustrates the ratio of US loans to the United Kingdom to repayments by the latter of the loans received.

The need to tightly control the defense budget and keep it in line with the overall national economy is a distinctive feature of all postwar British governments. This was reflected in the so-called consensus between political parties with regard to the main foreign and defense policy directions. Soviet scholars who studied the British colonial policy stated with regret that even the Labour Party, which promoted socialist ideas and was supposed to defend the interests of the working class against the Conservative Party, was infected with opportunism and anticommunist prejudice.[152] V. A. Ryzhikov wrote that the foundation of the foreign policy course of the British Labour government in 1945–1951 "was defined by attempts to strengthen the position of British imperialism in England and beyond its borders, serving the interests of British and US monopolies, hatred toward the USSR and socialist countries, the desire to strangle the national liberation movement that had emerged after the end of World War II in the countries of the British Empire and all over the world, fear of communist ideas, democratic forces and social progress all over the world."[153]

This claim is only partly justified. In Ritchie Ovendale's words, though the Labour Party pursued the ideals of socialism at home, it did not extend these to foreign policy. Attlee shared Churchill's perspective on foreign

Figure 3.7. Volume of Funds Received by the United Kingdom from the United States under Grants and Loans, Taking into Account Repaid Loans, 1945–1970 (US$ millions).

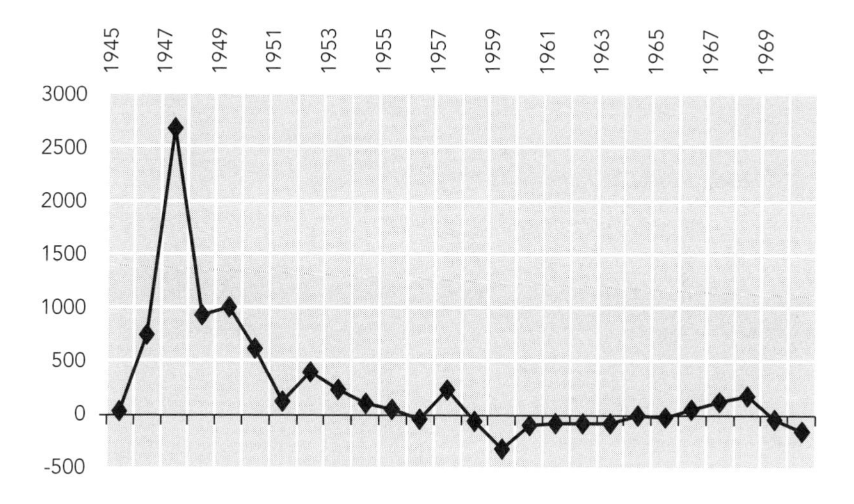

Source: Series U 75-186: U.S. Government Foreign Grants and Credits, by Country: 1945 to 1970, in *Historical Statistics of the United States: Colonial Times to 1970,* pt. 2 (Washington, DC: U.S. Department of Commerce, Bureau of the Census, 1975), 873–874.

policy issues, and everyone expected him to continue the course typical of the previous Conservative cabinet.[154] However, Attlee's Labour government laid the foundations for the transformation of empire under new conditions and in this way contributed to the defeat of the idea of empire. The Attlee government formulated the main directions of empire transformation: nation-building, a transfer of power, and constitutional reforms in the colonies. As David Fieldhouse wrote, "between 1945 and 1951 Attlee's governments dismantled Britain's imperial system in south Asia, accepted a fundamental change in the nature of the Commonwealth, and defined a process of political evolution which led directly to the dismemberment of empire in Africa, the Caribbean and the Pacific during the two decades after 1951." Fieldhouse also tried to answer the question of "how far such far-reaching changes were planned or whether they grew, unintended, from the pressure of circumstances." In general, he concluded that

"Labour's position on imperial issues was almost identical with that of most Conservatives and that their policies in office were shaped by circumstances rather than by principle." He also noted that "the Churchill government that came into office in 1951 could not put the clock back," even though "it attempted to slow up the extension of self-government" for Ghana.[155]

W. Roger Louis noted that Attlee regarded the system of international trusteeship created within the UN framework as a continuation of "the old story of British Imperialism," and quoted from a memorandum of British Prime Minster Clement Attlee in September 1945: "After the last war, under the system of mandates, we acquired large territories. The world outside not unnaturally regarded this as a mere expansion of the British Empire. Trusteeship will appear to most people as only old mandates writ large." At the same time, Louis believes that Attlee, "with the objections of a Labour politician who opposed Imperialism on ideological ground, sensed that the basis of power of the British Empire was being transformed," and realized the need to protect the Empire and the peoples inhabiting it.[156]

The need to maintain or transform the empire and preserve order in the dependencies posed a significant problem for the postwar governments, both Labour and Conservative. As Gillian Peele has noted, analyzing concurrences and differences in defense and foreign policy of the leading British parties, the Conservatives were characterized by a closed decision-making procedure, realism in defining priorities based on national interests, and the appointment of professional diplomats and officials to key positions in those areas. The Labour Party traditionally demonstrated a higher commitment to ethical and moral principles in politics, corresponding to the idealist paradigm in international relations; a significant degree of openness of decision-making procedure; and the appointment of popular politicians to key security and defense positions.[157] Labour governments were characterized by an active pursuit of policy to involve the colonial population in governance, protect its rights to land, develop democratic procedures, allocate funds for the development of colonial nations, develop the education system, and fight against racial discrimination and forced labor.

One of the most serious crises in relations between African countries and the United Kingdom broke out during the Labour government headed by Prime Minister Harold Wilson (1964–1970). Bitter differences were caused by the political system of Rhodesia and the apartheid practices supported by the local white minority. Meanwhile, researchers also note that the Labour Party seemed more concerned about the economic aspect of the problem, that is, sanctions. Conservatives were more pragmatic in evaluating the need to allocate money to support local forces loyal to the British and thoroughly weighed all the pros and cons of maintaining the British presence and choosing the forms of influence. Conservatives did not want the Commonwealth to become, like the United Nations, "a forum for expression of Third World grievances."[158]

Both Labour and Conservative parties regarded the Soviet threat in Europe and Asia in the 1940s to early 1950s as a primary one, and starting in the late 1950s, concern about the spread of communism in Africa was expressed more and more often. The Chinese threat was viewed primarily in the context of the security and defense of British interests in Southeast Asia and the Far East. The earliest official documents of the postwar period already emphasized the impossibility of maintaining full control in the colonies by military means only. Identifying the strategy with regard to dependent territories, British politicians paid significant attention to the issue of colonial development, going so far as to lay the foundations for a contemporary "development policy" that included political, economic, cultural, educational, and technical aspects. However, the humanitarian and educational projects also required significant spending to implement.

Ovendale, evaluating "what lay behind the British decision to abdicate in Africa" in 1960, wrote that many earlier theories about the reasons for the rapid dissolution of the empire were finally confirmed or refuted by government documents that were released in the 1990s after the 30-year rule had expired. Ovendale indicated that the "quit in Africa" position was caused by a number of international and national factors: (1) "the growing evidence that there would be disturbances in Africa similar to those in Asia at the end of the Second World War"; (2) the inability of Britain to prevent such a development and the concomitant possibility that the British presence would be associated with failure to preserve the order, which could push the political forces toward the communist bloc; (3) the failures of France and Belgium in conducting a colonial policy in Asia and Africa and the moral condemnation of their policy by world public opinion; (4) pragmatic profit-and-loss calculations in relation to keeping colonies; (5) "the need for the West to maintain a common front in Africa to prevent Soviet penetration"; and (6) "strategic calculations in relation to the protection of Middle Eastern oil." Like many other researchers, Ovendale pointed to the influence of the Suez Crisis, the launch of Sputnik, and the development of the Soviet nuclear program on the British defense strategy, as well as the fact that "the threat to Britain's position and influence in the world was political and economic rather than military."[159]

International Factors

The growing US influence in the world and tensions between the transatlantic partners were the most important international factors influencing the dissolution of the British Empire. According to Niall Ferguson, "the key to victory—and the key to the future of the Empire itself—lay, ironically, with the country that had been the first colony to throw off British rule," or "as one old Colonial Office hand already sensed...'the prize of victory [would] not be the perpetuation, but the honourable interment of the old system.'"[160] The anticolonial and anti-imperialist position of the United States was declared in the Atlantic Charter, signed on August 14, 1941, by President Franklin D. Roosevelt and Prime Minister Winston Churchill, despite differences between these two countries with respect to colonies and dependent territories.[161] The Soviet Union became a signatory to the Charter on September 24, 1941, together with European countries,[162] though Stalin had a skeptical attitude toward this document: "I actually thought that the Atlantic Charter was directed against countries striving for world domination, but now I see that it is directed against the Soviet Union," according to David Kaiser. Kaiser noted that "Anglophobia still dominated important elements of the American press, and the President was sensitive to any accusations of 'pulling British chestnuts out of the fire.'"[163] Louis wrote that many prominent Americans, including Roosevelt, believed that "national independence was the natural and desirable course in world affairs, though, in the case of the British, some were more willing than others to grant that self-government within the British Empire might be a satisfactory alterative."[164] President Truman in his speech "Restatement of Foreign Policy of the United States" of October 27, 1945, outlined the fundamental principles of foreign policy in an imperfect world after the end of the war. Seven of the twelve principles in one way or another stated the right of nations to free self-determination, the democratic principles of governance, and a ban on limited sovereignty and coercion.[165]

W. Roger Louis tried to identify the role of the United States in British decolonization during World War II. His assessment reflects the dual character and evolution of the US position. On the one hand, many American politicians and officials hoped that the empire would be destroyed during the war; however, by the end of the war British politicians had managed to prove that preservation of the empire would contribute to security and check the proliferating Soviet and Chinese influence. On the other hand, the negative attitude of the American public and many officials, as well as the growing dependence of the United Kingdom on the United States, prompted Britain to strive for reform of the empire and the development of colonies toward full independence.[166] The Suez Crisis became a serious test for the Anglo-American partnership. The unexpected concurrence of US and Soviet positions in the United Nations created an anti-imperialist front that France and the United Kingdom could not counter. British attempts to influence the United States in its anti-imperialist rhetoric and practice were not successful owing to US economic interests and the nation's colonial past.

Alan Burns's book *In Defence of Colonies: British Colonial Territories in International Affairs*, published in 1957, is a unique account of the emergence of an anti-imperialist alliance that crossed the space dividing political blocs. Burns spent 42 years working in the British Colonial Office; he was governor of the Gold Coast (Ghana) from 1941 to 1947 and permanent representative of the United Kingdom on the UN Trusteeship Council from 1947 to 1956. In the preface to his book, he noted that the Britons, under the influence of ever-growing international criticism with regard to the British rule in colonies, "are beginning to be a little ashamed of our position as a colonial Power and inclined to pay undue attention to the self-righteous attitude of other nations." The reason for writing the book, according to the author, was the lack of objectivity of the representatives of "anticolonial" nations, who, though they may be "personally friendly and reasonable in their approach to colonial questions, ... are generally bound by anti-colonial convention (or by their instructions) to an extent which does not allow them to deal with such questions publicly in an objective manner."[167] Burns cited the United States, the Soviet Union, India, Latin

American countries, and the United Nations as a whole as the most striking representatives of the anticolonial alliance. Characterizing the US position with regard to the colonial issue, he sarcastically noted that the difference between the United Kingdom and the United States was that the United States preferred to regard its colonies "as nothing more than 'possessions.'" Most Americans, he felt, did not understand the British contribution to the economic and political development of the peoples of the empire and the United Kingdom's help in preparing them for self-government.[168] Burns wrote that anticolonial feeling in America was "due almost entirely to an historical emotion, and is based to a large extent on a lack of knowledge and a convenient blindness to comparative facts. The anti-colonial attitude of the United States has been a great help to Communist policy in the past but there is fortunately a growing realisation of this fact in responsible American circles.... It is often said that the Communist leaders are trying to drive a wedge between the United Kingdom and the United States. There would be less chance of this wedge being effective if the British could feel that their American allies ... were not intent on weakening the British position by making colonial administration more difficult."[169]

Burns also believed that the Soviet Union was an empire, as it continued the policy of conquering new territories and exploiting those it had inherited from the tsarist regime. He also noted that in the Soviet Union, unlike in dependent British territories, all manifestations of nationalism and critique of existing order were suppressed.[170] He criticized the anticolonial position of India, stating flatly that the country had "assumed the leadership of the (non-communist) anti-colonial bloc."[171] Burns calculated the distribution of positions of 76 countries in the United Nations with regard to colonial issue: 8 Arab states, 13 Asian states, the Soviet bloc (9 countries), 20 Latin American states, and 26 "miscellaneous" states, a category that included the developed countries of Europe and North America. In his view, all regional groups were ready to support any resolution hostile to colonialism. Of the group of developed countries (the "miscellaneous" states), at least three countries, he thought, also had an anticolonial position.[172]

Prime Minister Harold Macmillan personally addressed President Eisenhower, asking the United States not to vote for the UN General Assembly resolution on immediate independence to colonies in December 1960. However, having abstained from voting at the direction of the White

House, US delegate Zelma Watson George stood up and applauded to welcome the resolution's passage.[173] The contradictions between partners were sometimes so pronounced that Macmillan in his diary ironically referred to Anglo-American affairs as "a very special relationship."[174]

In the Soviet literature, the US anti-imperialist rhetoric was often regarded as insincere, which led to an underestimation of US influence in the dissolution of the European empires after the war. However, an analysis of American and British official documents of the postwar period prompts agreement with the opinion expressed by the contemporary British historian Andrew Williams, who wrote that the liberal ideas of the American elite and its negative attitude toward British imperialism were not "just liberal hypocrisy" that "camouflaged for naked pursuit of national interest." On the contrary, US national interests stemmed from an understanding of the need for ethics in foreign policy.[175]

In the postwar period, the problems of constructing a new international security system inevitably forced US leaders to search for a balance among ideas of national liberation, the struggle against leftist and communist ideology, providing for US economic interests, and managing security on a global scale. Thus, the issue of a colonial legacy could not be disentangled from military-strategic issues. At the same time, the United States did not want its position on decolonization to be associated with the policies of the European empires. Acting in support of colonial independence, the United States did not want to accept responsibility for the transition period and had a pragmatic attitude toward military intervention, especially intervention in British dependent territories and the colonies of other European powers. The US position with respect to former European colonies in Africa and Asia was clearly outlined in declassified documents of the National Security Council of 1948–1949. A 1949 NSC document dealing with the possible capture of Hong Kong and Macao by Communist China noted that, taking into account the low strategic, economic, and military value of these areas for the United States, it would be "unwise for the United States to contribute forces for the defense of Hong Kong and Macao unless we are willing to risk major military involvement in China and possibly global war. Similarly, no United States military materiel should be provided or committed in advance of a re-examination of the situation at the time the British and/or Portuguese position is made known to us."[176]

The NSC reports of 1948–1949 dealing with the former Italian colonies in Africa (Eritrea, Libya, and Somaliland) express the US urge to emplace "unilateral British control," but Britain refused to accept the burden.[177] The NSC report of July 26, 1949, considered various options of voting at the UN General Assembly concerning the independence of Eritrea, Libya, and Italian Somaliland. It noted that US policy should be directed toward "the preservation of United States and United Kingdom strategic interests and positions." However, practical work aimed at a power transfer in Libya was to be undertaken, according to the document, by French and British administrations, while the United Kingdom was accorded responsibility for preparing Eritrea to unite with Ethiopia and Somaliland, a decision that was to be codified by the UN General Assembly. The US position was expressed as follows: "Our strategic interests will be protected until that time by the continuance of the British administration."[178] Following the UN decision in 1948, Libya (Tripolitania and Cyrenaica) officially gained independence on December 24, 1951. In November 1949, the United Nations made a decision to establish Italian trusteeship over Somaliland for 10 years, while Eritrea was attached to Ethiopia as a result of a referendum.

The United States maintained a similar position with regard to unrest in Malaya, where it had economic interests, as Malaya provided more than half of the natural rubber supplies used by the United States and more than one-third of its tin, and was also the biggest dollar-earner in the sterling bloc. A CIA memorandum noted that at that moment, there was no serious threat from communist forces in the region, the United Kingdom managed to maintain order, and US economic and strategic interests were protected.[179] Nevertheless, in some regions the British presence relied on US support—in Cyprus, Aden, Malaya, and Kenya, for example—or, in Niall Ferguson's words, where "British rule was essentially 'underwritten' by the US."[180]

The United States was forced to become more and more active in ensuring international security because of the Soviet and communist threat. Responsibility for maintaining the global order thus migrated from the United Kingdom to the United States. Niall Ferguson referred to this process as a "transfer of power," drawing a parallel with empire transformation as a transfer of power to the local forces, as the term is accepted in the British literature. However, the meaning of this phrase has a global context.[181] By the end of World War II, the United Kingdom was the only Western country that had a mass-trained army, fleet, and aviation capacity, as well as a worldwide network of military bases. This fact made the Anglo-American partnership in international security inevitable. The United States and the United Kingdom discussed the degree of responsibility and intervention in case of civil war in Greece and the situation in Yugoslavia, Turkey, China, the Middle East, and South Asia. All these situations were in a way related to confrontation with a leftist ideology promulgated by national liberation movements. Nevertheless, the United States did not plan to accept the role of world policeman alone, and insisted on sharing the responsibilities of demilitarization and de-Nazification of the Axis countries with both Western and Eastern bloc countries. In Ovendale's view, the Korean War demonstrated that the United States had chosen in favor of active participation in ensuring international security on a global scale.[182]

In considering the international factors that contributed to the rapid dissolution of the British Empire, we should also note the agreement of US and Soviet positions with regard to Southeast Asia in the first postwar decade. Alexey Bogaturov, a leading Russian international relations expert, has pointed out that both countries preferred to distance themselves from postcolonial conflicts. The positions of the two countries also concurred with regard to Chiang Kai-shek, who was rendering assistance to the growing national liberation movement in French Indochina. In both countries, the leadership was satisfied with the desire of nationalist forces to destroy the remains of European colonial power in the region. Later, the United States and the Soviet Union demonstrated restraint toward the government of the DRV under Ho Chi Minh. They had a similar policy with regard to Indonesia, and there was no confrontation with regard to the riots in British Malaya. In the first postwar decade, both countries practiced restraint in their approach to local revolutionary movements. Moreover, the Soviet influence and the military and economic assistance it supplied to revolutionary movements in the region were much less than deemed by Western analysts of that period.[183] Paul Kennedy noted that the growing popularity of communist ideas in the world was not a manifestation of the Soviet influence, though the Soviet Union could claim it as part of its victory.[184]

The final defeat of France in Indochina in 1954 and the subsequent active involvement of the United States in regional policy marked a new stage in the alignment of forces and interests in Southeast Asia between Western bloc partners and opponents of a bipolar world. British documents on the situation in Laos in 1961 reflect an increasingly restrained position of the United Kingdom as well as growing British distrust and anxiety about the overly aggressive behavior of the United States, which could draw SEATO countries into a full-scale military conflict.[185] US Secretary of State Dean Rusk, in response to the British decision to withdraw all troops from Southeast Asia and the Middle East in 1967, noted that this decision left him "profoundly dismayed," and that it represented "a major withdrawal of the United Kingdom from world affairs, and this was a catastrophic loss to human society." The British position with regard to the war in Vietnam was negative. The Labour Party rank and file evaluated it as "as the most immoral act since the Holocaust," though the official position of the Labour government was moderate. When asked why the United Kingdom did not adopt a tougher stance with regard to US policy in Vietnam, Prime Minister Wilson answered, "Because we can't kick our creditors in the balls."[186]

Soviet researchers traditionally consider the collapse of colonial empires to be conditioned by the overarching crisis of capitalism and inevitable revolutionary transformations. They wrote about the "liquidation of the colonial system as an integral part of world revolutionary process"; however, they also emphasized that "socialist revolutions … as such would not have caused a crisis and dissolution of the British Empire, if these processes had not been prepared by the development of characteristic contradictions. The criminal and shameful system of colonialism is collapsing under the weight of the crimes it has committed."[187] This suggests that Soviet and Western researchers identified some common reasons for the dissolution of the British Empire, though their conclusions and predictions differed. As a result, the studies of the 1950s to 1970s contained some conclusions important for understanding the nature of empire dissolution as an example of a clash between asymmetric antagonists. The main objectives of the military campaign were to prevent the victory of communist and pro-communist forces, because their coming to power would make it impossible to maintain traditional political and economic ties with a given country. The conclusions of Soviet historians generally were in accord with the fears of Western politicians, though this did not mean that the Soviet Union had a decisive influence on empire dissolution.

In fact, official documents from the 1940s and 1950s show that both the Soviet influence in the third world countries and the fears of Western countries concerning this influence were exaggerated, if substantiated. There are no data to confirm that the Soviet Union rendered significant military or material assistance to the national liberation movements in Asia, Africa, and the Middle East *in the first postwar decade.* The Soviet Union had to rebuild its economy under conditions much worse than those the United Kingdom faced. Control in Europe was associated with significant costs. The Soviet nuclear program consumed a lot of resources. Leaders of the newly independent states in Asia and Africa were inconsistent, and this did not contribute to strengthening the Soviet influence there. Hence the Soviet influence was expressed to a larger extent through the formation of an alternative ideology of liberation and anti-imperialism, especially using the UN platform. China, India, Indonesia, and Malaysia became prominent players in Southeast Asia, but they occupied a special position.

The presence of a multimillion-troop Soviet army in Eastern Europe and the Soviet nuclear weapon capability prompted the United Kingdom to abandon its customary isolationist thinking and contribute to the balance of power in Europe. By the mid-1960s, the United Kingdom had deployed in Europe three times as many troops as it had in Asia, the Middle East, and North Africa. The constant bargaining that went on between the United States and the United Kingdom over transatlantic obligations to counteract the Soviet threat in Europe illustrates this claim well.

Andrew Rothstein, a member of the British Communist Party and the author of *British Foreign Policy and Its Critics, 1830–1950*, published in Russian in the Soviet Union in 1973, wrote that British foreign policy was gradually made subordinate to US foreign policy, which was a consequence of "the policy of maximum hostility towards the Soviet Union." He listed the following events to confirm this claim: the Quebec agreement between the United States and United Kingdom in 1943 on a secret nuclear bomb development; a series of agreements with the United States on financial and economic issues in 1945–1947; the devaluation of the pound, carried out under US pressure in 1949; an agreement to place American bases with nuclear weapons on British territory; "restrictions in trade with Socialist countries introduced at American bidding in 1951 and being in place for many years; consent to rearmament of Western Germany in 1954, despite the fact that a minority in the House of Commons voted in favor; subsequent opportunity given to troops and tanks of Western Germany to hold exercises in the British territory; shameful support of American aggression in Vietnam."[188]

More clearly seen now is the contradiction between the United States and the United Kingdom with regard to the British colonies and the principles of international trade and financial system. The United States intended to replace the British pound with the US dollar as the major world currency, creating additional difficulties for British economic stability within the sterling zone and imperial preferences. US economic assistance to the United Kingdom was brought up earlier. This aid was made contingent on certain conditions being met, often as a price paid for the destruction of the imperial world order. In A. M. Rodriguez's opinion, the establishment of the Bretton Woods system contributed to the destruction of "the closed nature of imperial monetary systems of Britain and France" and prepared for the transition from colonial policy to neocolonialism.[189] The implementation of the Marshall Plan further strengthened the US influence through loans made to European countries. A similar view was expressed in the Soviet literature, framed as "the contradictions of imperialist powers in the struggle for sales markets and markets of resources." Niall Ferguson cited Hitler's words that it was "rival empires more than indigenous nationalists who propelled the process of decolonization forward." And the United States, in his view, is without doubt an empire. He believes that "the imperial renaissance might have led further if the United States and Britain had made common cause, for American backing was the sine qua non of imperial recovery."[190]

CONCLUSIONS

The main reason for the dissolution of the British Empire was the impossibility of preserving colonial rule and prewar forms of political and economic relations under the new historical conditions. The United Kingdom found it difficult to retain efficient control of its colonies and mandates; it had to change how it interacted with local elites and political forces that were not satisfied with a subordinate position. Thus, a major role in the dissolution of the British Empire was played by the rise of national liberation movements under the influence of the events of World War II and the anti-imperialist agenda of the two postwar world leaders, the United States and the Soviet Union. Liberal ideas—liberation and the absence of coercion—found proponents and opponents at both poles of the post–World War II world order. Many authors have noted the competing and often conflicting interests of the United States and the United Kingdom, especially with regard to European colonial possessions. The United States believed that independence should be granted as soon as possible, while European countries, especially the United Kingdom, preferred to execute a planned process over several decades. The United States openly expressed its anticolonial position and exerted pressure on European countries in order to destroy the remains of imperial possessions and create a new economic order based on the principles of liberalism and free trade.

It would be a mistake to consider the dissolution of the British Empire as a sign of military weakness without taking into account the political and economic aspects of the empire as a system. The British Empire was to a large extent an economic project, though its profitability is debatable. Nevertheless, it was impossible to protect economic interests without relying on military force. The imperial state in the colonies was founded on a combination of factors: (1) military power, to maintain order and subordination both within the empire and beyond; (2) political and economic factors, which brought certain benefits to local elites and the population; and (3) an ideological system emphasizing values, to justify the imperial order to elites and the population. World War II undermined the foundations of colonial dominion, and the colonial powers had to restore control. The postwar conditions were unfavorable for the preservation of imperial possessions, even by states with strong military and power resources. The United States, the new leader of the Western world, openly made known its anti-imperialist position and contributed, in rhetoric and in practice, to the demolition of the European empires. As a result, the United Kingdom was more concerned with safeguarding its economic and political interests in the new world order than with keeping an empire together.

The asymmetric conflicts sparked by national liberation movements in the British colonies hastened the transition from imperial to local control. Small wars in colonies were initially considered political rather than military enterprises. Their political feasibility was an important issue, and the United Kingdom wanted to reduce its efforts to keep the peace in regions going out of control. In the interwar period, the so-called inter-imperialist contradictions with regard to control, access to resources, and the use of colonial possessions represented the greatest challenge to security. After World War II, however, the Western powers had to coordinate their efforts as a growing communist influence in Asia and Africa became the overriding concern. Responsibility for control over the periphery and semiperiphery of the former British Empire was transferred from the United Kingdom to the United States.

Having lost its global leadership, the United Kingdom hoped for a Pax Americana to replace the Pax Britannica, and was concerned to prevent the rise of Russian communism. The United States reluctantly agreed to expand its sphere of authority and maintain order on a global scale, fearing that it would be accused of imperial behavior both at home and abroad. However, the persistent position of the United Kingdom with regard to its economic capabilities, defense spending, and army size helped push the United States into more active participation in ensuring international security, along with a conscious need to counteract the Soviet threat on a global scale. Yet even after the dissolution of the Soviet threat in the early 1990s, the effects of the transition from British power to American power would be felt in the continuation of asymmetric conflicts that echoed those of the Cold War years.

CHAPTER 4

The US War in Iraq, 2003–2011

The withdrawal of US troops from Iraq in December 2011 marked the end of one of the largest military operations of the United States in the global war on terror initiated by President George W. Bush. American military leaders and policy-makers often described the war in Iraq as asymmetric in terms of the strategies used by each participant, which provides formal grounds for analyzing it within the framework of asymmetric conflict theory. Unlike military analysts, however, we will focus on evaluating the political outcome of the conflict and whether US goals were achieved. I believe that the war in Iraq delivered a political defeat to the United States in the absence of a military defeat and hence is a perfect case for applying asymmetric conflict theory, which offers explanations for the paradoxical defeat of strong powers in wars against weaker adversaries. In the early 1990s, American military leaders talked about asymmetry in terms of the global military power superiority of the United States, which could not be undermined in "big" or "conventional" war. The United States is often considered to have repeated in the Iraq War the mistakes it made during the Vietnam War, the conflict that provided a springboard for asymmetric conflict theory.

An analysis of the war in Iraq through the lens of asymmetric conflict theory will focus on evaluating the results of the war, and specifically whether US goals in entering the war were met. This analysis has several stages: (1) looking at the war objectives and justification of the war cause;

(2) analyzing the course and outcome of the military operation, as well as the content and results of the reconstruction and democratization programs in Iraq; and (3) examining American public opinion toward the war. Put otherwise, this chapter evaluates the war in terms of the "just goals" claimed for it, the means used, and the results achieved.

The US administration had to present the war cause as legitimate to obtain congressional approval and ensure the support of both the US public and the international community, the latter represented by the United Nations and other influential international actors. Because national and international law regulates the right of sovereign states to use military force domestically and internationally, the exercise of military force today is not merely a military issue but also a legal and political one. For the United States, with its central place in the international relations system and its desire to be a role model, compliance with international norms is a necessity.

A close look at the course of a war makes it possible to pinpoint the components of success or failure of a military campaign and the relation between military and political victory. In this regard, the programs for reconstructing Iraq and facilitating democratization were a crucial part of the military campaign. They were carried out in accordance with international norms regulating occupation and were directed toward achieving long-term peace in the country and the region. These programs were aimed at "winning the hearts and minds" of the Iraqi people by laying the foundations for a modern, developed, democratic state to replace the destroyed autocratic regime.

Debates over the outcomes of the Iraq War are salted with references to the past experiences of the United States and other countries in small wars. Rarely, however, is the view expressed that the war was successful and justified, although assessments of the outcome and opinions about the lessons of the war vary greatly. It is interesting that President Obama began his January 24, 2012, State of the Union speech by labeling the war in Iraq a success, for his evaluation of the war during his first electoral campaign had been precisely the opposite. Here is the relevant part of his State of the Union speech:

> We gather tonight knowing that this generation of heroes has made the United States safer and more respected around the world. For

> the first time in nine years, there are no Americans fighting in Iraq. For the first time in two decades, Osama bin Laden is not a threat to this country. Most of al Qaeda's top lieutenants have been defeated. The Taliban's momentum has been broken, and some troops in Afghanistan have begun to come home. These achievements are a testament to the courage, selflessness and teamwork of America's Armed Forces. At a time when too many of our institutions have let us down, they exceed all expectations. They're not consumed with personal ambition. They don't obsess over their differences. They focus on the mission at hand. They work together.[1]

Because any discussion of what was achieved in Iraq is inevitably highly politicized, we will focus on objective criteria for assessing the war's outcome.

THE WAR'S OBJECTIVES AND LEGITIMATION OF THE WAR'S CAUSE

The war in Iraq—dubbed Operation Iraqi Freedom—was launched on the night of March 19, 2003, by international coalition forces, dominated politically, militarily, and materially by the United States. The accusations that the George W. Bush administration had laid against the Iraqi government served as grounds to start the war. Iraq was accused of developing and possessing weapons of mass destruction (WMDs) and maintaining contact with the international terrorist network al-Qaeda. The US-led military operation was aimed at overthrowing the Saddam Hussein regime and was a continuation of the war on terror declared by President Bush in 2001.

There is a widespread opinion that the notion of a forceful disarmament of Iraq was conceived as the result of neoconservative Republicans being in power; however, numerous facts testify in favor of bipartisan continuation of and support for the policy toward Iraq. The problem of Iraq disarmament became urgent in 1998, when Baghdad suspended cooperation with UNSCOM, the United Nations Special Commission on Iraq,[2] declared it an instrument of US military espionage, and forced UN inspectors to leave the country. In December 1998 the United Kingdom and the United States launched air strikes against targets in the north and south of the country with operation Desert Fox, in order to destroy facilities where biological, chemical, and nuclear weapons allegedly were being developed and produced.[3] The Clinton administration set the task of overthrowing the Saddam Hussein regime, and in December 1998 the US Congress adopted the Iraq Liberation Act, allocating US$97 million to support the political opposition in Iraq.[4]

Bob Woodward noted in his 2004 book *Plan of Attack* that even before the war the United States had effectively controlled a significant part of Iraqi airspace, and that President Bush was briefed on this before his inauguration in January 2001: "The United States enforced two designated no-fly zones, meaning the Iraqis could fly neither planes nor helicopters in these areas, which comprised about 60 percent of the country." Woodward then described the coalition's military control of Iraqi airspace:

> Operation Northern Watch enforced the no-fly zone in the northernmost 10 percent of Iraq to protect the minority Kurds. Some 50 U.S. and United Kingdom aircraft had patrolled the restricted airspace on 164 days in 2000. In nearly every mission they had been fired on or threatened by the Iraqi air defense system, including surface-to-air missiles (SAMs). U.S. aircraft had fired back or dropped hundreds of missiles and bombs on the Iraqis, mostly at antiaircraft artillery.
>
> In Operation Southern Watch, the larger of the two, the United States patrolled almost the entire southern half of Iraq, up to the outskirts of the Baghdad suburbs. Pilots overflying the region had entered Iraqi airspace an incredible 150,000 times in the last decade [1991–2001], nearly 10,000 in the last year. In hundreds of attacks not a single U.S. pilot had been lost.[5]

In February 2001, the United States and Britain carried out bombing raids to try to disable Iraq's air defense network. The bombings had little international support.[6] Thus, in retrospect, the issue of the use of military force against Iraq, as well as the desire to overthrow Saddam Hussein and change the regime in Iraq, comes into clear focus.

At the UN, the issue of Iraq was interpreted as a problem of disarmament and achieving efficient control over the disarmament process. It was considered important that such a process should involve the coordinated efforts of the entire international community. After the 1998 crisis in relations between the UN and Iraq, it took a year to settle the issue of modifying the verification and control regime. UNSCOM ceased its activities, and in 1999 a new UN commission was established, UNMOVIC, or the United Nations Monitoring, Verification and Inspection Commission. The new commission proceeded with the UNSCOM mandate under UN Security Council Resolution 1284 of December 17, 1999. The work of UNMOVIC was characterized by more active involvement of the UN Secretary-General, whom the chairman of the commission could address directly, thereby circumventing the Security Council.

At first Iraq rejected Resolution 1284 and demanded an unconditional lifting of sanctions, for, according to the Iraqi government, all demands in the field of disarmament had been met. UN Secretary-General Kofi Annan worked hard to find a compromise through a dialogue with the Saddam Hussein government, eliciting heavy criticism from the United States and its supporters. In the course of 2000 the new commission was formed, and another year was spent training inspectors in the special training programs. In July 2002, Hans Blix, the chairman of UNMOVIC, met with representatives of Iraq and Kuwait in Vienna, where the possibility of resuming inspections in Iraq was discussed. By mid-2002 conditions were in place for inspections to resume.

However, these developments did not align with the interests of the George W. Bush administration. For the US administration, the problem of Iraq had been transformed in the context of the war on terror and the war in Afghanistan. Bob Woodward confirmed that President Bush's close circle of advisers—National Security Adviser Condoleezza Rice, Vice President Dick Cheney, Secretary of Defense Donald Rumsfeld, and CIA Director George Tenet—tried to convince the president to go to war against Iraq after 9/11. Already by late November 2001 General Tommy Franks, combat commander of CENTCOM, had been requested by Rumsfeld to update the existing Iraq war plan, Op Plan 1003, which outlined an attack and invasion of Iraq designed to overthrow the regime of Saddam Hussein. Op Plan 1003 had been approved in 1996 and updated in 1998, and called for deploying a force of 500,000 for overall invasion and a preparation phase of seven months. The renewed plan called for a lower troop deployment, 400,000, and a shorter mission time of six months. Woodward also noted that at the time, the war in Afghanistan was seen as a victory: "The widespread prediction of a Vietnam-style quagmire had been demolished, at least for the time being, and Rumsfeld was in a buoyant mood."[7]

The Bush administration's readiness to launch a war against Iraq immediately after the terrorist attacks of September 11, 2001, is confirmed by the fact that a number of agencies, including the Department of Defense, the Bureau of Near Eastern Affairs of the Department of State, USAID, and the National Security Council, had begun planning the "postwar administration" of Iraq in the fall of 2001 and on into 2002.[8] Leading British journalist John Kampfner asserted in his book *Blair's Wars* that the decision on the United Kingdom's participation in the war was made during Prime Minister Tony Blair's visit to the Bush ranch in Texas in April 2002.[9] All subsequent actions by George W. Bush and Tony Blair in the runup to the war ought to be regarded as attempts to legitimize the planned military operation against Iraq at national and international levels.

The bellicose position of the United States led to a split in the UN Security Council with regard to the possible resolution of the crisis and worsened US relations with European and Arab countries. The United States believed that coercive action would convince Iraq that the international community was determined to finally resolve the problem of Iraq's disarmament. Moreover, a military operation was seen as preemptive, as was often stated by President George W. Bush in his speeches in the second half of 2002. (As an example, he devoted his whole speech before the UN General Assembly on September 12, 2002, to the Iraq problem.) Bush used the first anniversary of the 9/11 attacks to drum up emotional support from UN General Assembly delegates for his new coercive action in the war on terror. He called on General Assembly delegates to support decisive actions against a regime that personified all conceivable threats to peace—promoting terrorism, developing WMDs, engaging in human rights violations—and was also a source of instability and aggression in the region. Bush listed the violations committed by Iraq against UN and international law and called on General Assembly delegates to prevent "far greater horrors" in comparison to which "the attacks of September the 11th would be a prelude." Bush tried to prove that Iraq represented a threat to peace and security, and that if the UN did not approve military actions against Iraq, the United States would reserve the right to use force. Nothing was said about the use of force as such; however, the message was unequivocal. Bush stated that the United States "cannot stand by and do nothing while dangers gather," and that it "must stand up for our security, and for the permanent rights and the hopes of mankind"—and must do this even without UN support.[10]

US representative John Negroponte and Secretary of State Colin Powell repeated in their statements before the UN Security Council the argument about the need for war and its just nature. In their statements before the UN Security Council, American and British representatives said that Iraq's behavior discredited the role of the UN in conflict resolution, security, and the WMD nonproliferation regime and that it was necessary to demonstrate the UN's readiness to act decisively, which implied a military operation. However, most countries on the UN Security Council were in favor of a peaceful resolution to the Iraq problem.[11] The key arguments against the use of force were as follows:

- The use of force would be a violation of international law, for no evidence had been found to prove that Iraq was producing and storing outlawed weapons.

- The use of force would result in more casualties than an attempt at a peaceful resolution and would result in a humanitarian disaster in the country and the region.

- The use of force would destabilize the situation and create another source of instability and threat in the Middle East.

Three permanent members of the UN Security Council—Russia, France, and China—were ready to consider compromise solutions that took into account Iraqi demands yet continued the inspections. The proponents of this position hoped to benefit from economic cooperation with Iraq in the long run, after the lifting of economic sanctions, and hoped for cooperation in exchange for guarantees of debt repayment by Iraq. Most Arab states and the Non-Aligned Movement countries also supported peaceful resolution to the conflict. At the UN General Assembly on March 12, 2003, Mr. Lamba (Malawi) spoke on behalf of the African Group:

> The heavy consequences of war in Iraq will be felt very acutely, even in Africa.... Thousands, if not millions, of innocent lives will be lost in Iraq. The fragmentation of Iraq is not inconceivable. The spillover of the war could conceivably create a regional conflagration as the conflict transcends the borders of Iraq. In our global village today a backlash in various forms would destabilize the world even more, and New York or London would not be assured of any safety when the uncertainty of life leads to desperation. Africa considers the war against terrorism as a bigger threat to global peace.[12]

Two positions with regard to Iraq took shape among members of the UN Security Council, and the proponents of both tried to prove that they were pursuing the same objective, Iraqi disarmament, though the positions stemmed from different assumptions and different scenarios for solving the problem. The position that inclined toward a peaceful resolution of the crisis was founded on the presumption of innocence: until Iraq's possession of WMDs was proved, there were no legal foundations for the use of force. To solve the problem, cooperation with Iraq was needed, as well as continuation of the inspections, control over disarmament, and the coordinated efforts of the international community. The second position was based on a conviction that Iraq was guilty and that whatever steps Iraq took would be perceived as inadequate proof of its intent to cooperate, or as a new deception. The conviction that Iraq possessed WMDs and produced them was grounded in previous experience and did not require new evidence, so that continuing with inspections was meaningless. The use of force was to a large extent intended to punish Iraq rather than disarm it.

Under strong US pressure, the Security Council unanimously adopted Resolution 1441 on November 8, 2002, obligating Iraq to start fulfilling all disarmament demands without any preconditions or delays.[13] Bush expressed his approval of the resolution and stressed the language of the resolution: "Iraq is already in material breach of past U.N. demands." He also said that "the full disarmament of weapons of mass destruction by Iraq will occur.... The only question for the Iraqi regime is to decide how. The United States prefers that Iraq meet its obligations voluntarily, yet we are prepared for the alternative."[14] Thus, the only resolution of the Iraq problem that was acceptable to the US government entailed the use of sufficient force to destroy Iraq's military potential, overthrow Saddam Hussein's regime, physically eliminate the leader himself, and establish a regime that would correspond to US economic and geopolitical interests in the region.

An examination of the events of autumn 2002 and winter 2003 shows that proponents of the use of force were not prepared to consider the change in Iraq's position and the steps it had taken to resume inspections and fulfill its disarmament obligations. No doubt, all these actions were undertaken under strong pressure from the United States and United Kingdom, which were actively preparing for war and made clear their serious intent to initiate the engagement, whatever it took. Nevertheless, the legitimacy of the use of force in this situation remained a highly problematic issue. In accordance with the theory of a just war, a military operation can be considered just and legitimate if it meets the criteria of "jus ad bellum," or the right to go to war, and "jus in bello," the legitimate conduct of war. These concepts, important to the conduct of modern war, can be unpacked as follows:

1. The decision to go to war must be made by a legitimate authority. Here two general scenarios are possible: the decision can be made either at the level of a state or by the international community, represented by an authoritative international organization with *competence in such matters.*

2. The military operation must pursue a "just cause," which in accordance with existing tradition is defined as the right to self-defense or to provide assistance to other countries that have become victims of aggression.

3. The armed actions must stem from "peaceful intentions," which is a subjective expression of the just cause principle.

4. The military operation must be undertaken only as a "last resort," that is, it must be used only when all peaceful means of achieving the just goal have been exhausted.

5. The planned operation must anticipate a "reasonable hope of success."

The principles of humanity in warfare call for the observance of two further principles, that of proportionality and that of discrimination. Proportionality presumes the use of only the necessary and sufficient means to achieve the war objectives, while discrimination means that the fighting forces must distinguish between combatants and noncombatants. Furthermore, war participants must observe international humanitarian law regulating the behavior of belligerent parties and prescribing the rules of occupation and the responsibility of occupation authorities.

As the United States moved forward with the just war argument, a UN decision was needed to legitimate the military operation against Iraq. This decision could be made if a majority of Security Council members were convinced that Iraq had violated disarmament obligations and therefore concluded that there was a need for the forceful disarmament of Iraq. The UN Security Council could also approve a preventive operation to fight against international terrorism, if it was proved that a link between international terrorist groups and the Iraqi regime existed and that Iraq had indeed planned, supported, and carried out terrorist attacks. Another option was to legitimize the military operation at the national level by implementing a country's right to self-defense against aggression and initiating an engagement as a preventive act against the probable aggressive actions of Iraq.

In a September 2002 speech devoted to national security strategy, George W. Bush underscored the United States' readiness to fight regimes that supported terrorist organizations. He emphasized that this fight would be offensive and preemptive: "America will act against such emerging threats before they are fully formed.... We will disrupt and destroy terrorist organizations by defending the United States, the American people, and our interests at home and abroad by identifying and destroying the threat before it reaches our borders." He also pointedly declared the United States' readiness to go it alone: "While the United States will constantly strive to enlist the support of the international community, we will not hesitate to act alone, if necessary, to exercise our right of self-defense by acting preemptively against such terrorists, to prevent them from doing harm against our people and our country."[15]

Faced with a credible threat of war, Iraq agreed to serious concessions. On September 16, 2002, Iraqi foreign minister Tariq Aziz informed UN

Secretary-General Kofi Annan of the Iraqi decision to allow inspectors to return to the country without preconditions. On November 27, 2002, UN inspectors resumed work in Iraq before the end of the 30-day period indicated in Resolution 1441. In early December 2002, Iraq submitted a 12,000-page report to UNMOVIC on the disarmament obligations it had fulfilled. In the subsequent months prior to the military operation, Iraq submitted documents on its military programs and its destruction of chemical and biological weapons stockpiles, and UN inspectors were allowed to inspect Saddam's palaces, a condition previously rejected by the Iraqi government. UN inspectors were allowed to interview without witnesses the Iraqi researchers working for military projects and were permitted to carry out video recording of objects in Iraqi territory from aircraft.

From January to early March 2003, there were heated debates on Iraq at the UN Security Council. The majority of countries interpreted Iraq's actions as a willingness to cooperate and fulfill obligations. Only a small group of countries, led by the United States, insisted on ending the inspections and starting the military operation. These countries saw Iraq's actions as new attempts to deceive the United Nations. In this regard, the statement of Secretary of State Colin Powell at a meeting of the UN Security Council in February 2003 is illustrative: "Resolution 1441 (2002) was not about inspections. Let me say that again: Resolution 1441 (2002) was not about inspections. Resolution 1441 (2002) was about the disarmament of Iraq." He added, "There are no responsible actions on the part of Iraq. These are continued efforts to deceive, to deny, to divert, to throw us off the trail, to throw us off the path."[16]

On January 27, 2003, Hans Blix informed the UN Security Council of the work done by UNMOVIC in the course of 60 days after inspections had been resumed. The UNMOVIC report stated that inspectors had failed to find evidence of the production, storage, or development of banned weapons in Iraqi territory. The United States and the United Kingdom were unsatisfied with the UNMOVIC report and tried to use information provided by their intelligence services to prove that Iraq possessed WMDs. However, the UN Security Council did not deem such information sufficient to support a military operation. On March 7, another UNMOVIC report confirmed that no evidence of the presence of WMDs had been found in Iraq. After France announced its readiness to use its veto right if the issue of a military operation against Iraq was put to a vote in the UN Security Council, the United States abandoned further attempts to obtain UN support and directed its efforts toward forming a coalition of countries that would support the United States in war.

Opposition to a military operation existed not only on the UN Security Council but also in the United States domestically. The Bush administration was criticized for its fascination with military plans to the detriment of domestic economic and social problems. Gerard Powers, director of the Office of International Justice and Peace of the US Conference of Catholic Bishops, wrote: "Based on available information, there is no new evidence, no new precipitating event, no new threatening actions by the Iraqi government, no new reason to go to war that did not exist one, two, four, or even six years ago. It is entirely legitimate to ask, therefore: Why now? What is the basis for claiming a unilateral right to use preventive force to overthrow the Iraqi regime? What would be the consequences for Iraq, the Middle East, and international relations?"[17] General Douglas MacArthur's words from May 1951 concerning the war in Korea were frequently brought out to characterize the administration's pursuit of war with Iraq: "The wrong war, in the wrong place, at the wrong time, and with the wrong enemy."

To obtain democratic legitimacy, the Bush administration needed to overcome the opposition in Congress and ensure the support of a majority of Americans. However, when talking to American citizens, Bush gave a different interpretation of the Iraq problem, partially voiced in his statement to the UN General Assembly in September 2002. He cited the need to prevent terror attacks as the main objective of the military operation, rather than the need to disarm Iraq. Bush tried to prove that Iraq presented a real threat to US security and that a military operation was a necessary preventive strike against a potential aggressor. The main pressure point in public opinion and the political opposition that Bush activated was the American fear of terrorist attacks. Bush talked about a link between Saddam Hussein's regime and international terrorism, and claimed that Iraq armed terrorists with WMDs that it secretly produced.

The terrorism-backed perspective Bush gave to reasons for going to war is evident in his address to Congress on January 28, 2003, when he said, "Before September the 11th, many in the world believed that Saddam

Hussein could be contained. But chemical agents, lethal viruses and shadowy terrorist networks are not easily contained. Imagine those 19 hijackers with other weapons and other plans—this time armed by Saddam Hussein. It would take one vial, one canister, one crate slipped into this country to bring a day of horror like none we have ever known. We will do everything in our power to make sure that that day never comes."[18] Here Bush appealed to emotions, to the fear of a possible repetition of the events of 9/11 and of the use of biological weapons. He claimed that the US intelligence agencies had evidence gathered from records of secret conversations and statements of detainees proving that Saddam Hussein aided and armed terrorists, including members of al-Qaeda.

The speech had the effect of immediately increasing the level of support for the planned military operation. A comparison of opinion polls carried out by various agencies before the war indicates that in early January 2003, the level of support did not exceed 47 percent to 49 percent. After the speech of January 28, 2003, the share of Americans supporting the start of war went up to 58 percent, though this high figure was then followed by an (insignificant) decline in support. Another increase in the level of support was observed in the second half of February, after General Colin Powell's statement before the UN Security Council, when it was measured at 57 percent. It should be noted that the level of support differed according to how the question was formulated. A question that included a statement that the war against Iraq was a war against terrorism, against Saddam Hussein's tyranny and his military power, garnered more support than the question "Would you support or oppose the United States going to war with Iraq?" It is noteworthy that the question "Would you support or oppose sending your son or daughter to war to remove Saddam Hussein?" elicited a constant increase in positive responses, from 45 percent in September 2002 to 53 percent in early March 2003, with a slight decrease to 50 percent in mid-March 2003.[19] Americans were more ready to risk the lives of their children to defend the United States from Saddam's threat than they were to approve a war that could result in thousands of casualties on both sides. It is perfectly obvious that the main reason why the majority of Americans supported the war in Iraq was that it was regarded as a continuation of the war on terror. As the *New York Times* observer Michael Gordon noted, "for many Americans, protecting the nation against terrorists is a far more persuasive rationale for going to war than preventing Iraq from developing new weapons that could change the balance of power in the oil-rich Persian Gulf."[20] Still, more than half of Americans polled had a negative attitude toward starting the war without UN approval and without the support of the international community, and did not support going to war if the military operation would result in significant casualties among American military personnel and Iraqi civilians. Even before the start of the war, in March 2003, mass demonstrations by antiwar activists were taking place in large US cities.[21]

Opinion poll results demonstrated another crucial feature of American public opinion with regard to Iraq and Saddam Hussein: Saddam Hussein had been successfully demonized. Some experts believed that since the events of September 11, 2001, the American people lived daily in a climate of fear and anger that had been created and nurtured by the Bush administration in order to manipulate public opinion and convince Americans that waging war against Iraq would be beneficial. According to polls taken in December 2002, 66 percent of Americans believed that a connection existed between al-Qaeda and Saddam Hussein, and that Saddam was involved in the September 11 atrocities. An even greater proportion, 86 percent, thought that Saddam Hussein already had nuclear weapons or was on the brink of obtaining them, that he was the world's most brutal and ruthless tyrant, and was ready to use any weapons against the United States.[22] Peculiarly enough, from the start of the war in March 2003 through the rest of the year, this perception remained essentially unchanged, even though no evidence of any link was found, as reported by the CIA. Nevertheless, as the Program on International Policy Attitudes (PIPA) stated in September 2009, 69 percent of Americans still believed that Saddam Hussein was personally involved in the September 11 terrorist attacks.[23]

As part of the public discussion, the US Institute of Peace organized a symposium on December 17, 2002, at which four presenters, Gerard Powers, director of the Office of International Justice and Peace of the US Conference of Catholic Bishops; Robert Royal, president of the Faith and Reason Institute; George Hunsinger, professor at Princeton Theological Seminary; and Susan Thistlethwaite, president of the Chicago Theological Seminary, were asked to answer the question, "Would an invasion of Iraq be a 'just war'?" Referring to the experience of the Vietnam War, the symposium participants doubted both the need for war and the probability of success, and noted the following points:

- "A threat that is not clear, that is not direct, and that is not imminent cannot justify going to war. Measured by just war standards, the war proposed in Iraq fails completely of a sufficient cause." "Justifying preventive war in this way would represent a sharp departure from just war norms."

- "Just war tradition stipulates a reasonable chance of success, but the most probable outcome of an invasion of Iraq would be a long drawn-out bloody war."

- "An invasion would also wreak havoc on a civilian population already tortured by war and sanctions, clearly violating the non-combatant immunity stipulation."

- "Iraq is the test case for this muscular unilateralism. U.S. policy towards Iraq is based on three assumptions, each of which can be morally problematic: (1) the United States has a right to use preventive force against Iraq; (2) the objective of U.S. military action should be the overthrow of the Iraqi regime; and (3) the United States has a right to act unilaterally if others are not willing to do as it deems necessary."

- "U.S. credibility in justifying war on humanitarian grounds is weakened by the fact that some of its allies in the war on terrorism are themselves implicated in egregious human rights abuses and by the fact that humanitarian intervention" is selective.

However, there were those who thought it was necessary to act before it was too late, citing the failed attempt to "pacify an aggressor" with regard to Hitler before World War II.[24]

The most important task of the Bush administration was to obtain the permission of Congress to use armed forces in Iraq. One consequence of the Vietnam War was the adoption of laws restricting the right of the president to use armed forces outside the country and strengthening the power of Congress in resolving such issues. The War Powers Act or the War Powers Resolution, adopted in 1973, delimited the powers of the president as commander in chief and Congress as a legislative body at times when the armed forces of the country became involved in war. Under the provisions of this law, the president must obtain the approval of Congress for any use of armed forces.[25] On September 19, 2002, the White House forwarded a draft resolution to the House of Representatives and the Senate that would allow starting military action against Iraq, and would also allow using US armed forces "to restore international peace and security in the region." Discussion continued until early October 2002, and the Authorization for Use of Military Force Against Iraq, Resolution 114, was adopted on October 11 and entered into force on October 16, 2002. The US Senate voted 77 in favor and 23 against the resolution. Senator Hillary Clinton, who would later become US secretary of state, voted in favor.[26]

In accordance with Resolution 114, Bush was "authorized to use the Armed Forces of the United States as he determines to be necessary and appropriate in order to (1) defend the national security of the United States against the continuing threat posed by Iraq; and (2) enforce all relevant United Nations Security Council resolutions regarding Iraq.... [He shall inform Congress no later than 48 hours after exercising such authority and] shall at least once every 60 days, submit to the Congress a report on matters relevant to this joint resolution, including actions taken pursuant to the exercise of authority granted in section 3 and the status of planning for efforts that are expected to be required after such actions are completed."[27] It is important to note that this resolution cited the need "to restore international peace and security" and "enforce United Nations Security Council resolutions relating to Iraq," typical for UN Security Council resolutions, and specified authorization by Congress.

The Bush administration managed to ensure the necessary level of support from American society and government bodies to start the military operation. An "international coalition of the willing," comprising 34 countries that agreed to take part in the war, also was formed. On March 17, 2003, the United States presented Iraq with an ultimatum demanding that Hussein and his family leave Iraq within 48 hours. Iraq rejected the ultimatum, and on the night of March 19, 2003, the Combined Joint

Task Forces began military actions in Iraq. On March 20, Secretary of Defense Donald Rumsfeld stated: "The United States and the international community have made every effort to avoid war. Diplomacy and sanctions over more than a decade have not worked. And now, by rejecting President Bush's ultimatum, the Iraqi regime has chosen military conflict over peaceful disarmament.... This is not a war against a people. It is not a war against a country. It is most certainly not a war against a religion. It is a war against a regime."[28] On March 21, George W. Bush in a Presidential Letter informed Congress that US armed forces had been sent to Iraq, and referred to legislation that allowed him to start the war against Iraq, including the 1973 War Powers Act: "Consistent with the War Powers Resolution (Public Law 93-148), I now inform you that pursuant to my authority as Commander in Chief and consistent with the Authorization for Use of Military Force Against Iraq Resolution (Public Law 102-1) and the Authorization for Use of Military Force Against Iraq Resolution of 2002 (Public Law 107-243), I directed U.S. Armed Forces, operating with other coalition forces, to commence combat operations on March 19, 2003, against Iraq."[29] As Bush has said repeatedly, the war had two missions: disarming Iraq, and then transforming it into a "free and hopeful society."[30]

COURSE OF THE WAR AND THE EVOLUTION OF STRATEGIES

It did not take long for the coalition forces to crush the resistance of the Iraqi army. Baghdad was captured on April 9, and Tikrit was captured on April 15. On April 17, the remains of the government forces surrendered near Baghdad, and the Iraqi government ceased to exist. Saddam Hussein and Iraqi high officials fled and were put on the wanted list. Information about them was disseminated among coalition soldiers in the form of a deck of 55 cards; Saddam Hussein, the ace of spades, was wanted "dead or alive."[31] On May 1, 2003, on the deck of the USS *Abraham Lincoln,* which was returning to base after taking part in battle, President George W. Bush announced victory and the end of the military operation. In his speech, which quickly became known as the "Mission Accomplished" speech, after the banner displayed behind him, Bush acknowledged, "America's work in Iraq is far from done. If anything, securing a durable peace in Iraq will be harder than winning a military victory."[32] The coalition forces were to reconstruct what had been destroyed (cities, infrastructure, industrial objects), ensure security in the country, and prepare the country for the transfer of power to the local government in accordance with international law regulating an occupation regime. In the meantime, the international community severely criticized the looting of historical and cultural monuments in Iraqi territory that started after the Iraqi government fell.

Phases of the War

The US involvement in actions in Iraq can be divided into several phases, marked by important shifts in the military and political situation in Iraq and in US strategy.

Phase 1: March 20 to May 1, 2003. This phase of active military operation aimed at disabling Iraq's military power capability and overturning Saddam Hussein's government. This period ends with the coalition forces gaining control of a large part of Iraq and George W. Bush's "Mission Accomplished" speech on the flight deck of the USS *Abraham Lincoln*.

Phase 2: May 2, 2003, to June 28, 2004. This phase is marked by the start of guerrilla warfare by Ba'athist groups and supporters of Saddam Hussein. The struggle started in the area north and west of Baghdad known as the Sunni Triangle. Fighting for the city of Fallujah, in Al Anbar province, where Hussein had many supporters, was especially fierce. The United Nations Assistance Mission for Iraq (UNAMI) was established following Security Council Resolution 1500 of August 14, 2003. Security was precarious, however: on August 19, 2003, a suicide bomber attacked UN headquarters in Baghdad and killed 22, including Sérgio Vieira de Mello, the Secretary-General's Special Representative in Iraq. In total, more than 100 people were injured in the attack, and Kofi Annan ordered UN staff to withdraw from Iraq; staff did not return until April 2004, under greatly increased security.[33] In late October 2003, a terrorist attack was carried out against a delegation of the International Committee of the Red Cross, prompting this organization too to withdraw its representatives from Iraq. Partly in response to these attacks on nonpartisan aid organizations, on October 16, 2003, the UN Security Council adopted Resolution 1511, which authorized "a multinational force under unified command to take all necessary measures to contribute to the maintenance of security and stability in Iraq."[34] Paragraph 16 of the same resolution emphasized "the importance of establishing effective Iraqi police and security forces in maintaining law, order, and security and combating terrorism"—anticipating a handover of security responsibilities from coalition forces to the Iraqi police force.

In the second half of 2003, Shia opposition to occupation troops coalesced under the leadership of the religious figure Muqtada al-Sadr, and the Mahdi Army was created. In April, a Shia uprising started in central and southern Iraq. During this period, many representatives of Saddam's government were detained, and the former leader of Iraq was arrested on December 13, 2003. Shortly thereafter, in April 2004, the abuse of Iraqi detainees by American soldiers in Abu Ghraib prison became known to the general public.

On June 8, 2004, the UN Security Council adopted Resolution 1546. According to the provisions of this resolution, power was to be transferred to "[the] sovereign and independent Interim Government of Iraq by 30 June 2004," and the multinational coalition forces were to coordinate their stay and actions in Iraqi territory with the interim government. Coalition troops stayed in the country at the request of the new government and were given a UN mandate under Resolution 1546. This resolution also stipulated the right of Iraq to create its own security forces and armed forces.[35] The UN Security Council began to receive quarterly reports from the commander of the Multi-National Force–Iraq. The period ended with the termination of the occupation regime, when the Coalition Provisional Authority transferred sovereignty to the Iraqi Interim Government, led by Prime Minister Ayad Allawi.

Phase 3: June 29, 2004, to January 30, 2005. On October 12, 2004, the UN General Assembly gave the right to vote back to Iraq, thus restoring Iraq's official representation in the United Nations.

Guerrilla actions against coalition forces continued, however. In August 2004 the coalition forces inflicted a military defeat on the Mahdi Army, and Muqtada al-Sadr abandoned armed struggle. In late 2004 the coalition forces waged fierce battles for Fallujah, and by late November the city was captured. At around the same time, presidential elections took place in the United States, and Iraq was one of the key issues in the electoral campaign. Following George W. Bush's reelection to serve a second term, the period ended with general parliamentary elections in Iraq on January 30, 2005. During the elections the number of terrorist acts and attacks against coalition forces went up. The majority of the seats in the parliament,140, were won by the United Iraqi Alliance (Shia). Of the remaining seats, the Kurdistan Alliance took 75, the Iraqi List won 40, and the Sunnis won 17. Members of parliament were to elect the president and president's deputies, so that the latter could form a government.

Phase 4: January 31 to December 14, 2005. The Iraqi Interim Government was formed in April. Its tasks were to draft and adopt a constitution. As the US representative reported to the Security Council on May 31, 2005, "some 165,000 Iraqi soldiers and police officers have been trained and equipped. The Iraqi army has over 90 battalion-level units conducting operations. Some of those forces conduct independent security operations, and others operate alongside or with the support of the multinational force. The Iraqi battalions are out in the cities and rural areas of the country, and they are getting results. Iraqi police and military forces are shouldering the burden in 12 of Iraq's 18 provinces."[36] At the request of the Iraqi government, the United Nations extended the mandate of the Multi-National Force–Iraq to ensure security in Iraq. On October 15, a referendum was carried out to adopt a new constitution, and profound divisions between political factions became evident. The draft constitution was rejected in several Iraqi provinces (in Salah ad-Din by 82 percent, in Ninawa by 55 percent, and in Al Anbar by 97 percent).[37] This period saw continued fierce fighting with Sunni groups, and an increased number of terrorist attacks against Iraqi military and religious leaders.

Phase 5: December 15, 2005, to January 31, 2007. On December 15, 2005, new parliamentary elections were held. The Shiite United Iraqi Alliance was again victorious, winning 128 seats, while the Sunni Iraqi Accord Front won 58 seats and the Kurdistan Alliance took 53 seats. The elections were accompanied by extraordinary security measures. Some 150,000 Iraqi soldiers and policemen were delegated to guard the elections, and Iraq's borders were closed.[38] A new four-year, constitutionally based government took office in March 2006, and a new cabinet was installed in May 2006.

Observers often consider this phase as the start of the civil war, when coalition forces suffered the biggest casualties since the beginning of hostilities. In October 2006, George W. Bush for the first time publicly compared the war in Iraq to the war in Vietnam. During the congressional midterm elections of November 2006, the Republican Party lost its majority in both chambers. As the Baker-Hamilton Commission (the Iraq Study Group) report noted, many in the United States considered the results of the elections to be a referendum on progress in Iraq.[39] US Defense Secretary Donald Rumsfeld, who is considered one of the war initiators, resigned. The Iraq Study Group was appointed by Congress in March 2006, and in December 2006 its report was published. On December 30, 2006, Saddam Hussein was executed, following a ruling of the Supreme Iraqi Criminal Tribunal.

On January 10, 2007, George Bush announced the new US strategy in Iraq, according to which the American presence was to be increased by 21,000 troops, along with an increased term of deployment. The strategy "New Way Forward" was informally called the "surge."

Phase 6: February 1, 2007, to July 19, 2008. This phase saw the implementation of the surge strategy. From February to November 2007, the military operation Law and Order was carried out by American and Iraqi armed forces in Baghdad. In March 2007, during the visit of UN Secretary-General Ban Ki-moon to Baghdad, a mine exploded not far from where he was delivering his speech. The decision to increase the US military presence in Iraq resulted in strong disagreement between Congress and the president in spring 2007. George W. Bush vetoed a congressional resolution that linked the allocation of new war funds to a specific schedule of American troop withdrawal from Iraq.[40]

From June to August 2007, an operation was carried out to establish control over the city of Baqubah, north of Baghdad. An improvement of the situation in the province of Diyala allowed some US troops to be withdrawn from Iraq. In Al Anbar province, the American command concluded a contract with local elders on cooperation in the fight against al-Qaeda. Under this agreement, the United States armed the local Sunni militia, which brought criticism from the official Iraqi leadership. This period saw even more casualties among the multinational coalition forces, Iraqi security forces, and civilians. However, by late 2007 the wave of violence in Iraq had begun to subside. In March and April 2008, uprisings in Basra and Baghdad started again. The Mahdi Army forced the Iraqi government to start talks. In August a ceasefire was reached between the Iraqi government and Muqtada al-Sadr. Summer 2008 saw the lowest level of casualties during the whole war. According to official estimates, the reduced level of violence in Iraq was associated with three factors: (1) an increase in the number of US combat units; (2) the creation of local nongovernmental security forces that stood against al-Qaeda, which reduced violence in some parts of Iraq, especially in Al Anbar province; and (3) the conclusion of the ceasefire agreement in August 2007 with Muqtada al-Sadr.[41]

Phase 7: July 20, 2008, to December 18, 2011. After the troop surge and accompanying operations were completed, the coalition forces conducted operations in the northern provinces of Iraq against al-Qaeda fighters. In August to September there were large-scale terrorist attacks in Baghdad, Baqubah, and Mosul. Turkey bombed the Kurdish provinces of Iraq in response to the activities of Kurdish separatist groups in the north of Iraq.

In autumn 2008 the presidential elections in the United States ended with the victory of the Democratic candidate Barack Obama. On November 17, 2008, an agreement was signed between the Iraqi and US governments that specified a deadline for the withdrawal of all US combat units from Iraq. Under this agreement, the United States was to transfer complete control over military operations to Iraqi security forces and withdraw all combat units from cities and villages by June 30, 2009. The withdrawal of all American troops from Iraq was to be completed in December 2011.[42] On June 31, 2009, American troops withdrew from urban areas, a step that reinforced Iraqi sovereignty. In accordance with the agreement, the US strategic approach in Iraq changed from a conditions-based strategy to a time-based approach for drawing down US forces.[43] The Strategic Framework Agreement for a Relationship of Friendship and Cooperation between the United States of America and the Republic of Iraq outlined the main areas of interaction: technical, social, cultural, legal, and health-related. The need for joint combat operations against corruption was mentioned in the section on legal cooperation, along with joint combat against terrorism; drug trafficking, organized crime, money laundering, and the export of stolen archaeological artifacts.[44] The agreements entered into force on January 1, 2009.

The UN mandate sanctioning multinational forces in Iraq ended in late 2008. As of early 2009, the armed forces of four countries were still deployed in Iraq (the United States, Australia, the United Kingdom, and

Romania) with a total of 147,000 troops, 142,000 of them from the United States and 4,000 from the United Kingdom.[45] From July 2009, when all coalition forces were withdrawn, the United States was the only country retaining a military presence in Iraq (128,000 troops). On February 27, 2009, President Obama confirmed the change of strategy in Iraq and the time frame for US troop withdrawal.

The new US strategy in Iraq included the following components: (1) completion of combat operations and withdrawal of the main part of American troops from Iraq within 18 months (i.e., by August 31, 2010); (2) the retention of 50,000 US troops in Iraq to ensure stability; (3) the pursuit of "a regional diplomatic strategy" to solve the problem of refugees and to provide assistance to the Iraqi government in resolving political problems; and (4) a "principled and sustained engagement with all of the nations in the region, and that will include Iran and Syria."[46] The spring and summer of 2009 saw heated discussion in Congress about the economic recovery in Iraq and plans to terminate the US military presence. Until late 2011, the United States tried to fulfill a dual task—ensure security and reconstruct Iraq—to complete the actual missions announced at the beginning of war.

On January 31, 2009, Iraq held elections for provincial councils in all provinces except the three provinces of the Iraqi Kurdistan region, and Kirkuk province. On March 7, 2010, Iraq held a second round of national elections to choose the members of the Council of Representatives, which in turn would choose the executive branch of government. The Iraqi National Movement coalition, led by former prime minister Ayad Allawi, won the most seats (91), followed by Prime Minister Nouri al-Maliki's State of Law coalition (89 seats), the Kurdish bloc, headed by Kurdistan Democratic Party president Masud Barzani and Patriotic Union of Kurdistan president Jalal Talabani (57 seats), and the Iraqi National Alliance, led by Muqtada al-Sadr (70 seats), with the remaining 18 seats won by other smaller political and minority parties. On November 11, 2010, the Council of Representatives convened to elect Jalal Talabani to a second term as president of Iraq. Osama al-Nujayfi of the Iraqi National Movement coalition was elected parliament speaker. On December 21, 2010, the Council of Representatives approved President Talabani's nomination of Nouri al-Maliki for a second term as prime minister after al-Maliki proved able to secure the minimum parliamentary majority of 163 seats. The Council of Representatives also approved a majority of al-Maliki's Council of Ministers.[47]

Final phase? A new stage in the US military presence in Iraq started on August 31, 2010. It was associated with the end of combat operations and was symbolically indicated in the change of the mission title from Operation Iraqi Freedom to Operation New Dawn. The elimination of terrorist number one, Osama bin Laden, on May 2, 2011, was a crucial symbolic achievement of the Obama administration. On December 18, 2011, the last convoy of US soldiers pulled out of Iraq.

Evolution of US Strategies in Iraq

The strategies followed by the United States in the course of the Iraq War were of three general kinds. In chronological order, these were (1) an anticipated short-term military operation, which proved impossible to implement successfully as the struggle transitioned into guerrilla and civil war; (2) the 2007 troop surge in Iraq, under the working title (especially for press purposes) of "The New Way Forward," which took into consideration the findings of the Baker-Hamilton Commission (the Iraq Study Group) and its proposition to increase US troop presence; and (3) the strategy to bring to an end the American military presence in Iraq. These three strategies are associated with certain terms widely used by American politicians and military personnel during the corresponding periods of war. The Iraq War started as a continuation of the war on terror and went on as a COIN—counterinsurgency—operation; in the final stage, the war was regarded as a limited-presence or contingency conflict.

The war started with a clearly defined goal: to change the political regime and establish a new government, founded on democratic principles and representing the main population groups. This strategy was based on prewar assessments of the unpopularity of Saddam Hussein's regime among Shiites and Kurds. The operation was planned as a swift destruction of the unpopular regime, followed by a quick reconstruction of local civilian authorities, the de-Ba'athification of political life, and the creation of a foundation for the long-term cooperation of the United States with the new regime. The Iraq reconstruction plans implied the reconstruction of industrial structures and infrastructure destroyed during the war, so that Iraq could return to the global market as a large oil exporter.

A rapid normalization of the situation within the country and of Iraq's place in the international arena was regarded as a prerequisite for the country's ability to sustain itself on funds from oil exports. The United States hoped for beneficial terms of sale that would recognize its presence in Iraq and its role in regime change and postwar reconstruction. Thus, the prewar planning presumed massive funding of a short-term military operation and the allocation of limited funds for postwar recovery, which would allow the United States to derive immediate benefits from the military operation. Unlike the war of 1991, which to a large extent was a "no-contact" war of the new arms generation, Operation Iraqi Freedom relied on a massive military presence in the territory of the whole country, that is, an occupation regime, with its associated legal, political, economic, and military-strategic problems.

The initial course of the war in 2003–2004 followed the prewar planning script. The military defeat of the half-million-strong Iraqi army, the destruction of the political institutions of Saddam Hussein's regime, the insignificant resistance of the remaining Iraqi troops—all this dovetailed with the concept of a swift military operation backed by the significant material and technical superiority of the United States and its allies. The initial enthusiasm on the part of Iraqi society for regime change was broadcast by American mass media and inspired hope for a peaceful turn of events and the quick recovery of the political and economic life of the country. A few surprises followed: the emergence of a powerful Shia movement against foreign occupation by late 2003, and uprisings in central and southern Iraq, in addition to the prolonged struggle in the Sunni Triangle in central Iraq. The Kurdish north of the country was relatively quiet from the security standpoint. However, the United States had serious disagreements with Turkey about Kurdish autonomy in Iraq. American politicians expected that the situation would normalize after the transfer of power from occupation forces to the new Iraqi government. However, despite the military defeat of the Mahdi Army in August 2004, guerrilla operations and terrorist attacks against coalition forces and Iraqis cooperating with coalition forces continued to increase in number. Moreover, the development and adoption of the Iraqi constitution in 2005 not only did not lead to national reconciliation, but also demonstrated the strength of the conflicting tendencies in Iraqi society. Between May 1, 2003, when President George W. Bush announced

Figure 4.1. US Forces Fatalities by Year, 2003–2012

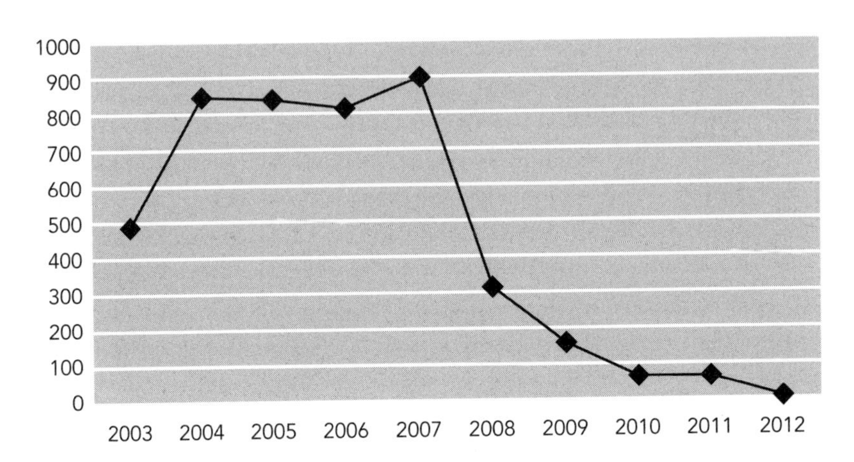

Source: "Iraq Coalition Casualty Count," iCasualties.org (http://icasualties.org/).

the end of the war in his "Mission Accomplished" speech, and December 15, 2005, 2,013 American servicemen died, or around half of all US troop fatalities sustained during the course of the war. The largest casualties were suffered by the US Army during the surge (see figure 4.1).

The beginning of civil war in 2006 was evidence of the coalition forces' failure to control the situation and ensure security in the country. In fact, the coalition forces failed to ensure security even in Baghdad, for terrorist attacks were successfully carried out in the most protected Green Zone of the city. In addition to military clashes, the number of terrorist attacks and instances of violence increased drastically, and the number of abductions went up. The abduction of people for ransom became a real business. According to the Brookings Institution's Iraq Index, a regularly updated index of information about Iraqi security issues, between 2003 and 2010, 312 foreign citizens were kidnapped in Iraq, of whom 60 were murdered, 149 were freed, 4 escaped from their kidnappers, and 6 were released as a result of special operations.[48] Some information agencies reported that in 2006, around 30 to 40 Iraqis were kidnapped every day in Iraq; however, in 2007–2008 the situation improved noticeably.[49]

There was a growing concern in US political circles and society that the war was turning into another Vietnam, a protracted, futile war that could be neither won nor lost. In the context of guerrilla warfare, the reconstruction of civilian and industrial infrastructure had become a black hole, absorbing millions of dollars allocated for recovery. The results of audits published by the US Government Accountability Office disclosed a serious problem with control over supplies of arms to Iraqi security forces. "Thus, DOD and MNF-I cannot fully account for about 110,000 AK-47 rifles, 80,000 pistols, 135,000 items of body armor, and 115,000 helmets reported as issued to Iraqi forces as of September 22, 2005. Our analysis of the MNSTC-I property book records found that DOD and MNF-I cannot fully account for at least 190,000 weapons reported as issued to Iraqi forces as of September 22, 2005."[50] These arms were often found in the hands of fighters and were used against American troops and allies.

The expectations of the US allies for a quick improvement in the situation and benefits from the postwar reconstruction of the oil-producing sector were not met. Iraqi oil production and export levels did not reach their prewar figures, even though Iraq ranks second after Saudi Arabia in terms of oil reserves. Starting in 2004, the European countries that had supported the United States as part of the coalition forces began to withdraw their troops from Iraq. After the terrorist attacks of March 11, 2004, in Madrid, the parliamentary elections were won by the Spanish Socialist Party, which was against the war. The new government, led by José Zapatero, announced the immediate withdrawal of Spanish troops from Iraq. The Netherlands withdrew its troops in 2005. Italy withdrew its forces in 2006 after the electoral defeat of the center-right coalition of Silvio Berlusconi, who had supported the war in Iraq. Prime Minister Romano Prodi called the war in Iraq a mistake. The United Kingdom reduced its military presence in Iraq from 46,000 troops in March 2003 to 7,200 troops in late 2006.[51] Significant resources of the United States' European partners were being drawn away by the ongoing military operation in Afghanistan, which was being carried out as a NATO operation.

Commentators at the time often voiced an unequivocal and desperate assessment of the situation in Iraq: it was chaos. The inability of the coalition forces to ensure security in Iraq nullified all efforts to reconstruct the country and contribute to its development. In the United States, the number of people supporting an immediate withdrawal of troops from Iraq went up. In March 2006, Republican former secretary of state James Baker and Democratic congressman Lee H. Hamilton were chosen to create a bipartisan congressional commission to consider the Iraq issue. The results of the Baker-Hamilton Commission activities were made public in December 2006. The word "consensus" was used three times in the short preface to the report. According to the commission, the United States needed consensus not only to develop the strategy in Iraq but also to implement it successfully: "Yet U.S. foreign policy is doomed to failure—as is any course of actions in Iraq—if it is not supported by a broad, sustained consensus." The report noted that "the situation in Iraq is grave and deteriorating" and that "there is no magic formula to solve the problems of Iraq."[52]

The report analyzed the situation in Iraq in four main areas: security, politics, the economy, and international support. The section on security evaluated the major sources of violence and the position of the coalition forces and the United States, the security forces, the police, and departmental security forces in Iraq. The authors suggested shifting the center of gravity in security to Iraqi forces and police. The training of local security forces was to become the main task of the American military presence in Iraq. The report also emphasized the need to render military and technical assistance to Iraqi armed forces and police. The main conclusion with regard to security was as follows: "Security efforts will fail unless the Iraqis have both the capability to hold the areas that have been cleared [of insurgents] and the neighborhoods that are home to Shiite militants. U.S. forces can 'clear' any neighborhood, but there are neither enough U.S. troops present nor enough support from Iraqi security forces to 'hold' the neighborhoods cleared."[53]

The review of the political situation assessed the main political forces in Iraq, their leaders, and their degree of participation in the legal political process and in armed struggle against each other and the coalition forces. In characterizing the various groups fighting the coalition forces, the authors of the report identified both uniting and dividing factors. The key objective in the political domain was to achieve national reconciliation that involved all the political forces in governance, including the Sunni technocratic elite that had been removed from power. Of the several possible forms of Iraqi political organization, preference was given to a federation, with the caution that a balance needed to be struck so that the federal bodies would be able to represent the interests of Iraq's main political, religious, and ethnic groups.

The economic review noted that the economy of Iraq was almost 100 percent dependent on the oil-producing sector and oil exports. It also noted that "Iraq's economy has been badly shocked and is dysfunctional after suffering decades of problems: Iraq had a police state economy in the 1970s, a war economy in the 1980s, and a sanctions economy in the 1990s." The main obstacles to the normal functioning of the country's economy were "insecurity, corruption, lack of investment, dilapidated infrastructure, and uncertainty."[54] The report noted that the Iraqi leaders aimed to redistribute control over the oil-producing sector and had concluded separate contracts with foreign companies on oil extraction, which aggravated discord in the country; one of the report sections was titled "The Politics of Oil." Inflation in the country was up to 50 percent a year, the unemployment rate had risen from 20 percent to 60 percent, and foreign investment was less than 1 percent of gross national product. The recovery efforts of various US agencies were nullified owing to the ongoing war and the never-ending attacks of fighters on infrastructure and the structural objects of the oil-producing sector, such as pipelines and refineries.

The report's analysis of international support in Iraq's recovery offered an estimate of the financial assistance available from developed countries. It also evaluated the degree to which countries in the region had a stake in continuing the war in Iraq. Iran was assigned the strongest negative impact; however, US attempts to start negotiations with Iranian officials had not paid off. The negative influence of Syria was also noted, as was the "passive and disengaged" position of Saudi Arabia and the Persian Gulf states. Turkey's position on the Kurdish issue—namely, its disapproval of the creation of a Kurdish autonomous region in northern Iraq, on the Turkish border—was considered a contradiction to the US position. The assistance of Jordan and Egypt was positively assessed, as those countries had helped train Iraqi security forces and police units and gave asylum to Iraqi refugees. Turkey's restraint with regard to Iraqi Sunnis was appreciated. The report noted that, with the exception of British forces in Basra and the southeast of the country, coalition forces played a limited role in Iraq; Britain also played an active role in working toward political settlement. The report briefly reviewed the degree of involvement of the United Nations, the World Bank, the European Union, and US nongovernmental organizations.[55]

The report offered a sobering assessment of the constantly worsening situation in Iraq: the increasing chaos in the country, the region, and the world, and the deterioration of the US security position as a result of the increased terrorist activity of various international groups. The report cited an Iraqi official as saying "Al Qaeda is now a franchise in Iraq, like McDonald's." According to the report's authors, "Iraq is a major test of, and strain on, U.S. military, diplomatic, and financial capacities."[56] As for future developments, the report clearly indicated that the war in Iraq could not be won by military means. Success completely depended on achieving national reconciliation in the country and on Iraq's ability "to govern itself, sustain itself, and defend itself." The US task was to render comprehensive assistance to the Iraqi people in this process.[57]

The report's 79 recommendations were divided into two parts, international and domestic tasks. A crucial feature was the synchronization of tasks for Iraq and the United States, or, in the language of the report, "Helping Iraqis help themselves." Recommendations with regard to resolving security problems included strengthening regional cooperation and achieving an international consensus on Iraq, the latter of which would entail greater diplomatic efforts to resolve the problems with Iraq's neighbors and the countries of the Greater Middle East: Afghanistan, Iran, Israel, Jordan, Kuwait, Saudi Arabia, and Turkey. Solving Iraq's internal problems would require a hierarchy of tasks. Any political settlement within the country had to be directed at winning the trust of Iraqis in the Iraqi government and at promoting national reconciliation, efficient governance, and amnesty for participants in the resistance. The problem of security was to be addressed by increasing the fighting capacity of the Iraqi armed forces and police; providing a clear schedule of US troop presence in or withdrawal from Iraq, with special heed paid to the US armed forces and defense budget; and augmenting cooperation between the intelligence agencies of the United States and Iraq. Economic tasks included the development of Iraq's oil-producing sector and coordinating US efforts aimed at reconstruction and efficient spending of funds.[58]

The Baker-Hamilton Report, the congressional debates, and the comments of leading analysts paved the way for a drastic revision of the war strategy. On January 10, 2007, President George W. Bush announced a new strategy for Iraq in his State of the Union address.[59] The strategy, known informally as the "New Way Forward," contained six key elements: devolving greater responsibility and independence to Iraqi authorities, protecting the Iraqi civilian population, fighting extremists, strengthening the political process aimed at national reconciliation, diversifying US efforts in Iraq's recovery, and promoting a regional approach to security. In this address and in a subsequent address to Congress on January 23, Bush argued that the United States could not lose the war in Iraq because it would mean a US defeat in the war on terror.[60]

The success of the military component of the "New Way Forward" is often linked to the name of American general, David Petraeus, who was appointed to command the multinational coalition forces in Iraq in January 2007. On assuming this position, General Petraeus told a meeting of the Senate Armed Services Committee that strengthening the security of the local civilian population and coordinating coalition efforts with those of Iraqi security forces and police ought to be the main objectives of the military operation in Iraq. He also pointed to the frequent change of government in Iraq (four governments in three and a half years) and the suffering of Iraqis. His remarks were largely concordant with the tone and recommendations of the 2006 Baker-Hamilton Report.[61]

Implementation of the comprehensive strategy had positive results. By the summer of 2008, significant improvements in stability had been observed, and that allowed some US forces to be withdrawn, control in Iraqi provinces to be transferred to Iraqi security forces, and planning to begin on the termination of a US military presence. The Obama administration systematically implemented the new US course in Iraq, moving from a conditions-based strategy to a time-based approach for drawing down US forces.[62] In February 2009, President Obama announced the change of strategy in Iraq, and all subsequent actions of the US government confirmed the intention to end at least one of the wars that were "inherited from the previous administration," as relentlessly repeated by the new president.[63] However, not everyone regarded the operation itself and the decision to start the troop withdrawal in a positive light. Some commentators believed that the surge had strengthened the hand of actors who would be able to continue a full-scale struggle for power and resources after the US departure from Iraq, and demanded that troop withdrawal represent a "responsible retreat."[64]

Starting in the spring of 2009, American political leaders and military personnel had to shift attention and resources from Iraq to Afghanistan because of a worsening situation there, the activation of the Taliban, and the Taliban's destabilizing influence in Pakistan. At first, American military personnel talked openly about the success of the counterinsurgency operation in Iraq and the intent to use this experience in Afghanistan. For instance, Admiral Michael Mullen, Chairman of the Joint Chiefs of Staff, made a statement at the Brookings Institution on May 18, 2009, in which he mentioned the undoubted success in Iraq and US troops in Afghanistan enthusiastically perceiving and understanding the new approach to warfare against Taliban fighters and al-Qaeda.[65] However, by the second half of 2009 the situation in Iraq and Afghanistan had shown that the apparent success of the reconciliation strategy in Iraq and the possibility of transferring the same strategy to Afghanistan were illusory.

THE WAR'S OUTCOMES

The main problems the Iraqi government and the missions of international organizations had to face together were generated by the military operations of the coalition, the civil war that followed, and the presence of international terrorist groups in Iraq. Some of these problems were inherited from Saddam Hussein's regime. For Iraq to become stable and peaceful, the following problems had to be resolved:[66]

- security problems created by various opposition movements and terrorist groups;
- reconstruction and the creation of infrastructure to ensure economic and political development;
- the integration of Iraqi society and elimination of discord and hostility between various groups to achieve stability in the country,

including conflict over access to power and the distribution of natural resources and related revenues;

- settling Iraq's conflicts with other countries in the region (Syria, Kuwait, Turkey, Iran);
- the problem of refugees and displaced persons;[67]
- improving living standards and medical services for the Iraqi population;[68] and
- Iraq's external debt.

Beginning with President Bush's "New Way Forward" strategy, there was a concerted attempt to shift the focus of the US presence in Iraq from security to reconstruction and development. Ironically, this strategy originated in the strengthened military presence of the surge. Furthermore, an analysis of funds allocated by the United States for reconstruction in Iraq indicates how the spending on civilian programs was insignificant compared with the spending on security programs and the training of Iraqi security forces. As of January 2009, Congress had allocated US$51.01 billion to Iraq reconstruction, of which $41.42 billion had been accepted for execution and $36.58 billion had been spent. From the total sum allocated for reconstruction in Iraq, around $18.4 billion was intended for the Iraq Relief and Reconstruction Fund 2, $18.039 billion for the Iraq Security Forces Fund, $3.569 billion for the Commander's Emergency Response Program, $3.737 billion for the Economic Support Fund, and $2.475 billion for the Iraq Relief and Reconstruction Fund 1.[69] Funds earmarked for the Emergency Response Program had in fact been fully spent on recruiting former fighters to Iraqi security forces in the program called Sons of Iraq.[70] The total size of the Sons of Iraq troops was around 99,000 in early 2009.[71] These troops were controlled by the coalition command. It was planned that all Sons of Iraq units would be gradually transferred to the control of Iraqi authorities and would be integrated into Iraqi security and police forces.[72] These troops predominantly consist of Sunnis (80%), with the remainder being Shias (19%).[73]

Beginning in 2005, the United States started pursuing the policy of tying the spending of funds allocated by the US government for the recovery of Iraq to the funds of the Iraqi government, mainly generated by oil sales. General Raymond Odeirno, commander of Multi-National Corps–Iraq, noted in September 2008 that "any money we spend must be tied to Iraqi spending and should be in a ratio of three to one. Three dollars of Iraqi money to every one dollar of U.S. money."[74] However, the Iraqi government not only should plan to spend certain funds for the country's recovery and development but also should want these funds to be purposefully spent. According to the US Government Accountability Office, in the period 2003–2008 the United States allocated US$10.9 billion for Iraqi reconstruction, and $9.5 billion was spent. In 2005–2008, the Iraqi government planned to spend the equivalent of $17 billion and actually spent $2 billion. The report referred to US officials' opinion that Iraqi managers lacked the skill level and authority to create plans and buy materials necessary to sustain Iraq's energy and water sector projects. Owing to high oil prices before 2008 and the fact that the budgeted amount had not been spent, the Iraqi government budget accumulated a surplus of $47.3 billion over 2005 to 2008.[75]

The United Nations Fund for Iraq was established, and total contributions to it amounted to US$1.36 billion as of December 31, 2008. In December 2008, the Paris Club wrote off almost US$45 billion of the total $52 billion Iraqi national debt within the program of assistance for Iraq's reconstruction. At a meeting of the UN Security Council on February 26, 2009, an Iraqi representative noted that an agreement had been reached with the governments of Greece and China to reduce the amounts that Iraq owed them by 80 percent.[76] In February 2008, Russia signed a bilateral agreement with Iraq on debt relief, reducing the amount owed from US$12 billion to $1.5 billion, without any preconditions.[77]

Despite an improving situation in Iraq, it still could not be called safe, which created problems for American civilian reconstruction representatives. The United States increasingly relied on private companies to ensure security. Thus, between 2006 and 2008, US$1.1 billion was spent to fund approximately 1,400 security contractors to provide security for US State Department employees in Iraq.[78] Security then was gradually turned over to private security companies, which deployed between 150,000 and 180,000 people in Iraq. The use of contractors to provide security quickly generated questions, which never completely died away. The New Hampshire congresswoman Carol Shea-Porter, referring to her experience visiting Iraq, noted at a meeting of the House of Representatives Committee on Armed Services that employees of private agencies protecting American government buildings often did not speak English, and she asked whether the security contractors in Iraq were reliable ("When I went to Iraq last time, we were using contractors to guard the bases. And some of the contractors in this particular group were from the continent of Africa, and I didn't even think that they even understood English, never mind understood what I thought they needed to know in order to properly defend our troops there. Is there a risk, an inherent risk, of having people besides Iraqis or US soldiers defending and protecting our bases, and have you looked at any of those contracts?").[79]

Inefficient spending was one of the most hotly debated aspects of Iraq's recovery. On March 25, 2009, the results of audits carried out by the Office of the Special Inspector General for Iraq Reconstruction were announced at hearings of the US House of Representatives Committee on Armed Services. According to the report of Inspector Stuart Bowen, some 15 to 20 percent of the funds from the International Reconstruction Fund Facility for Iraq had been "spent in vain," or between US$2.76 billion and $3.68 billion out of $18.4 billion.[80] More than 163,000 private contractors took part in Iraq's recovery and development projects.[81] (Though this statistic does not necessarily reflect the activities of civilian contractors, most private contractors in Iraq likely offered security services and were not civilians.)

Another problem standing between the US and Iraqi governments was the issue of detained Iraqis who were suspected of participating in military or terrorist actions against coalition forces. At the 6087th UN Security Council meeting in February 2009, the figure of 15,000 detainees was brought up.[82] That number reflected a significant reduction from December 2007, when there were 24,000 detainees in Iraqi prisons under the control of the coalition forces and another 26,000 in prisons under the control of Iraqi authorities. By July 2010 the number of detainees had fallen to 200 people in US custody. In total, according to the Brookings Institution's Iraq Index for 2011, nearly 90,000 individuals were detained by the United States during the seven years following the invasion.[83]

After control over security in Iraq was transferred to Iraqi authorities, US forces were to coordinate all military operations with Iraqi authorities. On a number of occasions, Iraqi prime minister Nouri al-Maliki severely criticized the actions of US troops that resulted in the deaths of Iraqi civilians. For instance, on April 2, 2009, al-Maliki gave an interview to the BBC after several large-scale terrorist attacks in Iraq. He acknowledged that the US announcement of combat unit withdrawal by June 30, 2009, had been followed by several terrorist attacks and a worsening of security in Iraq, but the Iraqi government did not plan to postpone the withdrawal of US troops.[84]

The loss of a professional class also hurt Iraq badly. As fighters from various ethnic and religious groups ramped up violence, thousands of people fled Iraq because of direct threats to their lives and property. According to various sources, up to 40 percent of professionals left Iraq—educated people who were unlikely to return to the country as long as it faced problems with security and survival. These individuals were representatives of the middle class, which is a necessary foundation for a democratic form of government. Without a professional middle class, the modern system of governance, education, health care, and other civic provisions is impossible. This social layer scarcely existed in the time of Saddam Hussein and was catastrophically reduced during the war. Andrew W. Terrill of the Strategic Studies Institute, in a work on the spillover effect of the Iraq War, referred to the 2008 International Crisis Group estimates that the number of refugees from Iraq exceeded the number of refugees during the Vietnam War, and that Iraq ranked second in the world in number of refugees. The number-one spot was held by Afghanistan,[85] where the United States had been waging war for more than 10 years.

The state of the economy in Iraq is still at a prewar level, according to some indicators, though the production of electrical energy exceeds that of the prewar period. According to the regularly updated Iraq Index of the Brookings Institution, access to essential resources in Iraq in 2009 looked as shown in table 4.1.

The sector of public services—education and health care—suffered the most from the war. According to the Brookings Institution, the number of Iraqi physicians registered before the 2003 invasion was 34,000, and around 20,000 left Iraq after the start of war, while only 1,200 physicians returned

Table 4.1. Estimated Availability of Essential Services in Iraq, 2008–2011

Service	February 2008	February 2009	February 2011
Sewage (% of population with access to sanitation)	8	20	26
Water (% of population with access to potable water)	22	45	70
Electricity (% with access to 12+ hours of power per day)	25	50	n/a
Fuel (% of population with needs met)	25	48	n/a
Public health (% of population with access to health services)	18	30	n/a
Housing (% of population with adequate housing)	25	50	n/a
Trash (% of population serviced)	18	45	n/a

Sources: Data for 2008 and 2009 from "Iraq Index," Brookings Institution, February 26, 2009, 43; data for 2011 from "Iraq Index," Brookings Institution, November 30, 2011, 25.

in 2007 and 2008. As of December 2008, around 16,000 physicians worked in Iraq. During the war, around 2,000 physicians were killed and 250 were kidnapped. In 2011, more than 20 percent of the 2,250 graduates of Iraqi medical training institutions planned to leave the country.[86]

According to the World Bank, Iraq ranked 164th out of 183 economies in the Doing Business 2012 report, and the indicators had steadily worsened over the preceding years (table 4.2). The World Bank's overall assessment of the economic situation was sobering:

> The level of economic freedom in Iraq remains unrated this year, because of the lack of sufficiently reliable data. The Iraqi economy has slowly recovered from the hostilities that began in 2003, but progress has been uneven, and the country faces continuing tension among different ethnic and religious factions. With its economic

Table 4.2. Global Rankings of the Business Environment in Iraq, 2008–2012

Indicator	Rank					Trend over past year
	2008	2009	2010	2011	2012	
Control of Corruption Indicator, World Bank Group	196 (202)	195 (202)	193 (202)			↑
Regulatory Quality, World Bank Group	172 (202)	169 (202)	173 (202)			↓
Country Credit Rating, Institutional Investor	161 (178)	144 (178)	136 (178)	144 (178)		↓
Index of Economic Freedom, Heritage Foundation and the *Wall Street Journal*[a]	N/A	N/A	N/A	N/A	N/A	
Political Risk Rating (ICRG), PRS Group	139 (140)	139 (140)	137 (140)	137 (140)	137 (140)	↔
Doing Business Ranking, World Bank Group	141 (178)	N/A	N/A	159 (183)	164 (183)	↓
Doing Business—Trading Across Borders	178 (178)	N/A	N/A	180 (180)	180 (180)	↔

Note: a. The Index of Economic Freedom, compiled annually by the conservative Heritage Foundation (US) and the *Wall Street Journal*, does not rank Iraq, though the comments contain an indication of the low development level of other countries of the region—Iran, Libya, and Syria. "Vast Oil Wealth Doesn't Translate into Economic Freedom, Index Finds" (Washington, DC: Heritage Foundation, January 13, 2009), http://www.heritage.org/press/newsreleases/Index09d.cfm.

> growth highly volatile, Iraq's ongoing economic reconstruction, though facilitated by high oil prices and foreign economic aid, has been fragile. Political instability and pervasive corruption continue to undermine the limited progress made over the past years. Operating well below potential, the Iraqi economy lacks effective monetary and fiscal policies. The weak state of the financial system, coupled with its limited role within the economy, also makes development of a much-needed dynamic private sector extremely difficult.[87]

In the Global Peace Index rankings compiled by the nongovernmental organization Vision of Humanity, Iraq ranked last from 2007 to 2010; only in 2011 did it give up this rank to Somalia and become next to last.[88]

Opinions on the War

When evaluating the situation in Iraq, it is important to include the results of Iraqi opinion polls. Regular polls conducted by the BBC and ABC beginning in 2003 allow close monitoring of changing public sentiments.[89] Iraqi society is divided with regard to the invasion, though a negative assessment prevails to some extent. A negative attitude toward the invasion was minimally present in 2004, only 39 percent negative versus 49 percent positive. The peak of the negative attitude was registered in August 2007 during the US troop surge (63% negative versus 37% positive). In February 2009, a negative assessment (combining the categories "somewhat wrong" and absolutely wrong") exceeded a positive assessment (combining the categories "somewhat right" and "absolutely right") by 14 percentage points. These percentages, and the predominantly positive attitude toward the invasion, suggest that Saddam Hussein's regime was not popular and that opposition sentiments, taken into consideration by US officials when planning the invasion, were to a large extent justified. Table 4.3 serves as an illustration.

Evaluation of the general situation beyond security issues also uncovered a division in society, but a positive assessment slightly prevailed. The gradual increase in positive views from 2005 to 2009 is noteworthy. In February 2009, a positive assessment exceeded a negative one by 18 percentage points, whereas in 2005 the ratio was reversed: 53 percent negative and 44 percent positive. Also of note is the serious decline in optimism in 2007, a result of the intensive military actions connected to the troop

Table 4.3. Iraqi Opinion on the Coalition Invasion, 2004–2009
Answers to the Question, "From today's perspective and all things considered, was it absolutely right, somewhat right, somewhat wrong, or absolutely wrong that US-led coalition forces invaded Iraq in spring 2003?"

	2004 (%)	2005 (%)	February 2007 (%)	August 2007 (%)	March 2008 (%)	February 2009 (%)
Absolutely right	20	19	22	12	21	19
Somewhat right	29	28	25	25	28	23
Somewhat wrong	13	17	19	28	23	28
Absolutely wrong	26	33	34	35	27	28
Refused/ Don't know	13	4	—	—	—	2

Source: Iraq Poll, February 2009.

Table 4.4. Iraqi Assessment of the Situation in Iraq, 2005–2009
Answers to the Question, "Now thinking about how things are going, not for you personally, but for Iraq as a whole, how would you say things are going in our country overall these days?"

	2005 (%)	February 2007 (%)	August 2007 (%)	March 2008 (%)	February 2009 (%)
Very good	14	4	3	7	20
Quite good	30	31	19	36	38
Quite bad	23	35	40	36	25
Very bad	30	30	38	20	15
Refused/ Don't know	3	—	—	1	2

Source: Iraq Poll, February 2009.

Table 4.5. Iraqi Assessment of Personal Situation (General Poll), 2004–2009 *Answers to the Question,* "Overall, how would you say things are going in your life these days? Would you say things are very good, quite good, quite bad, or very bad?"

	2004 (%)	2005 (%)	February 2007 (%)	August 2007 (%)	March 2008 (%)	February 2009 (%)
Very good	13	22	8	8	13	21
Quite good	57	49	31	31	41	44
Quite bad	14	18	32	34	29	19
Very bad	15	11	28	26	16	16
Refused/ Don't know	1	1	—	—	—	—

Source: Iraq Poll, February 2009.

surge. According to the opinion poll, in August 2007 the total negative assessment of the situation in Iraq was 78 percent, versus 22 percent positive (table 4.4).

Answers to the question about one's own prosperity showed a declining positive opinion from 2004 to 2007 and a rise in optimism from 2008. However, the share of Iraqis who assessed the situation as "quite bad" or "very bad" was stable and significant: 35 percent in 2009, little different from the low value of 29 percent in 2004 and 2005 (table 4.5).

A breakdown by the main population groups indicates greater optimism among Kurds and Shia in the polls of 2008 and 2009 (table 4.6).

The polls demonstrated a stable and negative attitude of Iraqis toward the coalition forces and their actions in Iraq. The maximum positive attitude was observed in 2005 (total of 37%), while the minimal negative assessment the same year, 59 percent, exceeded the positive assessment by 22 percentage points. A negative attitude on this dimension peaked in

Table 4.6. Iraqi Assessment of Personal Situation, by Community, March 2008 and February 2009

Answers to the Question, "Overall, how would you say things are going in your life these days? Would you say things are very good, quite good, quite bad, or very bad?" by community, February 2009 (March 2008 result in parentheses)

	Kurds (%)	Shia (%)	Sunni (%)
Very good	32(24)	25(14)	8(7)
Quite good	41(49)	46(48)	42(27)
Quite bad	23(20)	16(27)	23(38)
Very bad	4(7)	13(11)	28(28)
Refused/Don't know	—	—	—

Source: Iraq Poll, February 2009.

Table 4.7. Iraqi Assessment of Coalition Performance in Iraq, 2005–2009

Answers to the Question, "Since the war, how do you feel about the way in which the United States and other Coalition forces have carried out their responsibilities in Iraq? Have they done a very good job, quite a good job, quite a bad job, or a very bad job?"

	2005 (%)	February 2007 (%)	August 2007 (%)	March 2008 (%)	February 2009 (%)
A very good job	10	6	3	6	11
Quite a good job	27	18	15	23	19
Quite a bad job	19	30	32	35	30
A very bad job	40	46	48	35	39
Refused/ Don't know	5	—	1	1	1

Source: Iraq Poll, February 2009.

August 2007, with an 80 percent negative evaluation and an 18 percent positive evaluation. In February 2009 negative assessments exceeded positive ones by 39 percentage points (table 4.7).

The breakdown of poll results by community demonstrates that Kurds had the most optimistic attitude toward the situation (total of 63% positive in 2007 and 62% positive in 2009), whereas the lowest positive assessments were observed among the Sunnis (total of 6% in 2007 and 8% in 2009) (table 4.8). These results directly reflect the population groups that gained the most from the regime change (Kurds) and those that lost (Sunnis).

The polls demonstrated a low level of trust in US occupation forces among Iraqis. The total share of Iraqis who (variably) mistrusted American occupation forces ranged from 72 percent in 2003 to a maximum level of mistrust in August 2007 of 85 percent. In 2009, the share of Iraqis mistrusting American occupation forces exceeded the share of those who trusted them by 47 percentage points (table 4.9).

Table 4.8. Iraqi Assessment of Coalition Performance in Iraq, by Community, March 2008 and February 2009

Answers to the Question, "Since the war, how do you feel about the way in which the United States and other Coalition forces have carried out their responsibilities in Iraq? Have they done a very good job, quite a good job, quite a bad job, or a very bad job?" by community, February 2009 (March 2008 result in parentheses)

	Kurds (%)	Shia (%)	Sunni (%)
A very good job	23(19)	12(5)	3(-)
Quite a good job	39(44)	20(27)	5(6)
Quite a bad job	22(23)	33(36)	28(40)
A very bad job	14(13)	34(32)	62(53)
Refused/Don't know	2	1	2

Source: Iraq Poll, February 2009.

Table 4.9. Iraqi Level of Trust in US Occupation Forces, 2003–2009

	2003 (%)	2004 (%)	2005 (%)	February 2007 (%)	August 2007 (%)	March 2008 (%)	February 2009 (%)
Great deal of confidence	7	8	7	6	4	4	12
Quite a lot of confidence	12	17	11	12	11	16	14
Not very much confidence	20	23	23	30	27	33	28
None at all	52	43	55	52	58	46	45
Refused/ Don't know	9	8	5	–	–	1	1

Source: Iraq Poll, February 2009.

Table 4.10. Iraqi Assessment of Government Control, 2005–2009
Answers to the Question, "Who do you think currently controls things in our country; is it the Iraqi government, the United States, somebody else, or no one?"

	2005 (%)	February 2007 (%)	February 2009 (%)
Iraqi government	44	34	32
United States	24	59	53
Somebody else	17	4	9
No one controls things	6	3	3
Refused/Don't know	9	–	3

Source: Iraq Poll, February 2009.

Answers to the question about who controlled the situation in the country demonstrated a decline in the number of Iraqis who believed that the Iraqi national government exercised control: from 44 percent in 2005 to 32 percent in 2009. Over the same period there was an increase in the proportion of those who believed that the United States controlled the situation in Iraq, from 24 percent in 2005 to 53 percent in 2009 (table 4.10).

Assessment of the level of trust in national institutions demonstrated a higher level of trust in power institutions (police and army) than in the government, and the level of trust in power institutions kept rising. The level of trust in the army increased from 38 percent in 2003 to 73 percent in 2009; in the police forces, from 46 percent in 2003 to 74 percent in 2009. An insignificant decline in trust in these institutions was observed in 2007. It is not entirely clear what army and police were implied in the polls of 2003, as the army and police of Saddam Hussein's government had ceased to exist by then and the new forces were in the early stages of being formed (table 4.11).

The level of trust in the Iraqi national government changed over time. In 2005, after the transfer of power from the Coalition Provisional Authority to the Iraqi Interim Government, the level of trust was quite high (53%), whereas in 2007 the level of mistrust (61%) exceeded the level of trust (39%). In 2009 the situation was the direct opposite: 61 percent of respondents trusted the government and 39 percent did not. Polls from early 2007 and 2008 showed an almost equal division of society with regard to that issue.

One of the tasks that US leaders wanted to achieve in Iraq was a transformation of the political regime from tyranny (autocracy) to democracy. In this regard, it would be interesting to know the Iraqis' opinion on a preferred political regime for Iraq. The question that was asked, and the answers given, do not point to a univocal commitment of Iraqis to a democratic system. Furthermore, the question as it was phrased presented a definition of democracy that barely encompasses even the most rudimentary vision of this form of political system ("government with a chance for the leader to be replaced from time to time"). This definition of democracy is not inconsistent with two other of the suggested forms of political system, autocracy and an Islamic state. The poll results showed a similar degree of support for governance by a strong leader and governance by an Islamic state, though both fell short of the support recorded for even the

Table 4.11. Iraqi Level of Confidence in Different Institutions, 2003–2009
Answers to the Question, "I am going to name a number of organizations. For each one, please tell me if you have a great deal of confidence, quite a lot of confidence, not very much confidence, or none at all?"

	2003 (%)	2004 (%)	2005 (%)	February 2007 (%)	August 2007 (%)	March 2008 (%)	February 2009 (%)
Iraqi Army							
Great deal of confidence	13	18	36	24	23	28	37
Quite a lot of confidence	25	38	31	37	43	37	36
Not very much confidence	29	25	18	25	21	24	24
None at all	16	10	12	14	12	11	3
Refused/Don't know	17	9	3	–	–	–	–
Police							
Great deal of confidence	18	26	38	32	33	33	36
Quite a lot of confidence	28	41	31	32	36	34	38
Not very much confidence	30	20	18	16	17	20	20
None at all	15	8	12	20	15	13	5
Refused/Don't know	10	4	2	–	–	1	–
National government of Iraq							
Quite a lot of confidence	N/A	N/A	30	31	28	31	–
Not very much confidence	N/A	N/A	25	27	31	26	26
None at all	N/A	N/A	16	24	30	25	13
Refused/ Don't know	N/A	N/A	6	–	–	1	1

Source: Iraq Poll, February 2009.

impoverished definition of democracy. The poll results (table 4.12) demonstrate how answers changed from 2004 to February 2009.

The breakdown of answers to that question by community in 2009 shows a similar distribution of answers and the absence of significant differences. However, differences were evident in 2007. Among Kurds, "democracy" remained the most popular form of political system (66%), while Sunnis expressed greatest support for the idea of a "strong leader" (58%). Among Shiites, "Islamic state" (40%) and "democracy" (41%) were almost equally popular (table 4.13).

The polls of 2004 to 2009 show a low assessment of the state of the country (security, reconstruction, governance) by Iraqis, a low level of trust in created institutions, and a negative attitude toward the coalition forces. An especially negative attitude was recorded among the Sunnis.[90]

A poll conducted in February 2009 uncovered an increase among Iraqis with regard to security: 85 percent of all respondents described the situation as very good or quite good, up 23 percentage points from a year earlier; a total of 52 percent said that security had improved over the last year, up 16 percentage points from March 2008; only 8 percent said that it was getting worse, compared with 26 percent in 2008. Those who said their lives were going very well or quite well were 65 percent of the total, up 9 percentage points. There was a 14 percentage point increase, to 60 percent, in those who thought things would be better in Iraq as a whole a year later. The same poll revealed a negative assessment of the war itself and the actions of the United States, United Kingdom, and coalition troops. Fifty-six percent of respondents thought that the 2003 invasion was wrong (up 6%), while 42 percent said it was right (down 7%). Only 30 percent thought that the coalition forces were doing a good job, while 69 percent thought they were doing a bad job, more or less the same as a year earlier. Overall, 59 percent of those questioned thought that Britain's role was negative, 22 percent said it was positive; 64 percent said that the US role was negative, 18 percent said it was positive; 68 percent viewed Iran negatively, 12 percent viewed Iran positively.[91]

The higher assessments of the state of security most likely owed to the objective improvement in the situation, as well as to subjective factors, such as the expectation of complete withdrawal of foreign troops. The only legitimate reasons to retain a foreign military presence in the country

Table 4.12. Iraqi Preferences on Governance (General Poll), 2004–2009 *Answers to the Question,* "There can be differences between the way government is set up in a country, called the political system. From the three options I am going to read to you, which one do you think would be best for Iraq now?"

	2004 (%)	2005 (%)	February 2007 (%0	February 2009 (%)
Strong leader: government headed by one man for life	28	26	34	14
Islamic state: politicians rule according to religious principles	21	14	22	19
Democracy: government with a chance for the leader to be replaced from time to time	49	57	43	64
Refused/Don't know	4	3	—	3

Source: Iraq Poll, February 2009.

were to ensure security and to fight al-Qaeda. The optimistic assessments reflected the increasing confidence among Iraqis that the Iraqi army and police could ensure security in Iraq on their own.

Public opinion polls in Iraq in autumn 2010 demonstrated that in general, there were no noticeable improvements. Instead, assessments of the situation in the country were becoming increasingly pessimistic. The polls conducted by the International Republican Institute (IRI) showed that assessment of the economic situation almost equally divided Iraqis into optimists (49%) and pessimists (50%); 57 percent of respondents noted that things in Iraq were going in the wrong direction.[92]

An IRI poll carried out in April 2011 showed that in the most problematic—that is, northern—parts of Iraq, the Sunni or Northern Triangle, "the economic mood was fairly optimistic." A majority of respondents in three of the five northern provinces (Diyala, Ninawa, and Salah ad-Din) had answered

Table 4.13. Iraqi Preferences on Governance, by Community, February 2007 and February 2009

Answers to the Question, "There can be differences between the way government is set up in a country, called the political system. From the three options I am going to read to you, which one do you think would be best for Iraq now?" by community, February 2009 (February 2007 result in parentheses)

	Kurds (%)	Shia (%)	Sunni (%)
Strong leader: government headed by one man for life	12(25)	9(19)	20(58)
Islamic state: politicians rule according to religious principles	15(10)	26(40)	11(4)
Democracy: government with a chance for the leader to be replaced from time to time	71(66)	62(41)	65(38)
Refused/Don't know	2	3	4

Source: Iraq Poll, February 2009.

"good" or "very good" when asked to describe the current economic situation in Iraq: Diyala (58%), Ninawa (59%), and Salah ad-Din (54%). The other two northern province had less optimistic results. In Kirkuk, only 49 percent of responses were positive and 51 percent were negative, and in Anbar a pessimistic mood prevailed, with 22 percent positive responses and 75 percent negative. According to the analysis accompanying the published IRI poll results, "the positive mood about the economy did not carry over on the question of whether Iraq is headed in the right direction. Of the five provinces, only Diyala responded positively, with 60 percent responding that Iraq was headed in the right direction. The other provinces indicated the country was headed in the wrong direction: Anbar—86 percent wrong direction, 10 percent right direction; Kirkuk—57 percent wrong direction, 33 percent right direction; Salah ad-Din—45 percent wrong direction, 35 percent right direction; and Ninawa—47 percent wrong direction, 44 percent right direction. Diyala's relative optimism may be attributable to a decline in violence that occurred more recently compared to the rest of Iraq." When asked about "the single biggest problem facing their provinces, respondents were split among basic services, security, unemployment and other issues. The responsibility for the problems with electricity, security, unemployment and government corruption the majority has placed on Iraq's Prime Minister Nouri Maliki."[93]

On April 10, 2012, Martin Kobler, head of UNAMI, made a statement at a meeting of the UN Security Council. According to his briefing, many of the issues discussed above remained unsolved. According to Kobler, "in the first three months of 2012, a total of 613 civilians were killed and 1,835 injured, slightly less than in the same period of last year, but still horrific," and "there were still more than 1.3 million persons unable or unwilling to return to their places of origin." Baghdad hosted the largest number, with more than 300,000 refugees representing almost 60,000 families registered. Kobler: "The achievement of security and stability was and will remain a central priority for the national partnership Government, a principle clearly embodied in the Iraqi security forces' ability to take full responsibility for the country's security before and after the withdrawal for foreign forces." In his opinion, the domestic political impasse, continued terrorism and displacement, and the potential fallout from regional crises such as the one in Syria still posed obstacles for Iraq "as it marches on the road to full recovery." In particular, "the continued delays in convening a national conference underscores the urgent need for Iraqi leaders to summon the requisite political will and courage to work together to solve the country's problems through an inclusive political dialogue." Kobler pledged that UNAMI would remain steadfast in its commitment to help Iraqis address those challenges.[94]

The Cost of War

Servicemen and women from 38 countries took part in the military actions in Iraq. The largest number of non-US coalition troops (more than 1,000) were deployed by Australia, the United Kingdom, Georgia (whose troops were withdrawn in August 2008 because of the war in South Ossetia), Denmark, Ukraine, South Korea, Italy, the Netherlands, Romania, Poland, and Spain. Statistics on casualties among the coalition forces over the course of the war, according to the nongovernmental organization iCasulties.org,

Table 4.14. Iraq Coalition Military Fatalities by Year, 2003–2012

Year	US	UK	Other	Total
2003	486	53	41	580
2004	849	22	35	906
2005	846	23	28	897
2006	823	29	21	873
2007	904	47	10	961
2008	314	4	4	322
2009	149	1	0	150
2010	60	0	0	60
2011	54	0	0	54
2012	1	0	0	1
Total	**4,486**	**179**	**139**	**4,804**

Source: "Iraq Coalition Casualty Count," iCasualties.org (http://icasualties.org/Iraq/Index.aspx).

are presented in table 4.14. The largest numbers of casualties were sustained by the United States and United Kingdom.

A comparison of casualties in Afghanistan and Iraq shows that the overall number of casualties in the military operation in Afghanistan, which involved 22 countries, was 3,197 for the period 2001–2012. From 2009 on, however, the fatalities increased dramatically, from a total of 1,129 from the start of the war to 2009 to 3,197 by 2012. The greatest numbers of casualties were sustained by three countries: the United States, which lost 2,132 troops, the United Kingdom, which lost 433, and Canada, which lost 158.[95] According to the US Department of Defense, 4,489 US servicemen died in Iraq and 32,230 were wounded between March 19, 2003, and September 11, 2013. In Afghanistan the figures were 2,267 troops killed and 19,287 wounded between October 7, 2001, and September 11, 2013.[96]

A protracted military campaign and the rotation of troops are factors that often lead to post-traumatic stress disorders and increased rates of suicide among servicemen, and multiple deployments to a combat area are one reason for such psychological trauma. Of the 513,000 active-duty soldiers who served in Iraq between 2003 and 2008, more than 197,000 (68%) were deployed more than once, and more than 53,000 (31%) were deployed three or more times. According to official data, after a first deployment in Iraq 17 percent of soldiers were diagnosed with post-traumatic stress disorder. This figure increased to 18.5 percent after the second deployment and to 27 percent after the third.[97] On May 11, 2009, Sergeant John Russell shot five people, two officers and three soldiers, at a hospital for military personnel suffering from post-traumatic stress disorder on the US base in Baghdad, where he was undergoing a medical checkup. He was on his third deployment to Iraq.[98] The situation itself is not unique, though the number of victims killed in that episode was the highest over the course of the campaign. Defense Secretary Robert Gates stated in connection with this incident that the Pentagon should redouble its efforts to help servicemen suffering from stress. According to official statistics, in 2006 102 suicides were registered in the army, in 2007 115 suicides, and in 2008 140 suicides. In 2009, 244 suicides were registered among servicemen, rising to 300 in 2010 and continuing the rising trend from previous years.[99] Starting in 2009, the psychological stress that servicemen, their families, and their relatives experienced as a result of deployment to Iraq and Afghanistan was discussed at US congressional hearings.[100] Also in March 2009 the Department of Defense Task Force on the Prevention of Suicide by Members of the Armed Forces was established. In 2010, it presented 76 recommendations. Twenty-five of these recommendations focused on establishing Department of Defense units to address suicide prevention on a constant basis. The change of attitude toward suicide and its survivors was reflected in the core approach: "to establish a culture that fosters prevention as well as early recognition and intervention." This approach would include special attention to psychological support for servicemen and their families and "zero tolerance" policies regarding any discrimination against individuals with "emotional, psychological, relations, spiritual, and behavioral issues." Special programs would be designed to train servicemen and officers to be aware of the issue and to address it responsibly.[101] In November 2011, the Department of Defense Suicide Prevention

Figure 4.2. Losses among Iraqi Civilians, Iraqi Security Forces, and US Military Personnel, 2005–2011

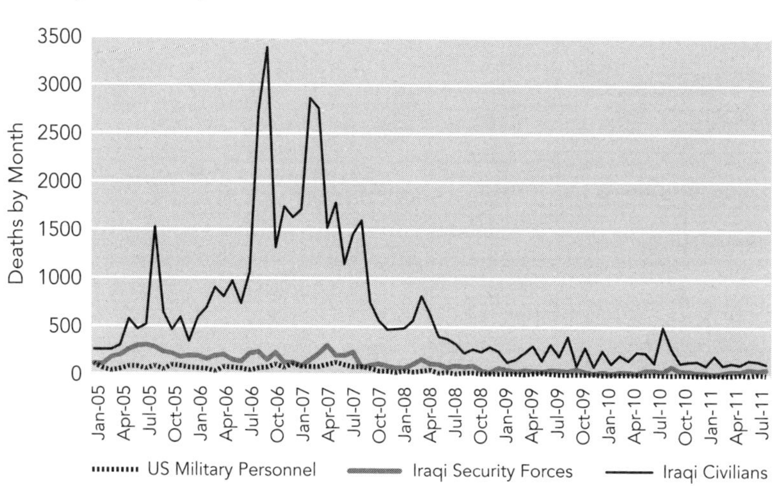

Source: iCasualties.org (http://icasualties.org/Iraq/IraqiDeaths.aspx).

Office was created to oversee the development and implementation of suicide and risk reduction programs within the department.[102]

According to data from iCasulties.org, there were 10,125 fatalities among Iraqi security forces and police and more than 50,000 civilian deaths from the start of warfare in 2003 to July 2011.[103] Iraqi civilian casualties significantly exceeded casualties among the military, though the exact figures are controversial and hard to establish. According to statistics provided by the Iraqi Healthcare Ministry, civilian casualties ranged between 104,000 and 223,000 people from 2003 to June 2006.[104] In November 2006, Iraqi minister of healthcare Al-Shemari publicly announced the figure of 150,000 civilian deaths and said that, in addition, there were three wounded per one dead in Iraq.[105] These figures gave rise to debates all over the world, for they exceeded any other estimates and came from a public official, based on a study supported by the World Health Organization. According to the data of the nongovernmental organization Iraq Body Count, the number of recorded deaths among Iraqis was 157,531 between January 2003 and December 31, 2011, including 114,212 civilian deaths, with the rest of the casualties accounted for by Iraqis fighting either for or against Western countries.[106] According to a widely cited study published in October 2006, almost 655,000 Iraqis died, or 2.5 percent of the population, which makes this war the most deadly conflict of the twenty-first century.[107] The highest casualty count comes from the US nongovernmental organization Just Foreign Policy; according to its estimates, more than 1.2 million people died in Iraq between March 2003 and August 2007.[108]

The proportional losses among Iraqi civilians, Iraqi security forces, and US soldiers can be compared using data from iCasualties.org. As figure 4.2 clearly shows, the war in Iraq resulted in a disproportionate number of civilian casualties. American public opinion and the citizens of European countries whose armed forces took part in the war were concerned about the number of dead troops, whereas the situation in Iraq could be called at least a humanitarian crisis and possibly a humanitarian disaster from the standpoint of the living conditions of the local population.

US expenses for the war in Iraq included spending on "military operations, base security, reconstruction, foreign aid, embassy costs, and veterans' health care." The total sum of these expenses approved by the US Congress for the war on terror from 2001 to June 30, 2008, amounted to US$864 billion, of which $657 billion, or 76 percent, went to the military operation in Iraq. According to the Congressional Budget Office, from $1.3 trillion to $1.7 trillion may additionally be allocated to the war on terror for the years 2009–2018,[109] which brings the total amount spent on the war on terror in the period 2001–2018 to more than $2.4 trillion.[110] The amount of $49 billion was allocated to programs for Iraqi stabilization and reconstruction beginning in 2003.[111] As Stuart W. Bowen Jr., inspector general in Iraq, noted in his February 2009 report *Hard Lessons: The Iraq Reconstruction Experience,* this represents "the largest relief and reconstruction effort for one country in U.S. history,"[112] though spending on the war has been vastly greater than spending on reconstruction.

The US budget deficit reached a record $490 billion by 2009, much of it associated with the war on terror initiated by President George W. Bush.[113] The world economic crisis doubled the budget deficit. The Obama administration had to deal with a budget deficit that grew to $1.752 trillion in 2009, and sought to reduce the budget deficit to $1.171 trillion in 2010 and to $912 billion in 2011.[114] Insofar as neither war was concluded, an increase in defense spending by 4 percent ($20.4 billion) was planned for

2009; additional funds were included in the budget to cover the withdrawal of US troops from Iraq and an increased military presence in Afghanistan ($75.5 billion for 2009, in addition to the $40 billion already allocated for 2009 and the $130 billion allocated for 2010). Policy-makers planned to increase the size of the army, raise the wages of servicemen and women, and increase spending on medical services to the military, including needed psychological assistance. The government fulfilled its pledges to the troops and veterans by increasing the budget for housing, providing compensation to the disabled, and increasing spending on health care and educational programs, all amounting to a total of $25 billion.[115] Payments to the military under medical programs increased almost three times on the budget line "Retiree Health Insurance Benefits," from $480 billion in 2000 to $1,220 billion in 2005, and the amount budgeted for "Veterans Disability Compensation" almost doubled, from $679 billion to $1,218 billion during the same period.[116]

In summing up the price of war in Iraq, it is appropriate to conclude with some observations by Ali A. Allawi, minister of trade and minister of defense in the Iraqi Transitional Government in 2005–2006. His book, *The Occupation of Iraq: Winning the War, Losing the Peace*, was published in 2007, but the author's assessments are still relevant:

> The law of unintended consequences broke out in Iraq with a vengeance. The USA invasion and occupation of Iraq broke a thick crust that had accreted over the country and region as a whole, and released powerful subterranean forces. The emergence of the Shi'a after decades, if not centuries, of marginalization was perhaps the most profound outcome, closely followed by the massive spur given to the drive of a Kurdish nation. On another level, the division within the world of Islam became far more pronounced. They are about to move on to an altogether different plane of mutual antipathy and internecine warfare.... It was the Bush Administration that acted as the unwitting handmaiden to history and denied, ignored, belittled and misunderstood the effects of what it had created....
>
> Iraq cannot, by any stretch of imagination, be seen as a model for anything worth emulating.

> America's "civilizing mission" in Iraq stumbled, and then quickly vanished, leaving a trail of slogans and an incomplete reconstruction plan. The billions that America had spent went unrecognised, and therefore not appreciated. Iraqis heard about the billions, like some memorable banquet to which only a few are invited. But what they experienced was the daily chaos, confusion, shortages, and the stark terrors of life. Death squads now compounded vicious attacks of terrorists. Opinions and divisions were hardening.... The corroded and corrupt state of Saddam was replaced by the corroded, inefficient, incompetent and corrupt state of the new order.
>
> Bush may well go down in history as presiding over one of America's great strategic blunders.... But it is Iraq and Iraqis who paid [for] most of the failed policies of their erstwhile liberators and their newly minted governors.[117]

DISCUSSION OF THE WAR'S OUTCOME IN THE UNITED STATES

Soon after the start of the war in Iraq it became clear that allegations about the presence of WMDs in Iraq were not confirmed, and the findings presented by the US intelligence agencies turned out not to correspond to reality. Claims that Iraq supported al-Qaeda and prepared terror attacks against the United States and the United Kingdom could not be confirmed. Thus, the tasks that the United States and its coalition partners set for themselves were not achieved, and the legitimacy of their actions was not affirmed. Moreover, policy-makers in both the United States and the United Kingdom faced the political consequences of the war in their own countries. George W. Bush and Tony Blair were accused of manipulating intelligence data and exaggerating the threat emanating from Iraq in order to win approval for military action, and of abusing their power and the trust of the members of Parliament or Congress and the general citizenry. In September 2003, the Committee on Intelligence of the House of Representatives completed its four-month-long examination of the data used by the Bush administration to secure congressional approval to go to war. The committee concluded that the materials used to support the war were "fragmentary," "outdated," and "circumstantial."[118] In May 2004, Secretary of State Powell acknowledged in his interview with ABC, an American news outlet, that the information the CIA claimed to have about the presence of WMDs in Iraq was "inaccurate and wrong and in some cases, deliberately misleading," and that he was "disappointed" and regretted his involvement in conveying incorrect information. He added, however, that this was the "best information that the Central Intelligence Agency made available" and that there had been a collective judgment of the intelligence community about the presence of WMDs in Iraq.[119] Thus, the main objective of the war, the disarmament of Iraq, turned out to be based on a fiction.

The war in Iraq, according to many analysts, worsened the situation in the region and gave rise to the emergence of al-Qaeda terrorist groups in Iraqi territory. Already in September 2003, Madeleine Albright, former US secretary of state, in an interview with *Time* magazine answered the question, "Has the war made the problem of terrorism better or worse?" by saying "The Administration immediately tied Sept. 11 to Saddam. They said, basically, that Saddam and Iraq were a hotbed of terrorism. While I had many criticisms of Saddam, that's not the way I saw it. But now Iraq is in fact a breeding ground for terrorists."[120]

The course of the US-led war and of the civil war in Iraq led to an even greater worsening of life for Iraqis compared to life under Saddam Hussein. The task of helping Iraqis in what the United States depicted as their struggle against tyranny and the creation of a free and prosperous society turned out to be not as simple as it had seemed to the war initiators. As was predicted, the war had extremely negative consequences for the political, economic, and humanitarian situation in the country and the region. In addition to the numerous human casualties, the war destroyed Iraq's infrastructure, including vital communications systems, the water supply, the electrical grid, transportation, and the foundations of economic activity and employment, as well as systems providing the population with food and vital resources and services. Creating a democratic regime is a complex process beset by controversy. Iraq's external debt, the state of its economy, and the life needs of its population make the future of the country dependent on international financial and political assistance. Stability, security, and democratic principles in domestic politics and peacefulness in foreign

Figure 4.3. Changing American Attitudes on Iraq's Connection to September 11 Terrorist Attacks, 2002–2008

Answers to the Question, "Do you think Saddam Hussein was personally involved in the September 11th, 2001, terrorist attacks on the World Trade Center and the Pentagon?"

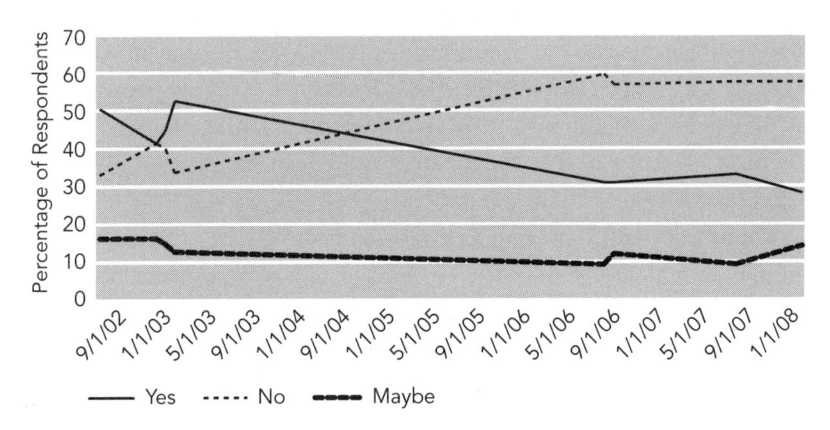

Source: CBS News poll, March 15–18, 2008 (http://www.pollingreport.com/iraq2.htm).

Figure 4.4. Changing American Attitudes on the Moral Justification of the Iraq War, 2006–2011

Answers to the Question, "Do you think the United States' action in Iraq has been morally justified, or not?"

Source: CNN/Opinion Research Corporation poll, December 16–18, 2011 (http://www.pollingreport.com/iraq.htm).

policy: that is what Western countries are hoping for, but such things are difficult to guarantee, especially if we consider the history, traditions, and current state of affairs of the country and of the region as a whole.

The war became possible because the Bush administration had created favorable domestic public opinion. Today, a decade later, it is possible to track the changing attitudes of Americans with regard to the key issues associated with the start of war and how the success of the campaign was generally assessed. Polls captured the change in opinion over time regarding the reasons for the start of the war and its legitimacy. Thus, answers to the question "Do you think Saddam Hussein was personally involved in the September 11th, 2001, terrorist attacks on the World Trade Center and the Pentagon?" reveal that in 2008, a minority of respondents believed that the former leader of Iraq had been involved in the terrorist attacks of 9/11 (figure 4.3).[121]

In 2003, 31 percent of respondents gave a positive response to the question, "Do you think the Bush administration deliberately misled the American public about whether Iraq had weapons of mass destruction, or not?," while 67 percent did not support this view. In February 2008, 53 percent of respondents believed that this was deliberate misinformation and 42 percent that it was not. In 2011, 57 percent believed that this was deliberate misinformation and 41 percent that it was not. The question about the morality of starting the war demonstrates a polarization of public opinion, with an insignificant predominance of war opponents (figure 4.4).[122]

The US success in achieving its stated objectives in going to war was evaluated by Americans increasingly negatively over the course of the polling period, though an almost equal division gradually emerged between those who believed that the war had contributed to an increase in US security and those who disagreed (figures 4.5 and 4.6).[123]

The most significant change in the American public's assessment of the Bush administration's policies in Iraq was observed between June 2003 and April 2008 (figure 4.7).

The party affiliation of respondents was evident in the responses to that question: the opinions of Republicans and Democrats were almost directly opposite. Table 4.15 illustrates this claim.

In April 2008, when asked, "Regardless of how you intend to vote, what would you prefer the next president do about the war in Iraq? Would you prefer the next president try to end the Iraq war within the next year or

Figure 4.5. Changing American Attitudes on the Iraq War and Long-Term US Security, 2003–2007

Answers to the Question, "Do you think the war with Iraq has or has not contributed to the long-term security of the United States?"

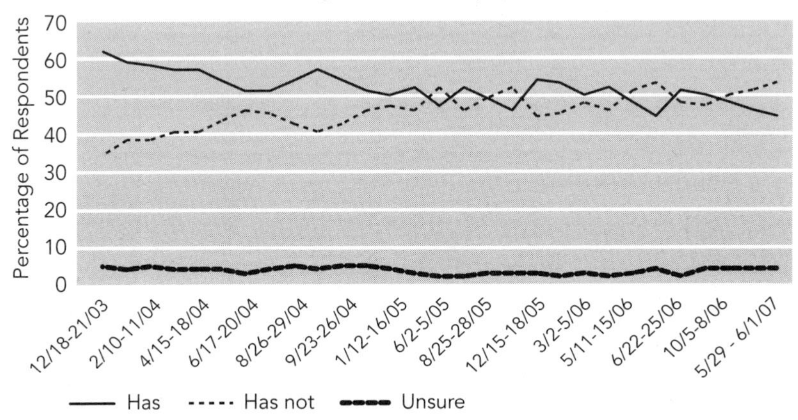

Source: ABC News/*Washington Post* poll, May 29–June 1, 2007 (http://www.pollingreport.com/iraq5.htm).

Figure 4.6. Changing American Attitudes on the Iraq War and the War on Terrorism, 2003–2007

Answers to the Question, "Do you think the war in Iraq has helped the war on terrorism, or has it hurt the war on terrorism?"

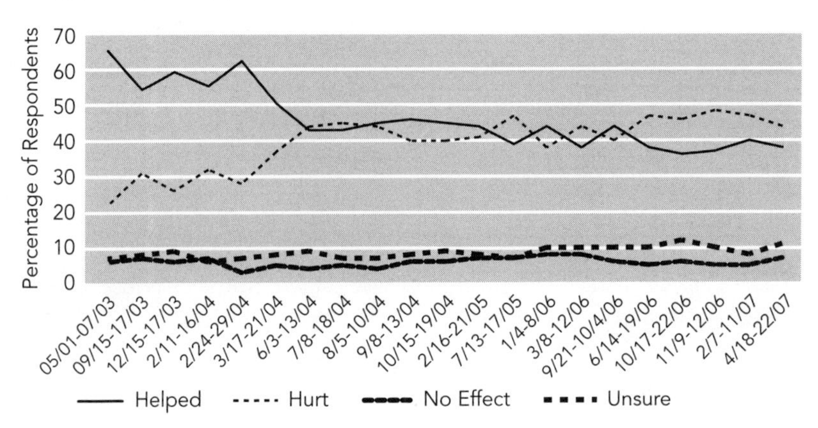

Source: Pew Research Center for the People & the Press, April 18–22, 2007 (http://www.pollingreport.com/iraq6.htm).

Figure 4.7. Changing Approval Ratings of the Bush Administration's Handling of the Iraq War, 2003–2008

Answers to the Question, "Do you approve or disapprove of the way Bush is handling the situation in Iraq?"

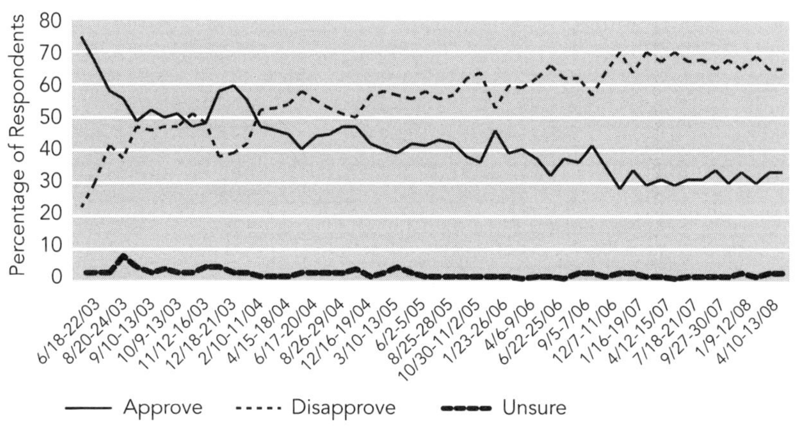

Sources: ABC News/*Washington Post* poll, September 4–7, 2007 (http://www.pollingreport.com/iraq3.htm); ABC News/*Washington Post* poll, April 10–13, 2008 (http://www.pollingreport.com/iraq2.htm).

Table 4.15. Approval Ratings of the Bush Administration's Handling of the Iraq War, by Party Affiliation, April 2008

Answers to the Question, "Do you approve or disapprove of the way George W. Bush is handling the situation with Iraq?"

	Approve (%)	Disapprove (%)	Unsure (%)
All adults	29	64	7
Republicans	66	26	8
Democrats	6	90	4
Independents	24	68	8

Source: CBS News/*New York Times* poll, April 25–29, 2008 (http://www.pollingreport.com/iraq2.htm).

two, no matter what, or continue to fight the Iraq war as long as they felt it was necessary?," 62 percent of all respondents spoke in favor of ending the war and 34 percent favored continuing it. Broken down by party affiliation, 89 percent of Democrats but only 26 percent of Republicans were in favor of ending the war, and 68 percent of Republicans but only 10 percent of Democrats favored continuing it (table 4.16).[124]

A divisive partisanship was also evident in the assessment of the war, its necessity, its justifiability, and the possibility of victory. Against the backdrop of a Republican administration and a Republican majority in both houses of Congress, Republicans more so than Democrats supported military action, and Republicans assessed President Bush's policies and plans for reconstructing Iraq more favorably than Democrats and independents did. As Charles A. Kupchan and Peter L. Trubowitz wrote in 2007, "The United States is in the midst of a polarized and bruising debate about the nature and scope of its engagement with the world. The current reassessment is only the latest of many; ever since the United States' rise as a global power, its leaders and citizens have regularly scrutinized the costs and benefits of foreign ambition."[125] The deep division in both American society and the political elite concerning the nature of the US engagement in Iraq grew as the war dragged on and its costs—both material costs and the cost in human lives—increased.

Over the course of the war, there was a significant change in the ratio of responses to the question concerning whether the decision to launch the war in Iraq had been a mistake ("In view of the developments since we first sent our troops to Iraq, do you think the United States made a mistake in sending troops to Iraq, or not?"). In March 2003, 75 percent of respondents believed that the United States had made the right decision in going to war, whereas in March 2010 only 41 percent thought so. The proportion of those who considered the war a mistake increased over the same period from 23 percent to 55 percent. Negative attitudes toward the war peaked in July 2007 (62%), April 2008 (63%), and the summer of 2009.[126]

The majority of Americans believed that the war had a generally negative effect on life in the United States, though from 2008 to 2011 the proportion of people thinking this way declined, from 63 percent in 2008 to 52 percent in 2011. The proportion of those who believed that the war had a positive effect remained practically unchanged, being 15 percent in 2008 and 16 percent in 2011. There remained a stable opinion that spending on the war in Iraq was the main reason for the economic problems of the United States (71%), with only 28 percent believing that the war and the economic crisis were not interconnected. It is interesting that Americans more optimistically assessed the effect of the war on life in Iraq, though this clearly contradicted all objective indicators: in 2011, 48 percent believed that "U.S. involvement in the war in Iraq has had a positive effect on life in Iraq generally," 28 percent responded that it "[has had] a negative effect," and 21 percent responded that it "hasn't had much effect."[127]

In September 2010, most Americans gave a negative response to the question, "Do you think the result of the war with Iraq was worth the loss of American lives and other costs of attacking Iraq, or not?" Seventy-one percent responded "not worth it," 23 percent responded "worth it," and 6 percent responded "unsure." Answers to the same question in

Table 4.16. Opinions on Continuation of War into Next Presidential Administration, by Party Affiliation, April 2008

Responses by Party Affiliation to the Question, "Regardless of how you intend to vote, what would you prefer the next president do about the war in Iraq? Would you prefer the next president try to end the Iraq war within the next year or two, no matter what, or continue to fight the Iraq war as long as they felt it was necessary?"

	End war (%)	Continue (%)	Unsure (%)
All adults	62	34	4
Republicans	26	68	6
Democrats	89	10	1
Independents	63	31	6

Source: CBS News/*New York Times* poll, April 25–29, 2008 (http://www.pollingreport.com/iraq2.htm).

August 2003 demonstrated the almost even split in society, with 46 percent responding "worth it," 45 percent responding "not worth it," and 9 percent responding "unsure." In the same poll, the party affiliation of respondents was considered, and the answers demonstrated the strictly negative attitude of Democrats toward the war (81% against vs. 14% for). Among independent respondents the figures were 67 percent against and 21 percent in favor, and among Republicans 42 percent were in favor and 49 percent were against.[128]

Answers to the direct question, "Do you favor or oppose the U.S. war in Iraq?" demonstrated a constant but not dramatic increase in the proportion of war opponents from 2006 to 2011. In June 2006, 38 percent of respondents supported the war and 54 percent were against it, but in November 2011, 31 percent of respondents supported the war and 68 percent were against it.[129] Americans' assessment of the success of the military operation in Iraq, however, changed significantly. In general, answers to this question reflected the successes and failures of American troops in Iraq, the division in society, and increased optimism with regard to the possibility of victory in Iraq and stabilization of the country with the coming of the Obama administration (figures 4.8 to 4.10).

Assessments of whether the situation in Iraq constituted a victory or a defeat caused difficulties in responses after the withdrawal of troops. In December 2011, 31 percent of respondents answered that "the outcome for the United States in Iraq" was a victory, 11 percent said that it was a defeat, and 54 percent described it as a stalemate. It is interesting to note the high and almost completely unanimous assessment of the US Army in Iraq: 96 percent of respondents said they "are proud of the U.S. troops who were stationed in Iraq," and only 3 percent said "no."[130]

Evaluating the outcome of the Iraq War from the standpoint of the objectives that the United States had set for itself, Americans negatively assessed the US success, though some polls demonstrated contradictory data. Thus, answering the question, "As a result of the United States' military action against Iraq, do you think the United States is more safe from terrorism, less safe from terrorism, or hasn't it made any difference?," most respondents said the war did not have an effect (figure 4.9).[131]

Opinions on the extent to which the United States had achieved its war objectives differed markedly in polls carried out by different agencies.

Figure 4.8. Changing American Attitudes on US Military Success in Iraq, 2006–2010

Answers to the Question, "Do you think the United States is winning or not winning the war in Iraq?"

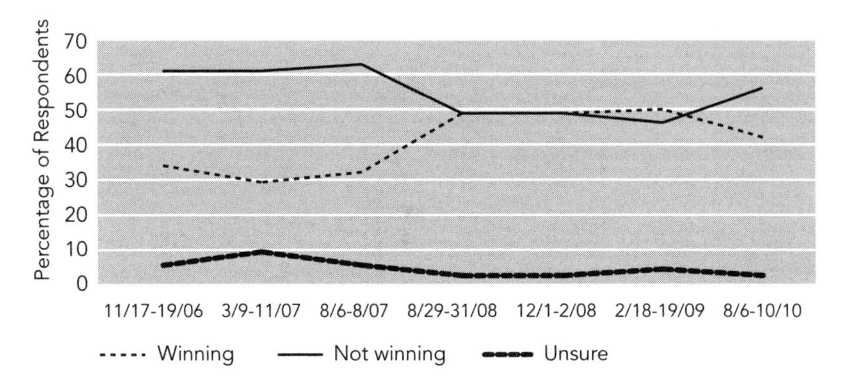

Source: CNN/Opinion Research Corporation poll, August 6–10, 2010 (http://www.pollingreport.com/iraq.htm).

Figure 4.9. Changing American Attitudes on US Safety from Terrorism as a Result of the Iraq War, 2007–2010

Answers to the Question, "As a result of the United States' military action against Iraq, do you think the United States is more safe from terrorism, less safe from terrorism, or hasn't it made any difference?"

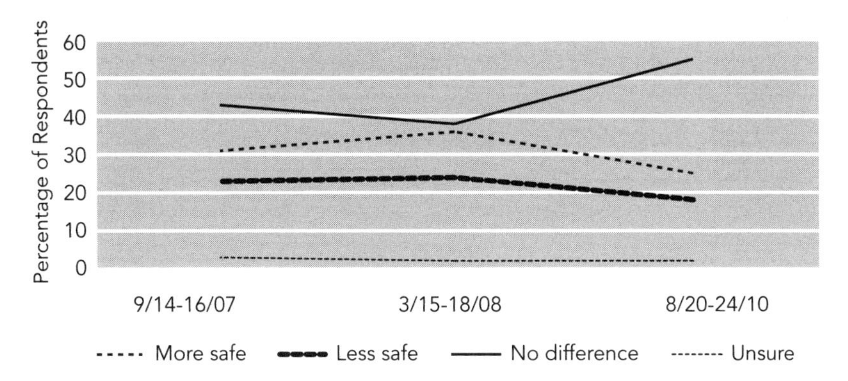

Source: CBS News poll, August 20–24, 2010 (http://www.pollingreport.com/iraq.htm).

Figure 4.10. Changing American Attitudes on US Goal Achievement in Iraq, 2010–2011

Answers to the Question, "Do you think the US has or has not achieved its goals in Iraq?"

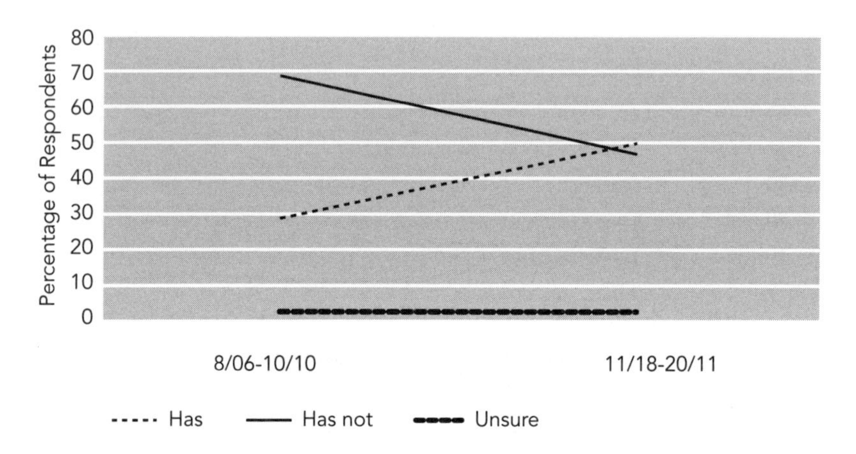

Source: CNN/Opinion Research Corporation poll, November 18–20, 2011 (http://www.pollingreport.com/iraq.htm).

Thus, the poll conducted by CNN/Opinion Research Corporation in November 2011 showed greater optimism in 2011 than in 2010, though still a deeply divided society (figure 4.10).[132]

However, the results of CBS News polls in August 2010 revealed a polarization of society rather than the large opinion gap noticeable in figure 4.10. The same poll also categorized responses according to the party affiliation of the respondents (table 4.17).[133]

Assessments of the situation in Iraq from the standpoint of constructing an "open democratic society" reflected a skeptical attitude of Americans and divided opinion with regard to American troops' responsibility for creating stability and security in Iraq. The polarization of opinion was evident in the assessment of whether "significant progress [has been made] toward restoring civil order in Iraq," with growing optimism manifested during the presidency of Barack Obama (figure 4.11).[134]

After the plan for US troop withdrawal was announced in early 2009, a majority of respondents believed that withdrawal of troops should be

Table 4.17. Opinions on US Military Success in Iraq, by Party Affiliation, August 2010

Answers to the Question, "Regardless of whether you think taking military action in Iraq was the right thing to do, would you say the United States has succeeded in accomplishing its objectives in Iraq, or has it not succeeded?"

	Has (%)	Has not (%)	Unsure (%)
All	41	51	8
Republicans	57	36	7
Democrats	36	57	7
Independent	34	56	10

Source: CBS News poll, August 20–24, 2010 (http://www.pollingreport.com/iraq.htm).

Figure 4.11. Changing American Opinions on Civil Order in Iraq, 2007–2009

Answers to the Question, "Do you think the United States is or is not making significant progress toward restoring civil order in Iraq?"

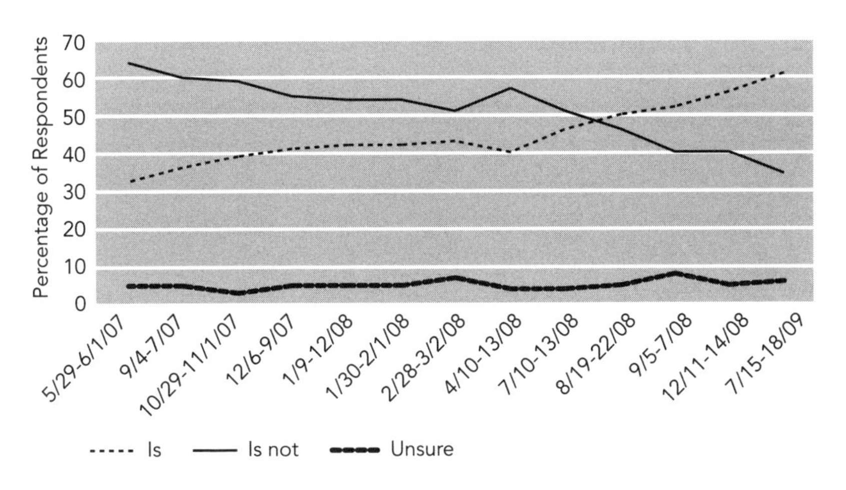

Source: ABC News/ *Washington Post* poll, July 15–18, 2009 (http://www.pollingreport.com/iraq2.htm).

carried out according to the schedule, regardless of whether the war objectives had been achieved. In June 2008, 52 percent of respondents believed that the troops should be withdrawn from Iraq as soon as possible, and 43 percent believed that the troops should stay until stability was established. This ratio almost mirrors the ratio between proponents (53%) and opponents (43%) of keeping the troops in Iraq registered in July 2004.[135]

It is interesting that in August 2010, a majority of respondents did not agree with the statement, "The U.S. should keep its troops in Iraq beyond 2011 if Iraqi security forces are unable to contain insurgent attacks and maintain order in Iraq": 53 percent responded "stick to timetable" and 43 percent responded "stay beyond 2011, if necessary."[136] Answers to a similar question asked in 2008, "Do you think the United States does or does not have an obligation to establish a reasonable level of stability and security in Iraq before withdrawing all of its troops?," showed that 65 percent of respondents believed the United States should stay and 32 percent did not.[137] Obviously, respondents understood that the problem was complex and could not be solved merely by withdrawing troops. After the withdrawal of US troops from Iraq, assessments of the future of Iraq with respect to constructing a democratic, peaceful, and stable society continued to be pessimistic.[138] Table 4.18 shows the breakdown of opinion.

One of the benchmarks for assessing Americans' attitude to the Iraq War is the war in Vietnam. The term "Vietnam syndrome" is often used as shorthand for the state of American society and politics as it coalesced during the war in Vietnam. The main manifestations of the syndrome are a negative attitude of society toward military interventions (anti-interventionism), a low level of trust in the military and political institutions of the country, a nationwide political crisis brought on by the loss of legitimacy of the foreign policy course, and an unwillingness of the political elite to initiate wars that could result in the political leadership losing legitimacy.

US society is considered to have put the Vietnam syndrome behind it during the Reagan presidency. The signs of such a turn included the recovery of trust in the armed forces, an end to the army's recruitment problem, and greater patriotism. The swift and successful war in Iraq in 1991 gave the United States confidence that the Vietnam syndrome had been overcome. Nevertheless, "Vietnam," as a cluster of symptoms around loss of legitimacy of the political leadership and lack of faith in abiding

Table 4.18. American Predictions of Outcome after US Troop Withdrawal, December 2011

Answers to the Question, "Now, thinking about Iraq, looking ahead to a time after US troops have left Iraq, how likely do you believe each of the following things is to happen: very likely, somewhat likely, somewhat unlikely, or very unlikely?"

	Very likely (%)	Somewhat likely (%)	Somewhat unlikely (%)	Very unlikely (%)	Unsure (%)
There will be all-out civil war	21	39	24	11	5
There will be more attempted terrorist attacks against the United States on our own soil as a result of our troops having left Iraq	12	33	30	22	3
The Iraqi government will achieve a stable democracy	4	34	32	28	2
Iraq will become more settled and less violent	7	28	30	32	3

Source: NBC News/ *Wall Street Journal* poll, December 7–11, 2011 (http://www.pollingreport.com/iraq.htm).

institutions, has become ingrained in the political vocabulary and is often hauled out in discussions of US participation in military operations abroad. The motto "no more Vietnams" has a lasting anti-interventionist appeal.

The lurking Vietnam syndrome as a backdrop to US involvement in foreign operations is shown in a remarkable cartoon by Daryl Cagle that was published on March 27, 2003 (figure 4.12), and appears in Donald Rumsfeld's memoirs, in the chapter on the start of the military campaign in Afghanistan in 2002. The cartoon shows Rumsfeld, secretary of defense under President Bush during the first few years of the Iraq War, as the driver of a car full of journalists pelting him with the question, "Is it Vietnam yet?"[139]

According to Henry Kissinger, secretary of state in the Nixon administration, "the experience of Vietnam remains deeply imprinted on the American psyche, while history has seemingly reserved for itself some of

Figure 4.12. Daryl Cagle, "Is It Vietnam Yet?" (March 27, 2003)

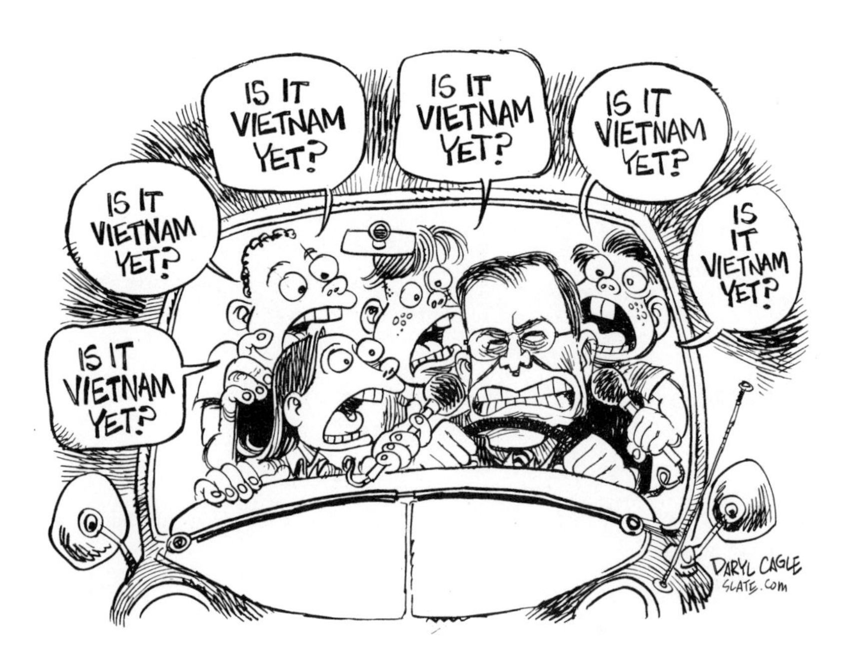

"Is it Vietnam Yet?" Cartoon by Daryl Cagle, March 27, 2003. Copyright 2003 by Daryl Cagle. Reproduced with permission.

its most telling lessons."[140] Melvin Laird, defense secretary in the Nixon administration, who held a negative view of interpretations of the lessons of Vietnam in contemporary American politics, wrote in 2005 that the legacy of more than 30 years of misinformation about the Vietnam War "[has left] the United States timorous about war, deeply averse to intervening in even a just cause, and dubious of its ability to get out of war once it is in one." Laird concluded this statement with a caustic remark: "All one need whisper is 'another Vietnam,' and palms begin to sweat."[141]

Numerous publications in the United States have been devoted to the lessons of Vietnam, with new papers, monographs, and memoirs appearing every year. In 1981 the leading international journal *International Security* published portions of a debate on the impact of the Vietnam War experience on US foreign policy.[142] The participants included such respected American analysts as Stanley Hoffman, Samuel Huntington, Thomas Schelling, Ernst May, and Richard Neustadt. They cited the following military and political lessons of Vietnam:

- The complexity of fighting a limited war, and the impossibility of achieving victory in such a war.

- "Democracies are not well equipped to fight lengthy limited wars" (Huntington).[143]

- Weak opponents impose on strong developed countries forms of warfare that make it impossible for the strong countries to win (guerrilla and terrorist strategies).

- Conventional armies cannot successfully participate in liberation wars and fight guerrilla wars, for this has a strong demoralizing effect on them (Hoffman).[144]

- Participating in limited wars in which one's opponents use guerrilla tactics drives military servicemen to commit cruel acts out of frustration, and leads to other forms of dehumanizing behavior (Schelling).[145]

- The fallacy of relying on split political forces with a simultaneous attempt to establish a new political system (state) (Hoffman, Huntington).[146]

- With Vietnam, Americans witnessed "the end of the era in which one could believe that a great industrial power is bound to win when it fights a small, poor, backward country" (Schelling). Schelling also pointed to the paradoxical lesson of Vietnam, namely, that "lessons are never learned."[147]

An important title among the numerous books on the subject is Robert McNamara's memoirs, *In Retrospect: The Tragedy and Lessons of Vietnam.* McNamara was defense secretary in the John F. Kennedy and Lyndon B. Johnson administrations. In later years, he concluded that the war damaged America terribly. He wrote, "We of the Kennedy and Johnson administrations who participated in the decisions on Vietnam acted according to what we thought were the principles and traditions of this nation. We made our decisions in light of those values. Yet we were wrong, terribly wrong."[148] McNamara was one of the initiators of the US mass military involvement in Vietnam, which led to an increase in the US military presence from 16,000 "military advisers" in 1963 to an army of half a million by the mid-1960s, and he was no doubt responsible for the expansion of that war and for the tens of thousands of American servicemen's lives lost. McNamara identified 11 reasons for US defeat in Vietnam. They can be generalized as follows:

- "Misjudgment of the geopolitical intentions of the US's adversaries and exaggeration of the dangers to the United States of their actions"; profound ignorance of the history, culture, and politics of the people in the area, and of the personalities and habits of their (Vietnamese) leaders.[149]

- Failure to recognize the limitations of modern, high-tech military equipment, forces, and doctrine in confronting an unconventional, highly motivated people's movement; underestimation of the power of nationalism to motivate a people to fight and die for their beliefs and values; failure to retain the popular support of the American people.[150]

McNamara also underscored the need to have the support of the international community when resorting to military actions and the importance of implementing such actions jointly with other countries, especially when there is no direct threat to the United States. Anticipating the trajectory of international relations in the twenty-first century, McNamara wrote that the United States should be cautious about participating in "limited wars," and noted that "our leaders—and our people—must be prepared to cut our losses and withdraw if it appears our limited objectives cannot be achieved at acceptable risks and costs."[151]

Analysts and politicians disagree on the reasons the United States went to war in Vietnam and Iraq. Most American analysts point out that in the first instance, the United States was motivated by the goal of combating communism in Southeast Asia and the expanding Soviet influence, which was an insufficiently well-defined task and did not correspond to the immediate national interests of the United States. In the case of Iraq, this was a fight against a real enemy, international terrorism, which had inflicted a tangible blow on the United States. Some analysts tend to lump the wars together, as if they were both started on the basis of false, untrue reasons that did not reflect the real national interests of the United States, and a substitution of reasons was discovered soon after the start of the war. In preparing for war, according to many analysts and public figures in the United States, the Bush administration successfully demonized the figure of Saddam Hussein and exploited the American fear of the dangers associated with international terrorism that had emerged after 9/11.

An exploration of American attitudes toward the Iraq War shows the differences of that period compared to the Vietnam War. During the Vietnam War, a significant share of the opposition was represented by young people, and for the war in Iraq, opposition was still high in the group less than 29 years old, according to polls taken in 2003 to 2005. However, a study by the Gallup organization in 2007 demonstrated that a sustained negative attitude toward the war in Iraq was observed only in the age group over 50 years. In senior age groups (more than 57 years old), up to 66 percent of respondents considered the deployment of American troops in Iraq a "mistake," while in the age groups from 18 to 49 years this assessment was supported by only 45 percent to 51 percent. A particularly notable difference is the higher level of support for the war from Americans with higher and postgraduate education, in contrast to the period of the 1960s to 1970s, when the highest support for the war was recorded among 30- to 39-year-olds.[152] Probably one of the reasons for the more tolerant attitude of young people to the Iraq War was the absence of the draft, which had existed until 1973 and was abolished because of the negative public perception of the US engagement in Vietnam. Criticism of the Iraq War among senior Americans may be connected to their having witnessed or participated in the Vietnam War or having grown up under the influence of the Vietnam syndrome.

American analysts sometimes refer to a problem that emerged during the Vietnam War and manifested itself again with the Iraq War: the deepening division in American society and increasing isolationism. In the course of the Iraq War, more and more Americans expressed the desire that the United States pay more attention to its own problems and not try to solve all the world's problems. Charles A. Kupchan and Peter L. Trubowitz in their article, "Grand Strategy for a Divided America," cited Dean Acheson, former secretary of state, as saying that 80 percent of the job of foreign policy was the "management of your domestic ability to have a policy."[153]

Comparisons with the Vietnam War were being made even before the start of the war, and after the war became protracted, analogies with the Vietnam War became constant. In 2004, the Strategic Studies Institute of the US Army War College published a report by Jeffrey Record and Andrew W. Terrill titled *Iraq and Vietnam: Differences, Similarities, and Insights*. The authors believed that it was impossible to compare the two wars, as all historical events are unique. They also emphasized that these two wars had more differences than similarities. The only possible analogy, in their point of view, related to the process of nation-building in the context of a foreign culture, as well as the problem of gaining and retaining the support of American society for a protracted war against guerrilla fighters.[154]

Some publications draw parallels between these wars. The first such work was published by the American historian Clark C. Smith in 2004 under the title *Vietnam ... in Iraq: Reflections on the New Quagmire*. Smith argued that the second war in Iraq was a concession to the pro-Israel lobby in the United States that sought to eliminate Saddam Hussein, who had rendered support to terrorist anti-Israeli groups. However, the start of the war went according to a scenario that American strategists had not foreseen. The mission was not accomplished, the struggle dragged on, and the quagmire in Iraq, according to Smith, was poised to cost much blood and money to both the United States and Iraq. Smith thought that the war in Iraq would not end while George W. Bush was in power,[155] and his prediction proved true.

The problem of Iraq occupied the leading place in the party struggle, while the problem of confrontation of parties and the role of the Congress in terminating wars drew the special interest of American researchers. William G. Howell and Jon C. Pevehouse, authors of the book *While Dangers Gather: Congressional Checks on Presidential War Powers* and the article "When Congress Stops Wars," considered cases in postwar US history in which Congress entered into a confrontation with the president and suspended the funding of military operations abroad, authorized the withdrawal of troops, or banned the use of troops, even contrary to international agreements. As an example, they cited the period of war in Vietnam from 1964, when Congress adopted the Gulf of Tonkin Resolution to allow the start of full-scale involvement of American troops in Vietnam, to the annulment of this resolution in 1970. After the Paris Peace Accords were signed between North and South Vietnam, Congress took additional measures that put an end to funding of a US military presence in Southeast Asia, including Cambodia, Laos, and North and South Vietnam. Howell and Pevehouse also referred to the extraordinary measures taken by Congress when, after the defeat of South Vietnam in 1975, a ban was imposed on using American troops pursuant to the Paris Peace Accords but contrary to the position of President Ford. Emphasizing the crucial role of Congress in the start and termination of wars, the authors said that "reports of Congress' death have been greatly exaggerated."[156] However, the problem of party consistency in initiating or terminating wars no doubt is intricate and should be considered in the context of checks and balances, rather than serving as proof of the peacefulness or aggressiveness of one party or another.

One of the constantly discussed questions with regard to Iraq has to do with losses among American servicemen, or the "cost of war." Contrary to the simplified idea that antimilitary feelings are generated by the fear of loss of lives of American soldiers, there are authoritative opinions that the experience of Vietnam demonstrated the presence of another interconnection, one that is confirmed by contemporary public opinion studies.

Melvin Laird has observed that "the American public will tolerate loss of life, if the conflict has worthy, achievable goals that are clearly espoused by the administration and if their leadership deals honestly with them."[157] This opinion is shared by Christopher Gelpi, professor of political science at Duke University, and Peter Feaver. They relied on public opinion studies to prove that the need and importance of the objectives set for the war are more important to Americans, and that Americans are ready to tolerate losses for the sake of achieving certain objectives.[158]

John Mueller has put forward the idea that the number of casualties in wars has a significant influence on Americans' approval of continuing or terminating military actions. He used the wars in Korea, Vietnam, and Iraq to argue his point. It seems likely that the level of military technology development leads to reduced "tolerance to losses." According to his estimates, by early 2005, when casualties among American troops in Iraq reached 1,500 people, the degree of approval or condemnation of war was similar to that of 1968, when around 20,000 American soldiers died in Vietnam. Mueller believes that the war in Iraq was considered by many Americans as "something of a humanitarian venture," and quotes the words of Francis Fukuyama that "a request to spend several hundred billion dollars and several thousand American lives in order to bring democracy to ... Iraq" would "have been laughed out of court."[159] Mueller believes that an "Iraq syndrome" had already developed by the end of 2005, that is, a negative attitude on the part of Americans toward an ongoing war whose objectives are unclear or seem difficult to achieve by military force.

More than 57,000 Americans died in Vietnam, and almost 4,500 Americans died in Iraq, so it is possible that the critical threshold has not been achieved. The desire of politicians to make the war as "international" as possible and to obtain the approval and assistance of the United Nations in the course of reconstruction was taken into consideration. However, for Americans, any military actions revolve around the question of tax money spending. Americans are concerned about budget deficits, the redistribution of budget monies in favor of war at the expense of social programs, inflation, crisis, the possibility of an increased tax burden, and a weakening of the dollar exchange rate, and all these concerns became decisive arguments in favor of electing the "peace party." Only the Democrats could claim to be such a party during the 2008 elections.

Another obvious aspect for comparison is the active participation of the United States in "nation-building" in Iraq. As for the lessons of the Vietnam War, Kissinger, McNamara, and others warned about the danger and low productivity of US involvement in nation-building. They noted that democracy is the result of internal development and that the involvement of US troops in maintaining the stability of such regimes is not productive. McNamara's book presents two main lessons of the Vietnam War: "We do not have the God-given right to shape every nation in our own image or as we choose," and "External military force cannot substitute for the political order and stability that must be forged by a people for themselves."[160]

By mid-2006 it had become clear that there was a stalemate in Iraq. Whatever was done would not lead to a quick improvement and stabilization of the situation. Moreover, the withdrawal of US troops could result in an even greater worsening of the situation than maintaining a US presence in a situation where a positive outcome is not guaranteed. In 2005–2006, many articles and books devoted to Iraq were published. In *Squandered Victory: The American Occupation and the Bungled Effort to Bring Democracy to Iraq*, Larry Diamond stated that the greatest sin of the United States was not that it had launched that war, but that it had entered the war unprepared, with limited knowledge of the language and culture of the nations it had come to free from dictatorship.[161] In *Losing Iraq: Inside the Postwar Reconstruction Fiasco*, David Phillips provided a negative assessment of the success of postwar nation-building in Iraq and stated that the main problem consisted of numerous mistakes made in the very process of postwar nation-building by American military and policy-makers, which turned a "decisive and potentially historic victory" into a defeat.[162] Both Diamond and Phillips were involved in the development of programs for postwar reconstruction and democratization in Iraq.

James Dobbins, director of the RAND International Security and Defense Policy Center, wrote in his paper "Who Lost Iraq?" that the American public's perception of the war in Iraq was to a large extent predefined by the Vietnam syndrome and a negative attitude toward military actions against guerrilla movements. In his view, there were attempts in the United States to find who was guilty of the war, and accusations were primarily directed against President Bush and his administration, as well as against Congress, which had approved the start of war; the intelligence services; the Department of Defense; and the Department of State. According to Dobbins, "above all, Americans should accept that the entire nation has, to one degree or another, failed in Iraq. Facing up to this fact and drawing the necessary lessons is the only way to ensure that it does not similarly fail again."[163] The military analyst Anthony Cordesman noted in his 2003 book *The Iraq War: Strategy, Tactics, and Military Lessons* that many problems were associated with excessive politicization of the activities of the intelligence services, insufficient readiness for war, and lack of coordination of coalition actions.[164]

Debates demonstrated the division of American society with regard to the war in Iraq and an urgent need to find a way out. A crucial problem that occupied the minds of many US analysts and policy-makers in 2005–2006 was a search for an answer to the question, what to do in Iraq? A roundtable discussion involving leading analysts, those who had taken part in elaborating the strategy at the start of war and Iraqi reconstruction, was published under this title. The main task, in their point of view, was to search for specific steps, in order to prevent Iraq slipping into a full-scale civil war. The main task for the United States was to prevent Iraq from turning into a foreign policy fiasco.[165]

The recommendations in the 2006 Baker-Hamilton Report echoed the debates about an "Iraqization" of the war, analogous to "Vietnamization" of the earlier war, which implied shifting the burden of fighting from American troops to Iraqi forces. Many American authors pointed out that such wars should not be fought by the American army and that protecting the new regime should become the task of Iraqis themselves. However, the problem of training local forces and transferring power to them became complex. One of the main tools used to reduce the number of attacks against the coalition forces was engaging former Sunni fighters to serve in Iraqi security forces in the Sons of Iraq program.

Stephen Biddle, senior fellow for defense policy of the Council on Foreign Relations, in his paper "Seeing Baghdad, Thinking Saigon" regarded the policy of Iraqization from the standpoint of its relation to the lessons of the Vietnam War and efficiency. He thought that the premature withdrawal of American troops could lead to the start of genocide, and so US troop withdrawal from Iraq should be postponed until a sustainable compromise was found between the domestic factions competing for power and resources. However, the US desire to prepare the local armed forces as quickly as possible may have contributed to the growing antagonism between these groups. The only solution to the problem, Biddle thought, was to retain a US military presence in Iraq until Sunnis, Shias, and Kurds reached a compromise and managed to ensure their security on their own. This understanding of the problem meant that accelerated Iraqization of the war was a counterproductive and dangerous strategy.[166]

Melvin Laird assessed the situation in Iraq and expressed a similar judgment on the basis of his experience dealing with the end of war in Vietnam. Contrary to the opinion of those who believed that the presence of US troops in Iraq only strengthened the civil war, Laird wrote that the United States had not lost the war in Vietnam, but had failed—partly because of party disagreements—to preserve a continuity of policies with regard to the regime in South Vietnam. The United States ceased to render necessary assistance, and in the end this resulted in the victory of the communists, and then in thousands of casualties in Vietnam and a worsening of the US image in the world. This outcome, Laird thought, was the key lesson of Vietnam that the United States should pay close attention to in its new war. He thought that the United States could not withdraw from Iraq without having ensured permanent support for local forces that strived for democracy and that the United States should render assistance to the new regime, if it did not want Islamists to celebrate victory, which would happen once American troops were withdrawn. Withdrawal of troops would mean a US defeat in Iraq and a betrayal of the interests of US allies and adherents of democracy in Iraq and the region. If the United States withdrew its troops and under the pressure of interparty disagreements ceased to support Iraq, then an "Iraq syndrome" would likely emerge, to become part of national narrative, along with the Vietnam syndrome.[167]

The issue of which strategies to use to combat insurgents also attracted significant attention. Andrew F. Krepinevich in "How to Win in Iraq?" proposed that a prerequisite of success was ensuring the safety of Iraqis rather than hunting down insurgents. He also pointed to the need to have a clear war strategy and explicitly defined objectives, and sarcastically cited the words of George W. Bush: "As the Iraqis stand up, we will stand down." In his view, this was "a withdrawal plan rather than a strategy."[168] Colin

Kahl, in "How We Fight?," discussed the need for the American military to strictly follow the laws of war in Iraq, in order to maintain a high morale level among the troops and win the trust of the Iraqis. It was especially important, he thought, to observe the immunity of noncombatants, that is, to protect civilians from possible assault during warfare or nonselective attacks. He wrote, "Tracking how U.S. operations affect Iraqi civilians is not simply a moral and legal imperative; it is vital to the United States' national interest."[169] To appreciate the importance of this issue, we need only recall an extremely negative experience of the Vietnam War, when the United States violated international law, and the scandal in Abu Ghraib prison in the summer of 2004, caused by the cruel treatment of Iraqi prisoners of war, as well as the Bush administration's rejection of the International Criminal Court. In 2006, several long articles were aggregated and published in book form as *Vietnam in Iraq: Tactics, Lessons, Legacies, and Ghosts*,[170] which continued the discussion about what lessons the United States had learned in the Vietnam War and whether Iraq would be another Vietnam.

Debates in the United States about whether the war in Iraq was a "new Vietnam" elicited a wide range of opinions, from unequivocal agreement to appeals not to turn Iraq into "another Vietnam." Some of the disagreement no doubt owed to different understandings of what is meant by "the lessons of Vietnam." For some analysts, the key lesson of Vietnam was the need to limit US attempts to transform the life of other nations in its own image, while for others the key lessons were mistakes, unmet promises, inconsistency, and the betrayal of allies' interests. War strategies were widely discussed, especially in the cities, including strategies to be used against insurgents that rely on terrorist tactics. Based on these discussions, it seems that the memory of Vietnam as an unsuccessful US attempt to bring freedom to Southeast Asia and stop the spread of communism is the most crucial measure for assessing American military actions abroad. The war in Iraq gave American policy-makers, analysts, and society an opportunity to once again dredge up the memory of one the most painful periods in US history and try to answer the question, how well were the lessons of Vietnam learned?

Toward the end of the Iraq War, analysts tried to summarize the results of the war and the reconstruction efforts. Some of them noted the emergence of a spillover effect[171] of the war across the region. Andrew W.

Terrill, professor at the Strategic Studies Institute, US Army War College, identified the following signs of a spillover effect: (1) refugees and displaced persons fleeing Iraq in large numbers for neighboring countries, (2) cross-border terrorism, (3) an intensification of separatism and sectarian discord among Iraq's neighbors, fueled by the conflict in Iraq, and (4) transnational crime. These problems are impossible to resolve quickly, and their resolution inevitably depends on the willingness and ability of Iraqi authorities to tackle them. As Terrill notes, some of the problems are direct consequences of the war in Iraq, and these consequences will last. For instance, the problem of refuges was not reduced to quantitative parameters; it also included the need to give these people an opportunity to return to their homeland and to provide them with the things needed for them to resume a normal life.[172]

The war in Iraq is an example of a war initiated by the United States in pursuit of the lofty goal of "liberating the nation from tyranny and building a democratic society" but in the course of which the United States once again faced the "ingratitude" of the local population; its leaders stood accused of imperial ambition and misunderstanding the traditions and values of the nation they strived to liberate. The comparison with the outcome of the Vietnam War stems from the negative outcome of the warfare (the lack of a victory), the large number of fatalities, the considerable political and material costs, and the decline in US prestige in the world. All these are components of political defeat in an armed conflict.

As Thomas E. Ricks, *Washington Post* reporter and author of the Pulitzer Prize–winning book *Fiasco: The American Military Adventure in Iraq*,[173] noted in a March 7, 2009, interview with CNN, that the decision to launch the war in 2003 was "the worst foreign policy decision in U.S. history." In his view, Iraq in the nearest future would be able to become neither a stable and democratic state nor a reliable regional ally of the United States.[174] Ricks wrote that 2009 would be the most difficult year in this long war. He cited the words of Shawn Brimley, a former Canadian infantry officer who worked as a defense analyst at the Center for a New American Security: "In many ways the entire war was a huge gamble, risking America's future power and prestige on a war that, at best, is likely to be inconclusive." "Bush's gamble," as Ricks dubbed it in another book, would "force Obama into a series of his own gambles and trade-offs—between war and domestic needs, between Iraq and Afghanistan, between his political base and his military."[175] In May 2009, CNN broadcast an interview of the British correspondent Nic Robertson with Zabiullah Mujahid, a representative of the Taliban movement in Afghanistan. Mujahid observed that the war in Afghanistan could turn out to be "Vietnam" for Americans who shifted the center of gravity of the struggle against al-Qaeda from Iraq to Afghanistan and Pakistan.[176] This prediction was fully confirmed, and for the Obama administration, ending the war in Iraq and setting a deadline for the withdrawal of troops from Afghanistan became the most important and costly tasks from the financial and political standpoints.

The start of troop withdrawal from Iraq prompted an active discussion of the similarities and differences between the wars in Afghanistan and Iraq, as well as between both wars in Iraq. Such comparisons were thoughtfully articulated in the book *War of Necessity, War of Choice: A Memoir of Two Iraq Wars*, by Richard N. Haass, president of the Council on Foreign Relations, who took part in decision making with regard to the first and second wars in Iraq and served as a consultant for the reconstruction effort in Afghanistan.[177] At a Brookings Institution gathering celebrating the book's release, Haass formulated its main thesis as follows: "I would argue that this [the Iraq War] was not simply a war of choice, it was not simply a preventive war, but ultimately it was a bad choice, and it was a bad choice badly implemented, adding insult to injury. It was a bad choice, again, because not only did the United States have many other options, but also options that I thought were preferable and far less costly. And I'm thinking about both the direct costs of this war and the indirect costs of this war, the distraction cost, and the opportunity cost."[178] Haass called the war in Vietnam a "war of choice," thus provoking the military personnel present in the room to ask about treaty obligations between the United States and South Vietnam that had required the United States to help its allies.[179] Meanwhile, President Obama in a statement on June 4, 2009, in Cairo called the war on terror a war "of necessity" rather than a war "by choice."[180]

Zbigniew Brzezinski, national security adviser under President Jimmy Carter, published a review of Haass' book in the journal *Foreign Affairs*. Titled "A Tale of Two Wars," the review was timed to accompany the book's release. Brzezinski gave a generally positive review, noting ironically that "once a war's outcome is known, the difference between necessity and choice is brutally simple. The ex post facto verdict of history is inevitably derived from a simple maxim: nothing fails like failure, and nothing succeeds like success." However, he noted, "until the outcome of a war becomes known, the difference between necessity and choice is rather ambiguous."[181]

CONCLUSIONS

The asymmetric conflict model turns a spotlight on the defeat of the stronger party in an armed conflict and the various asymmetric relations obtaining between the belligerent parties that could explain such an outcome. If we consider the Iraq War to have ended in political defeat for the United States, a superpower, we will also want to know the grounds for deeming it a defeat. The main measure of success or failure in any war is whether the war objectives were achieved. Failure to accomplish the war goals through military means and the subsequent termination of military operations can be considered indications of defeat. The deterioration of the situation as a result of war can also be laid to the failure of the military campaign.

Evaluating the outcome of the war from the standpoint of established objectives leads to several conclusions. For example, we may conclude that Iraq was not disarmed because no WMDs were found on its territory. Similarly, links between Saddam Hussein's regime and al-Qaeda were not destroyed because such links were never found. In the meantime, however, the war contributed to al-Qaeda's penetration into Iraq—a development counter to the US war goals. Tyranny was not transformed into a free and prosperous civic society because the war, the occupation regime, and the presence of foreign troops in the country exacerbated the divisions and tensions within Iraqi society. Protracted warfare and the activities of terrorist groups aggravated the economic crisis in the country and the overall state of Iraqi society. This conclusion is supported by statistical data and public opinion polls conducted in Iraq.

The war in Iraq started with declarations about its just cause; however, during the war it became evident that the principles of entry into a war (jus ad bellum) had been violated, as had the principles of the just conduct of war (jus in bello). International humanitarian law regulating the treatment of prisoners of war was violated. The civilian population was unprotected from combat operation; that is, the principle of discrimination between combatants and noncombatants was violated. The war continued for years and was terminated without a clearly achieved, positive outcome. Violations of international humanitarian law by American servicemen (including the cruel treatment of prisoners of war, the murders of civilians, and the torture of prisoners) drew the attention of the US national justice system. Individuals found guilty of perpetrating abuse in Abu Ghraib prison have been convicted, and other cases of war crimes are being investigated. These investigations have been accompanied by public debate over the context in which the incidents took place, which negatively affected the moral standing of the United States globally. In May 2009, President Obama banned the publication of photographs of abuse of Iraqi prisoners by American servicemen as evidence in court, which caused a heated national debate. Many analysts and representatives of the military expressed the opinion that this was the right decision, for publishing the photographs could have cost the lives of American soldiers who served outside the United States.[182]

The manifest war objectives aimed to achieve a strategic goal, that of enhancing US security and protecting the country from possible terrorist attacks. Indeed, after 9/11 there have been no comparable attacks on US territory, and this fact is perhaps the only argument of any weight put forward by the war's proponents. However, the war in Iraq contributed to the rise of terrorist groups in other regions and the execution of major terrorist attacks against US allies from among the coalition of the willing. Most Americans—ordinary citizens, analysts, and politicians—do not support the attempts of Bush administration representatives[183] to prove that the war strengthened US security. Summarizing the range of opinions, we can state that the majority believes that this war should not have been started. The war failed to ensure US security and to improve the image of the country in the world and in the Middle East. As American politicians and analysts have noted on numerous occasions, the war led to an increase in anti-American sentiment in the world, stimulated the rise of terrorist groups in the Middle East and Greater Middle East, in Asia and in Europe, and had spillover effects.

The reasons for the US political defeat in Iraq may be attributed chiefly to strategic, domestic, and international factors. Strategic factors included the difficulty of waging a victorious war with a fortiori false objectives; the difficulty of winning guerrilla warfare under occupation conditions; an insufficient understanding of the region and the country, which led to poor prewar planning; and underestimation of the potential for resistance of the local forces and overestimation of the regime's unpopularity. The main forms of struggle against the occupation forces in Iraq were guerrilla warfare and terrorist strategies. American military and political analysts widely apply the term "asymmetric warfare" to the war in Iraq. Though today the US strategy in Iraq tends to be considered successful, it is a relative success. The coalition forces managed to avoid an unequivocal military defeat at the hands of local insurgents after almost a decade of war, but they also failed to ensure an explicit victory. The situation is regarded as stable, but it can hardly be called secure. Nevertheless, a relative success allowed the coalition forces and the United States to withdraw their combat units and leave only military specialists, who are training Iraqi security forces and police units.

Public opinion and the war's effect on the economy and the political life of the United States are the domestic factors that led to the perception of a US defeat in this war. The war in Iraq, unlike the Vietnam War, did not cause massive and open protests of Americans, but the description of the state of society given in a Gallup report seems apt here—"cornered." Americans realized that the war could not be lost, as that would mean acknowledging the victory of the radical ideologies and movements against which it had been launched. Neither could the war be won by military means, as it proceeded from the combat phase to the stage of winning the hearts and minds of a nation liberated from a dictator. The withdrawal of American troops was predetermined, though security in Iraq was the most serious problem that the coalition forces tried to resolve. Few doubted that the withdrawal of US troops would further aggravate the political struggle in Iraq and possibly the resumption of the civil war. The events of 2012 and 2013 in Iraq confirmed these fears.

A protracted war without clear positive results activated the mechanism of checks and balances built into the US political system. The Democratic Party, which had consistently opposed going to war, won the midterm congressional elections in 2006 and the presidential election in 2008. Debate over the impossibility of withdrawing American troops ended, and deadlines for troop withdrawal were established. This meant that even if the

United States failed to achieve its reconstruction objectives by the time of complete troop withdrawal, the troops were still to leave Iraq. After troop withdrawal, the United States has had limited levers with which to influence the situation in Iraq, other than its political and economic ones. It is hard to predict to what extent the United States will manage to retain the commitment of Iraqi leaders to a bilateral partnership. Anti-American sentiments are strong in Iraq and in the region in general.

Funding the Iraq War siphoned large amounts from the federal budget. The war led to the highest level of national debt in US history—more than a trillion US dollars. The global economic recession of 2008–2009 focused attention on the state of the US economy, which defines the degree of US influence in the world no less than its military power does. In a situation of economic crisis, the government had to cut spending, and thoroughly checked how the money of taxpayers allocated to war and reconstruction was spent. Audits demonstrated that significant funds had been spent in vain, and that control over contractual obligations in Iraq to acquire supplies and execute recovery projects had been unsatisfactory. As a result of the audits, dozens of legal proceedings were initiated, and plans were developed to modify how the allocation and spending of funds from the federal budget were determined and controlled. The economic effects of the war for the United States also will be long-lasting, for the withdrawal of fighting forces does not end the expenses associated with war. The government must ensure the removal of troops and equipment from Iraq and fulfill its obligations to the half-million American servicemen and women who served in Iraq. In addition, to stabilize the situation in Iraq and the region after troop withdrawal, the United States must allocate significant development funds for Iraq and other countries in the region. During President Obama's term in office, the volume of international assistance doubled.

Among the international factors that contributed to the political failure of the United States in Iraq was the international community's ambiguous assessment of the war's objectives. The United States failed to convince UN members of the need to go to war before the start of the war. After the war began, criticism of US actions only became stronger. The coalition of the willing gradually broke down. The most consistent US partner, the United Kingdom, withdrew its troops in late April 2009, one month before the deadline. Few shared the opinion of

Prime Minister Gordon Brown, who claimed that the war in Iraq was a success. Opposition groups in the United Kingdom noted that the losses outweighed what had been achieved in Iraq.[184] The BBC report on the withdrawal of British troops from Basra cited British politicians who said that "the effects of the operation through the invasion were not high enough to earn respect." According to British politician and diplomat Lord Ashdown, "the Army is broken as a result of Iraq and Afghanistan." Sir Jeremy Greenstock, Britain's ambassador to the UN at the time of the start of the Iraq War, noted that the United Kingdom would in the future be unlikely to join military actions that were not authorized by the United Nations and lacked broad international support.[185]

The regional factors that negatively affected the probability of US success in the Iraq War included the influence of other countries in the region that directly or indirectly rendered moral, military, technical, and financial support to anti-American forces. The course of the war was significantly shaped by international terrorist groups that had a direct interest in the war and in a US defeat. In this way, the global war on terror initiated by the United States turned into a global war of terrorist groups against the United States and its allies. The outcome of this global war has not yet been determined, but a victory for the developed countries enmeshed in it is not obvious. Aggravation of the situation in Afghanistan and Pakistan demonstrates that the war on terror is far from over and its outcome is not certain. According to the nongovernmental organization NationMaster, Iraq ranks first in the world in fatalities sustained as a result of terrorist attacks for the period 1968–2006. It is followed by the United States, India, Pakistan, Israel, Colombia, Russia, Lebanon, Algeria, and Afghanistan.[186]

Thus, the war in Iraq of 2003–2011 will make it into the history books as yet another confirmation of the asymmetric conflict theory. In the early 1990s, American military analysts talked about asymmetry in terms of the global military power superiority of the United States, which could not be undermined in a big or conventional war. Today, the war in Iraq is often regarded as a repetition of the mistakes made in the Vietnam War, the conflict that provided the foundation for the development of asymmetric conflict theory. The Iraq War proved once again that military superiority does not guarantee military victory. The absence of victory, a protracted engagement, and the use of guerrilla and terrorist strategies by the militarily weaker adversary made terminating the war difficult, especially under the conditions of occupation of or maintaining a military presence in the foreign territory. Numerous factors that determine the political defeat of a great power in a war against an incommensurately weaker adversary were evident in this conflict.

CONCLUSION

Analyzing Asymmetric Conflicts Using the Model

This book has defined asymmetric conflict as a struggle between belligerents vastly incommensurate in force strength and resources. In a general sense, asymmetric conflict has been known and reported since antiquity. In the modern era, following the convulsions of two world wars in the twentieth century, it has become a more acutely observed phenomenon. The political background is the restraint shown by modern titans in waging "big wars" against each other, taking their antagonisms instead into the realm of economic, cultural, and ideological competition. There remain, however, small wars, limited engagements that enmesh protagonists of different capabilities, and here the paradoxical outcome in the postwar era has often been the political defeat of the militarily superior party. This phenomenon has become so pronounced that it has drawn the scrutiny of researchers, and a theoretical model is coalescing that considers the interaction of multiple quantitative and qualitative factors as a means of characterizing such conflicts. From the theoretical discussion at the beginning of the book, followed by two detailed case studies of great powers caught in asymmetric warfare, we can draw certain conclusions about the elements of the asymmetric conflict model and their relative weight in specific cases. This short discussion first recapitulates the findings of the previous chapters, then suggests how they could be applied in future research.

Over the past 15 to 20 years, the term "asymmetric conflict" has become widely used, though often quite nonrigorously, to describe quantitative asymmetries in resources or qualitative asymmetries in struggle strategies and the status of antagonists; the US military in particular has understood asymmetric conflict in terms of differences in military power and capabilities. More generally, researchers have explored the concept of asymmetry in several instances from the post–World War II era: in the French defeat in Indochina, the U.S defeat in Vietnam, the Israeli defeat in Lebanon, and the Soviet defeat in Afghanistan. As the concept has developed, researchers have moved beyond the basic quantitative and qualitative differences of belligerents to studying asymmetric strategies. Further efforts to develop the concept of asymmetric conflict have drawn on methods used in political analysis to identify correlations among a limited number of variables that represent the key characteristics of an armed conflict. Such variables usually condition each other's expression, making it difficult to predict the outcomes of future conflicts.

Such a multifactor analysis of dynamic relations between unequal adversaries in armed conflicts was proposed in chapter 1 as a theoretical model. The factors in the analysis, particularly factors that could explain the paradoxical political defeat of a great power by a weaker challenger in an armed conflict, are of three kinds: (1) internal or endogenous factors, or those having to do with the domestic characteristics of the belligerent parties (including both quantifiable resources and political, economic, and moral characteristics); (2) international or exogenous factors, or those that are defined by external influences on the participants in an armed conflict or on the development of the conflict (such as a great power coming to the aid of a weaker party, or international opinion weighing negatively on the conduct of the stronger power); and (3) tactical and strategic factors, or the manner and method in which the struggle is carried out—guerrilla actions versus fixed fighting units, for example. How such parameters assort themselves and influence other factors in any given case is an enduring challenge in asymmetric conflict analysis.

Attempts to measure asymmetry, to characterize it rigorously, typically begin with an inventory of possible constituent features or manifestations of such a relationship in databases on war. However, database records often do not catch all of the relevant information, and the measurements may be tricky. Simply establishing a casualty level sufficient to call a dispute an armed conflict has proved inconsistent over the years. Checking one database against another is laborious but occasionally must be done; the first two chapters of this book take advantage of multiple databases to more securely ground the theoretical arguments. A further salient use of multiple databases is to acquire more views of basic data and more angles, perhaps complementary ones, on the conflict of interest. To undertake a multifactor historical analysis using the asymmetric conflict model, researchers must construct and work with a matrix of relationships and possible correlations among variables, and developing a good data set is a crucial first step in beginning such a study.

To explore the asymmetric conflict model, this book uses the experiences of the two most powerful countries of the Western world: Great Britain during the dissolution of the empire and the United States after the Cold War. Both countries bore the burden of global influence, and both present themselves as liberal democracies striving to disseminate their political model around the world. Although their war experiences are different and stand in contrast to one another, certain parallels can be drawn, though history is not shy about revealing the irony of coincidences and contradictions. The United States welcomed and encouraged the dissolution of the British Empire, as the United States was the first colony to free itself from the imperial burden through armed struggle. The self-liquidation of the "global British policeman," however, in part a result of Britain's weakened state after World War II, forced the United States to actively expand its sphere of responsibility in world affairs in the 1950s and 1960s. During the Vietnam War, interestingly enough, British military experts advised their American colleagues on how to efficiently tackle guerrilla fighters.

The dissolution of the British Empire was largely historically predetermined. The reasons for the relatively swift and probably inevitable dissolution included the economic weakening of Britain during World War II and the consequent depletion of its human, material, and financial resources available to maintain order in its vast imperial territory. Official documents from that era show that Britain had to reduce its armed forces, limit its military presence overseas, and dilute the assistance it had previously provided to loyal forces in the colonies; it simply did not have sufficient resources to sustain the previous level of control.

Whether and to what degree each of the military campaigns in the colonies ended in political defeat for Britain is occasionally a matter of debate. However, it seems undeniable that the imperium ceased to exist long before the empire came to an official end. In fact, an important component in the defeat of the empire was the defeat of the idea of empire, and here the crucial actors bringing influence to bear on Britain were the United States, the Soviet Union, and continental China. Despite the sincere intent of the metropolis to pave the way for a gradual transition of its colonies to self-governance, the desire of local nationalist forces for immediate liberation from colonial bondage was insurmountable. The revolutionary ideology of China, combined with an anti-imperialist struggle against Japanese occupation, proved more attractive to Britain's colonies in Asia. Nor should the role of the Soviet Union be overlooked: for the first time in its short history, the world's new superpower demonstrated its military and moral power in a persistent struggle against fascist bloc countries, becoming in the course of these efforts an ally of the leading Western powers, and one of equal standing. The Soviet isolation ended and its ideology gained weight and gravitas, as reflected in the growing popularity of leftist and communist parties all over the world and in the Soviet Union's contribution to the postwar world structure.

Thus, the British Empire suffered a political defeat in its real struggle against three powers of the new world: the Soviet Union, the United States, and the national liberation movements in the colonies, which were united in their common intent to eradicate the colonial order completely. The Commonwealth of Nations format to which the British Empire transitioned failed to provide the same degree of control over the resources, politics, and economy of the former colonies that had existed in the times of the empire. Here I will mention that the term "neocolonialism," widely used in the Soviet literature, needs to be correctly interpreted. In accordance with Marxist logic, Soviet literature characterized the new relations between the former metropolis and its dependencies as another form of exploitation that benefited developed countries at the expense of the developing world. It is true that the former colonies were tied economically, financially, and often strategically to the former metropolis. However, to maintain the loyalty of its former colonies, the United Kingdom had to allocate substantial funds to its development policy. Today, there are debates in the United Kingdom about the profitability of the colonial system as a whole, about whether the imperial holdings ensured economic growth and a high standard of living in the metropolis or whether, on the contrary, the system of imperial dominance was associated with too many problems. One of these problems is constantly in the news today: the United Kingdom faces many difficulties as a society accepting millions of immigrants from its former colonies, and some of these difficulties are so pervasive as to influence national elections.

Applying asymmetric conflict theory to analyze the reasons for the dissolution of the British Empire, a structural change that was accompanied by military actions in all of its dependent territories, shows that economic factors associated with the dying empire and the influence of anti-imperialist ideology were of great importance in the postwar period. For Britain, the best outcome of warfare in the colonies was a minimal political victory consisting of offering independence and not letting the country fall under the influence of a leftist ideology. A maximum political victory, which can be described as preservation of British control and the gradual transfer of power to local forces under the control of British authorities, was often abandoned for the sake of a lesser victory. (On a slightly different note, a lasting result of its experiences with small wars in the colonial empire is that Britain now tends to be rather reserved with respect to the kinds of military force it uses, and it maintains a relatively small army.)

The US war in Iraq confirms the applicability of the asymmetric conflict concept. After the end of the Cold War, American strategists used the term "asymmetric" in reference to the incommensurability of US power and that of its potential adversaries, and believed that the main threat to US security came from rogue states that might try to use weapons of mass destruction against the United States. In initiating a war on Saddam Hussein's regime, which was a continuation of President George W. Bush's "war on terror," begun after the attacks of September 11, 2001, the United States sought to present its actions as a just war, in this way hoping to legitimize this and other "last resort" actions against international terrorist organizations in developed and developing countries. The military, economic, and financial might of the United States led many to believe that the superpower would make short work of a war against Saddam Hussein's regime, with victory inevitable. This thinking was also supported by the successful and swift war against Iraq in 1991. However, soon after the start of the Iraq War the international coalition forces faced strong resistance from local forces in Iraq. Attempts to create a new political order in Iraq only aggravated the tensions between different factions in the country, intensifying the struggle for power and resources and leading to full-scale civil war in 2005–2006. Paradoxically, the war on terror stimulated the rise of terrorist organizations in the region and the world, which now engage in conducting limited conflicts and guerrilla actions against the developed Western countries.

The United States launched the war in Iraq for the sake of the ideals of democracy and liberation, but the war had an opposite effect, strengthening the global image of the United States as a bully, an empire that exercises its right to the unilateral use of force. This image is far from the democratic ideals that the United States wanted to defend in Iraq. Reviewing its strategy in Iraq forced the United States to examine its experiences fighting in Vietnam against forces much poorer in military might and economic resources, as well as Britain's experiences with underpowered skirmishes in its own dominions. The new strategy that was developed in late 2006 and implemented in 2007 was oriented toward a scheduled withdrawal of US troops from Iraq. In the 2006 midterm congressional elections, the Democratic Party won a majority of seats in both the House of Representatives and the Senate, and this political change of horses prepared the ground for a change of policy in Iraq. The victory of the Democratic presidential candidate Barack Obama extended the political defeat of the Republican Party, the "party of war," and its representatives. However, the internal logic of asymmetric conflicts suggests it would be difficult to terminate such a war without an obvious political defeat of the superpower. Despite strong criticism of the war during the electoral campaign, the Obama administration became a prisoner of the situation, or more precisely a casualty of the asymmetric confrontation factors that dictated the modus operandi. The actual withdrawal of troops took place later than Obama had promised as a candidate and in strict accordance with decisions made during the Bush administration.

The Iraq War is without doubt an example of a US political defeat. None of the goals laid out for this war was achieved. Nor were the undeclared war objectives, namely, establishing control over the oil-producing sector of Iraqi economy and creating a loyal regime in the strategic region, attained. To this day, Iraq has failed to reach its prewar level of oil production. Funds allocated from the US budget and by private investors to reconstruct the infrastructure in this sector of economy often did not reach their target. Instead, the oil sector infrastructure became a constant target of attacks by guerrilla fighters. Moreover, the system of corrupt private contractors and local authorities came into being and will be hard to remove; international statistics place Iraq among the world's most corrupt nations. Many American analysts find it hard to acknowledge the political defeat of the military campaign in Iraq, which they take as tantamount to acknowledging the victory of al-Qaeda and its ideologists, against whom the war was waged in the first place.

The factors in the British and US political defeats can be grouped as follows:

1. Economic factors

In addition to domestic financial difficulties, the dissolution of the British Empire was accompanied by the collapse of the sterling zone and the devaluation of the pound sterling and its replacement by the US dollar in the global market. In works by British economists and historians and in official documents of the period, the currency devaluation and its fall from a world standard are named as significant problems in the postwar history of the empire.

During the Vietnam War, the United States also faced an acute financial crisis, one that forced the Nixon administration to end convertibility between the US dollar and gold in 1971. Beginning in 1973, the dollar exchange rate became determined by market mechanisms. The Iraq War was the main reason for the US domestic debt reaching an almost record level of $1.7 trillion in 2009. The war's expenses also contributed to an unprecedented economic crisis in the United States that started in 2008 (and indeed, became a global economic crisis). Because the US dollar is an important world currency, in 2008 the leading powers began discussions on the need to develop new principles for the global monetary system.

2. Domestic factors

There is an established view that postwar Britain's defense and foreign policies were upheld by a bipartisan consensus, and historical documents do suggest that both Conservative and Labour governments used careful and balanced actions to expand Britain's military presence abroad. For Britain, the front of struggle against leftist ideology shifted to Europe, where a significant share of its armed forces was deployed. Documents from that era show that the consensus was the result of a difficult economic situation in the country after the war, and that room for maneuver was strictly limited by available resources and a still vast zone of global responsibility. In this instance, what is widely regarded as thoughtful foreign policy really arose from domestic exigencies.

In the United States, domestic factors were manifested in the disagreement of the main parties over military policy. The Vietnam War came to an end as a result of pressure exerted by Congress and by war opponents in the United States. The Iraq War did not inspire the same level of opposition as the Vietnam War but did lead to the 2008 electoral defeat of the Republican Party, which had initiated the war. However, the process of war termination carried out by the Obama administration demonstrated that inertia of a political system can result in policy continuity despite declared party disagreements. In other words, squabbling parties that disagreed over foreign policy had much in common in the day-to-day functioning of realpolitik and ended up pursuing a similar course, the chronologically second mover following the plans of the first mover.

3. International factors

The international factors that significantly conditioned the defeat of the British Empire included an anti-imperialist ideology, which was supported by the Soviet Union, the United States, and national liberation movements. The newly created international relations system reflected these ideas in UN charter documents, and then in the implementation of the regulations in the documents. International factors included US influence and the special form of the Anglo-American interaction, especially in their competition and disagreements in the areas of ideology, economy, finances, and politics. The Soviet influence was to a large extent indirect, exercised as it was through the emerging Eastern bloc in Europe and expressed in anti-imperialist rhetoric and the support of national liberation movements.

The United States' failures in Iraq can also be explained by the influence of international factors. A negative attitude toward the war in many countries, especially in the Middle East, contributed to the collapse of the coalition of the willing and the rise of anti-American sentiments. Iraq became the place where international terrorist groups fought the United States. The pressure of global public opinion became a crucial factor in the change of policies in Iraq and helped place the United States on a trajectory to end the war.

Another important factor that influenced the outcome of asymmetric conflicts in the post–World War II period was the strengthening of norm- and value-oriented approaches in international politics, in conjunction with a tighter linkage between national and international politics. The concepts of justice and legitimacy were for a long time associated with the system of norms inherent to a closed social system, but later they became the foundation of international politics. Democratic values and principles became foundational to the postwar system of international relations. Their gradual and consistent implementation tore the veil from the contradictions in the norms, principles, declared values, and real conditions dividing very diverse actors that are still united under the world system of international relations.

⋆ ⋆ ⋆

Thus, the asymmetric conflict phenomenon to a large extent refutes established perceptions about the consequences of power domination in armed conflicts between unequal adversaries, rather than determining the winner and loser. Secondarily, this phenomenon confirms the presence of stable interconnections between participants in international relations, a tight linkage between domestic and foreign policy, and a hierarchy of military and political domains of contemporary states.

Asymmetric conflict is a crucial part of past and present international relations, and an appropriate analytical model helps us better understand such conflicts. To apply asymmetric conflict theory, it is first necessary to formulate a complex matrix of possible variables whose weight in the model and impact on the outcome of the conflict are difficult to predict.

Nonetheless, the model can be usefully applied, provided that certain conditions are observed. As a historical analysis of asymmetric conflicts demonstrates, the asymmetric conflict phenomenon finds expression in the interconnection of multiple variables. It is virtually impossible to identify any single factor whose presence would guarantee a manifestation of the phenomenon. The asymmetric conflict phenomenon is an equation with many unknowns, rather than a strict matrix of dependencies and determinants.

The asymmetric conflict model proposed in this book is one that can be used in applied analysis; it is not a theory establishing any rigid correlation of variables. However, relying on this model, one can develop and apply the concept so long as it is understood as an aggregate of asymmetric characteristics of an armed conflict that follow the basic asymmetry in power, resources, and status. The weaker party to a conflict will always attempt to change the balance of power and relations within the system in order to expand its authority, increase its resources, and elevate its status. This should be taken into account when analyzing specific examples of armed conflicts between asymmetric antagonists to better understand the logic of the struggle and predict the outcome. The asymmetric conflict phenomenon emerged from a historical analysis of completed wars, and is of interest for historical research. In political and strategic analysis, the model and the concept of asymmetric conflict can be used instead to make assumptions about, and prepare for, possible paradoxical developments that run counter to expectations based solely on the military and resource dominance of one of the parties.

APPENDIX

List of Armed Conflicts from the COSIMO Database Used in the Study

Following is the list of armed conflicts with at least 25 casualties, selected from COSIMO database of conflicts, for the years 1945–1999, the period covered in the study. Asymmetric conflicts are highlighted in light gray. Asymmetric conflicts with great power involvement are highlighted in dark gray. A key to country abbreviations is provided at the end of the table, as is a description of the variables used in the COSIMO publications.

Name	Start	End	Duration (years)	Items and issues	Intensity	Direct participants	External/ Indirect participants[1]
1. India II (partition)	1942	1948	6	2, 3	4	AND (MUSLIM)// AND (HINDU)	//UKI (3)//
2. Greece (civil war I)	1944	1945	1	4	3	AND (ELAS), AND (EAM)//GRC, UKI, AND (EDES)	—
3. Morocco (independence)	1944	1956	12	2	4	AND (MOR), AND (BERBER)//FRA, AND (SETTLERS)	USA (1)//
4. Iran (Kurds I)	1945	1946	1	2, 3	3	AND (KURDS)// IRN, IRQ	USR (2)//
5. Iran–USSR (Azerbaijan)	1945	1946	1	1, 5, 7	3	IRN//USR, AND (TUDEH), AND (DEMOCRATIC PARTY)	USA (1), UKI (1)//
6. Philippines (Luzon, HUK)	1945	1954	9	4	3	AND (PKP), AND (HUK)//PHI	USA (2)//
7. Algeria (independence I)	1945	1946	1	2, 6	3	AND (ALG), AND (ULEMAS), AND (MESSALIS)//FRA	—
8. Indonesia (independence)	1945	1949	4	2	4	AND (INS)//UKI, NTH, AND (KNIL)	IND (1), USR (1), USA (1)//
9. Indochina Ia	1945	1954	9	2, 6	4	FRA, RVN, LAO// DRV, AND (PATHET LAO), AND (KHMER ISSARAK)	USA (2), UKI (2)//CHN (2)
10. China (civil war)	1945	1949	4	4	4	AND (KUOMINTANG)// AND (KPCH)	USA (2)//USR (2)
11. Eritrea I (annexing)	1946	1952	6	3	3	AND (ERITREANS)// ETH//SUD, EGY	ITA (1)//
12. Israel I (independence)	1946	1948	2	2	3	AND (ZIONIST.)// AND (PALEST), UKI	USR (1)//EGY (3), SYR (3), LIB(3), SAU (1), AL (1)
13. Bolivia (teachers' strike)	1946	1952	6	4	3	AND (MNR)//BOL	—

Instruments of initiator	Instruments of affected party	Initiator	Minimum no. of victims/ casualties	Maximum no. of casualties	Outcome
AND (MUSLIMS): 10M–	AND (HINDU): 10M–	AND (MUSLIM)	500,000	800,000	M2, T2, P3
AND: 10M–, 5L–	GRC:10M–, 5L–	AND (ELAS), AND(EAM)	16,000	16,000	P2, M1
AND: 10A–, 10B–, 10N–	FRN:10G–, 10I–, 10M–,10F+	FRN	100	1,000	P3, M2, P16
AND: 10M–, 8A+	IRN:10M–	AND (KURDEN)	25	100	P11, P2, M3
AND: 10O–, 10N–; USR: 5B+, 5F–	IRN: 2B+, 1D+ , 5G	IRN: 2B+, 1D+, 5G–	25	100	M3, T4, P11, P1
PHI: 10M–, 10G–, 10F+, 10R–	AND: 10M–, 10N–, 10O–	PHI	9,000	9,000	M2, P3, P11, P15
AND: 10A–, 10F–, 10N–	FRN: 10G–, 10M–, 10F+	AND (ALG)	1,500	45,000	P2, M3, P11
INS: 10D–, 10M–, 10N–, 10Q–	NTH: 10M–, 10G–, 5F–	AND(INS)	5,000	100,000	M2, P3, T3, P9, P16
FRA: 5G–, 10M–	"DRV: 5G–; LAO:10B–, 10M–"	FRN	95,000	600,000	P1, T1, M3
CHN: 10M–, 10A+	AND (KPCH):10M–	CHN (KUOMINTANG)	1,000,000	2,000,000	P6, M3, T1
AND: 10M–	ETH: 2B+, 9A–	AND (ERITREANS)	100	1,000	P4
AND (ZION.): 10B–	"AND (PAL.): 10B–; UKI: 10M–"	AND (ZIONIST), AND (PAL	1,000	2,000	P2, P3
AND: 10A–, 10D–, 10F–, 10M–	BOL: 10M–, 10I–, 10H–, 10F+	AND (MNR)	2 000	3,500	M2, P5, P3

Name	Start	End	Duration (years)	Items and issues	Intensity	Direct participants	External/ Indirect participants[1]
14. Greece (civil war II)	1946	1949	3	4, 6	4	AND (DSE)//GRC, UKI (BIS 47)	USR (1), ALB (2), BUL (2), YUG (2)//USA (2)
15. Afghanistan–Pakistan (Paschtunistan I)	1947	1963	16	1, 6, 3,	3	AFG, AND (PASHTUN PEOPLE)//PAK	USR (2)//USA (2)
16. China–Nationalist China	1947	1947	0	5	3	AND//CHN (KMT)	—
17. Indonesia (Darul Islam separation attempt)	1947	1991	44	3	3	AND (DAR–UL–ISLAM)//INS	MAL (1)//USR (1), USA (2), CHN (1)
18. India IV (Kashmir I)	1947	1949	2	2, 3, 1	4	AND (MUSLIMS)// AND (HARI SINGH)	PAK (3)//IND (3)
19. Paraguay (coup d'état)	1947	1947	0	5	4	AND (PFR)//PAR	//USA (1), ARG (1)
20. Malagasy Republic (independence)	1947	1960	13	2	4	AND (MAG)//FRA	—
21. Costa Rica (exiled people)	1948	1949	1	5	3	AND//COS	NIC (3)//
22. Yemen–United Kingdom (Aden I)	1948	1963	15	2, 1	3	YAR//UKI (ADEN)	—
23. India V (Hyderabad)	1948	1948	0	3, 5	3	IND//AND (HYDERABAD)	—
24. Malaya (independence)	1948	1960	12	2, 4	4	AND (MRLA)//UKI, AND (MAL), AUL, NEW	CHN (2)// THI (2)
25. Israel II (Palestine war)	1948	1949	1	1, 3	4	JOR, EGY, SYR, LEB, IRQ//ISR, AND (HAGANA), AND (LEUMI), AND (STARS)	CZE(3)// UKI(3)

Instruments of initiator	Instruments of affected party	Initiator	Minimum no. of victims/ casualties	Maximum no. of casualties	Outcome
AND: 10D–, 10M–	GRC: 10H–, 10M–	AND (DSE)	44,000	160,000	M3, P1, P9, P11
AFG: 3A–, 5C–, 5B–, 5F–, 4K–	PAK: 1D+, 1E+, 1C–, 1A+, 4K–	AFG	100	1,000	M1, T5, P1, P2
AND: 10G–	CHN: 10H–, 10N–, 10E+	AND	4,000	4.000	M3, P7, P11
AND: 10K–, 10N–, 10O–	INS: 10H–, 10M–	AND (DARUL ISLAM)	4,000	4,000	M3, P1, P8
AND: 10A–, 10B–, 10F–	AND: 10E–, 10G–, 10M–	AND (MUSLIMS)	1,500	10,000	M1, T1, P1
AND(PFR): 10M–	PAR: 10H–, 10M–	AND (PFR)	28,000	28,000	P9, P11, M3, P15
AND: 10M–	FRN:10M–, 10F–	AND (MAG)	5,000	80,000	P3, M3, P16
AND: 5F–, 5D+	COS: 5C–, 5H–, 5F–, 2B+	AND	1	25	M3, P9, P1
YAR: 2B+, 1D–	UKI: 5F–, 1D–	YAR	100	1,000	M2, T5, P9, P11
IND: 10H–, 10M–	AND: 10N–, 10O–	IND	2,000	10,000	M2, P3, P9
AND(MCP): 10A–, 10D–, 10O–	MAL, UKI: 10M–, 10A+, 10R–	AND (MRLA)	12,500	13,000	M3, P11, P17, P15
"JOR: 5G–;CZE: 5B+"	ISR: 5F–	JOR, EGY, SYR	8,000	20,000	M3, T5, P2

Name	Start	End	Duration (years)	Items and issues	Intensity	Direct participants	External/ Indirect participants[1]
26. Burma/ Myanmar (minorities)	1948	1999	51	3, 4	4	AND (KAREN), AND (MOJAHIDS), AND (WHITES), AND (REDS), AND (BAN)//BUR	CHN(2)// USA(1)
27. Colombia (Violencia I)	1948	1953	5	5, 4	4	AND//AND, COL	—
28. Burma (Chinese troops)	1949	1961	12	4	3	AND (KMT)//BUR, CHN	USA(2), TAW(2)//
29. Israel–Arab States (cease-fire)	1949	1956	7	1	3	EGY, JOR, IRQ, SYR, LEB//ISR	
30. India VIII (Kashmir II)	1949	1964	15	1	3	PAK//IND	—
31. India X (Nagas)	1950	1964	14	3	3	AND (NAGAS)// IND	—
32. Tunisia (independence)	1950	1956	6	2	3	AND (TUN)//FRA	—
33. China (Tibet I)	1950	1951	1	5	3	CHN//TIB	—
34. Somalia–Ethiopia (border)	1950	1961	11	1	3	AND (SOM)//ETH	ITA (1)//
35. Indonesia (South–Moluccas)	1950	1965	15	3	4	AND (KNIL)//INS	—
36. Korea II (Korean War)	1950	1953	3	4, 5, 6	4	PRK//ROK	USR (2), CHN (3)//USA (3)
37. Egypt (1st Suez crisis)	1951	1954	3	2	3	EGY//UKI	—
38. Kenya (independence, Mau-Mau)	1952	1956	4	2	4	AND (KEN)//UKI	—
39. Sudan (independence II)	1953	1955	2	5	3	AND (ARMY), EGY//AND (SUD), UKI	—
40. British Guyana (independence)	1953	1966	13	2	3	AND (PPP, GUY)// UKI	—
41. China–Taiwan (Quemoy I)	1954	1954	0	1, 4	3	CHN//TAW	//USA (2)

Instruments of initiator	Instruments of affected party	Initiator	Minimum no. of victims/ casualties	Maximum no. of casualties	Outcome
AND: 10D–, 10M–	BUR: 10H–, 10M–	AND (KAREN), AND (KOMM)	40,000	60 000	M4, P2
AND: 10M–, 10B–, 10F–	COL: 10H–, 10K–, 10M–	AND	80,000	300, 000	P17, P11, M1
"AND (KMT): 5G–; USA: 2D+, 6A–"	BUR: 2B+, 5H–	AND(KMT)	100	1,000	M3, P11
EGY: 4B–, 5H–, 2B+	ISR: 2B+, 5F–	EGY, JOR, IRQ, SYR, LEB	1,000	2,000	M1, T5, P2
PAK: 5A–, 5B–, 2B+	IND: 1D–, 5A–, 5B–	PAK	2,000	10,000	P2
AND: 10B–, 10N–, 10O–	IND: 10G–, 10M–, 10F+	AND (NAGAS)	300	1,000	P17
AND: 10A–, 10F–	FRN: 10G–, 10A+, 10F+	AND (TUN)	100	1,000	P3
CHN: 5L–	TIB:2B+,5L–	CHN	1,000	2,000	P4, M2, T3
SOM: 2B+	ETH: 1D+	SOM	1,000	2,000	P2
AND: 10N–, 10O–	INS: 10M–	AND (KNIL)	5 000	10,000	P1, P8, M3, T3
PRK: 5L–	ROK: 5L–	PRK	1,500,000	2,000,000	M1, T5, P1
EGY: 5K–, 4I–, 2B+, 1D+	UKI: 5A–, 5D–, 5F–, 1D+, 5D+	EGY	100	1,000	P16, P17
AND: 10A–, 10B–, 10O–	UKI: 10H–, 10J–, 10M–, 10G–	AND (KEN)	10,000	10,745	M3, P1
AND (ARMEE): 10M–, 10O–	AND (SUD):10F–	AND (ARMEE)	300	350	P3, M3
AND: 10F–, 10A–, 10A+, 2B+	UKI:10G–, 10H–, 10A+, 10E+	AND (PPP,GUY)	100	1,000	P3, P5
CHN: 5F–	TAW:5A–, 5F–	CHN	1	25	P1, P2, P15, M1

Name	Start	End	Duration (years)	Items and issues	Intensity	Direct participants	External/ Indirect participants[1]
42. Guatemala I (intervention)	1954	1954	0	4, 6	3	AND//GUA	HON (2), NIC (2), USA (3)//
43. Cyprus I (independence)	1954	1960	6	2	3	AND (EOKA)// UKI, AND (ISLAND TURKS)	GRC (2)//TUR (1)
44. China–India (Aksai Chin)	1954	1962	8	1	3	CHN//IND	—
45. Oman (Imam–Sultan conflict)	1954	1971	17	5, 7	3	AND (OMA), UKI// AND (IMAM)	//SAU(2), EGY(1)
46. China (Tibet II)	1954	1959	5	4, 3	4	AND (TIBETIANS)//CHN	—
47. Algeria (independence II)	1954	1962	8	2	4	AND (ALG)//FRA// AND (SIEDLER)	MOR(2), TUN(2)//
48. Nicaragua–Costa Rica (exiled people I)	1955	1956	1	4, 8	3	AND (RIGHT-WING EXILES), NIC//COS	//USA(2)
49. Turkey–Syria (border)	1955	1957	2	2	3	TUR//SYR	USA (2)//USR (2), EGY (3)
50. Cameroon (independence)	1955	1967	12	2	3	AND (CAO)//FRA, CAO	—
51. Cambodia (border)	1956	1970	14	6	3	USA, RVN, THI, AND (LON NOL)// KHM	—
52. Poland (October uprisings)	1956	1956	0	4,5	3	AND (KPLB), POL//AND (COMMUNIST PARTY–STALINIST), USR	—
53. Jordan (Arab Legion)	1956	1957	1	5	3	AND (NUWAR, ARMY)//JOR	SYR(3), USR(1)// SAU(2), USA(2)

Instruments of initiator	Instruments of affected party	Initiator	Minimum no. of victims/ casualties	Maximum no. of casualties	Outcome
AND: 5F–, 5G–	GUA: 2B+, 5L–	AND	1	25	M2, P5, P3, P15
AND: 10B–, 10D–, 10N–, 10O–	UKI: 10G–, 10I–, 10M–, 10F+	AND (EOKA)	359	621	P17
CHN: 1D–, 1C+, 1D+, 5B–, 5F–	IND: 1D–, 1C+, 1D+, 5B–, 5F–	CHN	9	1,000	P2, T5
OMA: 10M–, 5H–, 2B+	AND:10K–,10O–	AND	100	1,000	M2, T3, P11
AND: 10F–, 10M–, 10O–	CHN:10G–, 10H–, 10M–, 10R–	AND (TIBETIANS)	65,000	65,000	M3, P2, P11
AND: 5K–, 5L–, 10A–, 1D+	FRN:5L–, 10G–, 4L–, 1D+, 5D+	AND (ALG)	100,000	190,000	M3, P3, T2
AND: 5G–, 2C–	COS: 2B+, 5L–	AND (RECHTE EXIL.)	1	25	M3, P1, P9
TUR: 5B–, 5F–	SYR: 2B+	TUR	1	25	M1, T4, P1
AND: 10B–, 10P–, 10M–	FRN: 10M–, 10F+	AND (CAO)	100	1,000	M3, P3
USA: 5F–, 4B–, 4I–, 5G–, 10K–	KHM: 2B+, 2F+, 10E+	USA, RVN	25	30	P7, P15
AND: 10E+	USR:4C+, 5E–	AND	53	53	P15, P17
AND: 10A–, 10F–, 10K–	JOR: 10E–, 10G–	AND (NUWAR, ARMEE)	25	100	M3, P11

Name	Start	End	Duration (years)	Items and issues	Intensity	Direct participants	External/ Indirect participants[1]
54. Sri Lanka (Ceylon) (Tamils I)	1956	1958	2	3, 4	3	SRI//AND (UNP), AND (BUDDIST MONKS), AND (SINHALESE PEOPLE)//AND (TAMILS)	—
55. Morocco (French troops)	1956	1958	2	2, 6	3	MOR//FRA	—
56. Hungary (revolt)	1956	1957	1	5	3	AND (KPLB), AND (BEV)//AND (COMMUNIST PARTY–STALINIST), HUN, USR	
57. Cuba (revolution)	1956	1959	3	4	4	AND (CASTRO)// CUB	DOM (2)// USA (2)
58. Egypt (Suez war)	1956	1957	1	1, 2, 6	4	ISR, UKI, FRA// EGY	//USR (2)
59. Honduras–Nicaragua (border I)	1957	1957	0	1, 7	3	HON//NIC	USA (1)//
60. Israel III (border)	1957	1967	10	1, 6	3	EGY, SYR, JOR// ISR	USR(2)//
61. Morocco–Spain (attempt at expansion)	1957	1958	1	1, 2	3	AND (AOL), MOR//SPN	//FRA(3)
62. China–Taiwan (Quemoy II)	1958	1958	0	1, 4	3	CHN//TAW	USR (1)//USA (2)
63. Tunisia (Sakiet)	1958	1958	0	6	3	FRA//TUN	–
64. Tunisia (Remada)	1958	1958	0	6	3	FRA//TUN	–
65. Iraq (Mossul revolt)	1958	1959	1	5, 4	3	AND (NATIONALISTS, SHAWWAL)//IRQ	SYR (2), EGY (1)//USR (2)
66. Lebanon (first civil war)	1958	1958	0	5	3	LEB, AND (MARONITES)// AND (UNF)	IRQ, USA, UKI(3)//SYR(2)

Instruments of initiator	Instruments of affected party	Initiator	Minimum no. of victims/ casualties	Maximum no. of casualties	Outcome
SRI: 10H–, 10M–	AND: 10F–, 10B–, 10Q–, 10O–	SRI	500	500	M2, P2, P9, P11
MOR: 1D–, 1D+	FRN: 1D–,1D+,5D+	MOR	100	1,000	P16, P17
AND: 10A–, 10G–, 10M–, 8A–	AND: 5G–,5L–	AND (BEV)	10 000	32,000	M2, P1, P11, P15
AND: 10B–, 10D–, 10N–, 10M–	CUB:10E–, 10G–, 10M–	CUB	2 000	5,000	P5, M2, P3
EGY: 4B–, 4K–, 5B–	UKI: 10M–	ISR, UKI, FRN	3,230	10,000	M1, P17, P8
HON: 5D–, 2B+, 5F–, 1D+	NIC: 5B–, 5G–, 5C–, 1D+	HON	25	100	T5, P17
EGY: 2B+, 2B–, 5B–, 5C–, 5F–	ISR: 2B+, 2B–, 5B–, 5C–, 5F–	EGY, SYR	1,000	2,000	M1, T5, P2
MOR: 5F–, 1D+	SPN: 5A–, 5F–, 1D+, 5D+	AND (AOL),MOR	25	100	T3, P17, M1
CHN: 2E+, 5A–, 5F–	TAW:5A–,5F–	CHN	100	1,000	P2, M1, P1, T5
FRA: 5F–, 4L–, 1D+	TUN: 1D–, 1A–, 5H–, 1D+, 1E+	FRN	69	69	P1, P2, M1, P16
FRA: 5F–, 5A–, 1D+, 5D+	TUN: 5H–, 5C–, 1E+, 1D+, 4A–	FRN	305	305	P16, P17, M1
AND (NATIONALISTEN): 10K–	IRQ: 10M–	AND (NATIONALISTEN)	2,000	2,000	M3, P11
LEB: 10L–, 2B+, 10M–	AND (MILIZEN):10M–, 10O–	LEB, AND (MARONITEN)	1,000	2,000	M3, P5, P1

Name	Start	End	Duration (years)	Items and issues	Intensity	Direct participants	External/ Indirect participants[1]
67. Rwanda–Burundi (independence)	1958	1964	6	3, 5	3	AND (RWA)//BEL	—
68. Nepal II	1959	1961	2	1	3	CHN//NEP	—
69. Malawi (independence)	1959	1964	5	2	3	AND (NAC)//UKI, RHO	—
70. Dominican Republic I	1959	1962	3	4, 5	3	AND, CUB, VEN// DOM	USA (1), ECU (1), COL (1), PER (1), BOL (1)//
71. South Africa (Sharpeville)	1960	1960	0	4,5	3	AND//SAF	AFRICAN. STATES (1)//
72. Mali–Mauritania (border)	1960	1963	3	1	3	MLI//MAA	MOR (2)//
73. Nepal III	1960	1960	0	5	3	AND (CONGRESS)// NEP	IND(1)//CHN (1)
74. Spain (Basque autonomy)	1960	1999	39	3, 4	3	AND (ETA)//SPN	AND (2)// FRA (2)
75. Venezuela (guerrilla)	1960	1969	9	4	3	AND (MIR, PCV)// VEN	CUB (2)//USA (1)
76. Indonesia (West Irian II)	1960	1969	9	1, 6, 7	3	INS//NTH//AND (PAPUANS)	USR (2)//
77. India XII (Goa II)	1961	1961	0	2	3	IND//POR	—
78. Cuba (Bay of Pigs)	1961	1961	0	4	3	AND//CUB	USA (3)//
79. Tunisia (Biserta)	1961	1963	2	2, 6	3	TUN//FRA	—
80. Iraq (Kurds I)	1961	1970	9	3	4	AND (KURDS)// IRQ	TUR (1), IRN (2)//SYR (3), USR (2)

Instruments of initiator	Instruments of affected party	Initiator	Minimum no. of victims/ casualties	Maximum no. of casualties	Outcome
AND: 10B–	BEL: 10F+, 2B+	AND (RWA)	100	20,000	P2, M1
CHN: 5F–, 1D+, 5B–, 4A+	NEP: 1D+, 5B–, 1D–	CHN	1	25	T1, P17
MAW: 10A–, 1D+, 10N–	UKI: 10H–, 10A+, 1D+, 10F+	AND (NYASA)	52	52	P3
AND: 2C–, 1C–, 1D–	DOM: 2B+, 2C–, 5K–, 1H–	AND	25	100	P3, P5, P10, M3
AND: 10A–	SAF:10G–, 2D–	AND	74	74	P2, P11
MLI: 5F–, 1C+	MAA: 1D+, 1C+	MLI	25	100	M1, T3, P3
AND: 10B–, 10O–	NEP: 10H–, 10M–	AND (CONGRESS)	25	100	P1
AND: 10A–, 10B–, 10O–	SPN: 10M–, 10E–, 10G–, 10H–	AND (ETA)	500	500	T5, P2
AND: 10B–, 10M–	VEN: 1C–, 10M–, 10G–, 10C+	AND (MIR, PCV)	600	600	P9, M3, P1
INS: 5A–, 1D+	NTH: 5A–, 1D+	INS	100	1,000	P3, T5, M1, P16
IND: 5A–, 5G–	POR: 5F–, 2B+, 5D+	IND	40	40	M2, T3, P3
AND: 5F–, 2C–, 6A–	CUB: 5B–, 2B+, 5F–	USA, AND	100	1,000	M3, P9, P1
TUN: 1F+, 5F–, 5H–, 2B–, 1C–	FRN: 5F–, 5A–, 2A–, 5D+, 4C+	TUN	1,100	1,100	M1, P3
AND: 10N–, 10O–	IRQ: 10A+, 10F+, 10M–	AND (KURDEN)	25,500	50,000	M1, P17, P10

Name	Start	End	Duration (years)	Items and issues	Intensity	Direct participants	External/ Indirect participants[1]
81. Angola (independence)	1961	1974	13	2	4	AND (MPLA), AND (UNITA), AND (FNLA)//POR	USA (1), CHN (2), ZAI (2), USR (2), CUB (2)/SAF (2), RHO (1)
82. Cuba (Cuban Missile Crisis)	1962	1962	0	6	3	USR, CUB//USA	—
83. Gabon–Congo (Soccer Revolt)	1962	1962	0	1	3	GAB//CON	—
84. Brunei (uproar)	1962	1962	0	2, 5	3	UKI, AND (SULTAN, BRU)// AND (NKNA), AND (SUPP)	//INS(2)
85. Somalia–Ethiopia (Ogaden I)	1962	1964	2	1	3	SOM//ETH, KEN	ARABS (1), CHN (1), USR (2)//
86. China–India (war)	1962	1963	1	1	4	CHN//IND	—
87. Yemen AR (civil war II)	1962	1968	6	5	4	AND (REP), EGY// AND (ROY)	USR (2)//JOR (2), SAU (2), IRN (2), UKI (3)
88. Morocco–Algeria (Tindouf I)	1963	1963	0	1, 7	3	MOR//ALG	
89. Cyprus II (civil war)	1963	1964	1	3, 5	3	CYP, AND (EOKA–B)//AND (TMT)	GRC (2)//TUR (3)
90. Yemen PR–Oman (Dhofar–uproar)	1963	1979	16	2, 6, 5	3	AND (PFLOAG), YPR//OMA	CHN (2), SAU (2)//JOR (3), EGY (2), UKI (3), IRN (3)
91. Zanzibar (massacre)	1963	1964	1	5	3	AND (ASU)//TAZ	—
92. Rwanda (Bugesera–invasion)	1963	1964	1	3, 5	3	AND (TUTSI)// RWA	BUI (2)//

Instruments of initiator	Instruments of affected party	Initiator	Minimum no. of victims/ casualties	Maximum no. of casualties	Outcome
AND: 10A–, 10B–, 10F–, 10M–	POR: 10M–, 10R–	AND (MPLA), AND (UNITA)	60,000	90,000	P3, M2
USR: 5B–, 3A–, 1F+	USA: 1D–, 5H–, 1F+	USR	1	25	P16, P1
GAB: 10B–, 10I–, 2B+	CON: 10B–, 10I–, 2B+	GAB	1	25	P1
UKI: 10M–, 10G–, 10E–, 10H–	AND: 10D–, 10M–, 10N–, 10O–	UKI	25	100	M3, P3, P11, T4
SOM: 10B–, 10D–, 5L–, 2B+, 1D+	ETH: 5L–, 10H–, 2B+, 1D+	SOM	2,000	2,000	P2, M6
CHN: 5A–, 5F–, 5D+, 1D+	IND: 5A–, 5F–	CHN	1,000	4,500	M2, T3, P2
AND(REP): 10K–, 10M–, 1E+	AND (ROY):10M–, 1E+	AND(REP),EGY	100,000	100,000	P5, M1, P17
MOR: 5A–, 5G–, 5F–, 1D+, 1C–	ALG: 5A–, 5F–, 2B+, 1D+, 1C–	MOR	100	100	M1, P1, P2, T5
CYP: 10G–, 10R–, 10M–	AND: 10B–, 10N–, 10O–	CYP, AND (EOKA-B)	100	1,000	M1, P2, P8
YPR: 5F–, 5K–	OMA: 5F–, 10C+	AND (PFLOAG)	1,000	2,000	M3, T5, P11
AND: 10K–	ZAN: 10E+	AND (ASU)	4,000	4,000	M2, P5
AND: 5F–, 5K–	RWA: 5L–	AND	14,000	14,000	M3, P2

Name	Start	End	Duration (years)	Items and issues	Intensity	Direct participants	External/ Indirect participants[1]
93. Guinea-Bissau–Portugal (independence)	1963	1974	11	2	3	AND (GNB, PAIGC)//POR	USR (1), CUB (2), GUI (2), CHN (1), SEN (2)//
94. Malaya–Indonesia (Sarawak/Sabah)	1963	1966	3	1, 2	4	UKI, MAL//AND (SUPP), AND (SAYA), INS	AUL (3), NEW (3), CAN (2)// USR (2)
95. Laos II (civil war)	1963	1975	12	4, 6	4	AND (PATHET LAO)//LAO, AND (MEO)	DRV (3)//THI (3), USA (3), RVN (3)
96. Sudan (civil war I)	1963	1972	9	3, 5	4	AND (ANYA NYA)//SUD	ISR (2)//USR (3), EGY (3)
97. Panama (Canal I)	1964	1967	3	1	3	PAN//USA	—
98. Zaire (Stanleyville—hostages)	1964	1964	0	5, 3	3	AND (CNL)//ZAI	//USA (2), UKI(2), BEL (3)
99. Zaire (civil war)	1964	1965	1	3, 5	3	AND (CNL)//ZAI	CON (2), BUI (2), USR (2)// USA (2)
100. Mozambique (independence)	1964	1975	11	2	4	AND (FRELIM), AND (FRLM)//POR	//SAF (3), RHO (3), USA (2)
101. Indochina II (Vietnam War)	1964	1973	9	4, 5, 6	4	AND (FNL), AND (VIETMINH), DRV// USA, THI, RVN	USR (2), CHN (2)//ROK (3), AUL (3),PHI (3), NEW(3)
102. Yemen PR (independence)	1965	1967	2	2	3	AND (NLF), AND (FLOSY), EGY//UKI (ADEN)	//USR (3)
103. Peru (guerrilla)	1965	1966	1	4	3	AND (ELN, MIR)// PER	—
104. India XV (Ran of Ketch II)	1965	1969	4	1, 6, 7	3	PAK//IND	—
105. Zambia–Rhodesia (border)	1965	1987	22	1, 2, 4	3	RHO, SAF//ZAM, AND	—
106. Thailand (communism)	1965	1980	15	4,5	3	AND (CPT)//THI	DRV (2), CHN (2)//USA (3)

Instruments of initiator	Instruments of affected party	Initiator	Minimum no. of victims/ casualties	Maximum no. of casualties	Outcome
AND (GNB): 10A–, 10M–	POR: 10M–	AND (GNB, PAIGC)	2,000	15,000	M2, P5, P3
"UKI; MAL: 5A–, 5B–, 5C–»	"AND; INS: 10M–, 10N–, 5F–»	AND	740	1,000	P3, P9, M4, T3
AND: 10N–	LAO: 10N–, 10A+	AND (PATHET LAO), DRV	100,000	100,000	M1, P12, P15
AND: 10N–, 10M–	SUD: 10M–, 10H–, 10K–	AND	100,000	500,000	M1, P4
PAN: 10A–, 1C–, 2B+, 1D+, 1A+	USA: 5D–, 1D+, 1A+	PAN	1	26	P1, P16, T5
AND: 10B–, 10M–	ZAI:5F–,10M–	AND (CNL)	1,000	1,000	P1, M3
AND: 10N–, 10O–	ZAI:2B+	AND (CNL)	20,000	100,000	M1, T5, P2
AND: 10N–, 10M–, 1D+	POR: 10M–, 1D+	AND (FRELI), AND (FRLM)	25 000	30,000	M2, P3, P16
USA: 5L–, 5M–, 2E+, 5A+, 5B+	DRV: 10M–, 10N–, 10O–, 5M–	USA	1,215 992	2,000,000	T3, P3, M2
AND: 10A–, 10B–, 10N–, 10O–	UKI: 10H–, 10M–	AND (NLF), AND (FLOSY)	425	425	M2, P3
AND: 10B–, 10D–, 10M–	PER: 10G–, 10M–	AND (ELN, MIR)	500	500	P11, M3, P1
PAK: 5A–, 1D+, 1D–, 1E–, 5F–	IND: 1D–, 1D+, 1E–, 1D–, 5F–	PAK	100	1,000	T1, M1, P17
ZAM: 10H–, 5A–	RHO: 5F–, 4A–	SAF, RHO	100	1,000	P3, T5
AND: 10B–, 10N–, 10O–	THI: 10G–, 10M–	AND (CPT)	5,000	5,000	M3, P11, P1

Name	Start	End	Duration (years)	Items and issues	Intensity	Direct participants	External/ Indirect participants[1]
107. Dominican Republic II (intervention)	1965	1965	0	6, 4	3	AND//DOM	//USA(3)
108. India XIV (Kashmir III)	1965	1965	0	1, 2	3	AND (AZAD–REBELS)//IND	CHN (1), PAK (3)//
109. India XVI (Kashmir IV)	1965	1970	5	1, 2	4	PAK, AND (MUSLIMS)//IND	CHN (1)//
110. Zaire (Katanga—mercenaries)	1966	1967	1	5	3	AND (MERCENARIES)// ZAI	POR (2)//
111. Mozambique (border)	1966	1974	8	1, 2	3	POR//ZAM, TAZ, MAW	—
112. Namibia II (SWAPO)	1966	1990	24	2, 4	3	AND (SWAPO)// SAF	USR (2)//USA (2)
113. Bolivia (Che Guevara, Mar. 23, 1967–Oct. 10, 1967)	1967	1967	0	4	3	AND//BOL	CUB (2)//ARG (2), USA (2)
114. Egypt–Israel (confrontations)	1967	1973	6	1, 6	3	EGY, SYR//ISR	USR (2)//
115. Egypt–Israel (Six-Day War)	1967	1967	0	1, 6, 7	4	EGY, SYR, JOR, AND (PLO)//ISR	ALG (3), KUW (3), USR (2)// USA (1)
116. Eritrea III (civil war)	1967	1993	26	3	4	AND (EPLF)//ETH	EGY (2)//CUB (2), USR (2)
117. Nigeria (Biafra secession)	1967	1970	3	5, 3, 7	4	AND (BIAFRA)// NIG	POR (2), FRA (2), SPN (2)// USR (2), UKI (2), CZE (2)
118. Thailand–Cambodia III (border)	1968	1969	1	1	3	THI//KHM	—
119. CSSR (Prague Spring)	1968	1968	0	4, 6	3	USR, DDR, POL, HUN, BUL//CZE	—
120. Northern Ireland	1968	1999	31	2, 3	3	AND (IRA)//UKI, AND (ORANGE ORDER, RUC, B–POLICE)	—

Instruments of initiator	Instruments of affected party	Initiator	Minimum no. of victims/ casualties	Maximum no. of casualties	Outcome
AND: 10N–, 10M–, 1D+	"DOM; USA: 10M–, 1D+, 10D+, 5F–"	AND	3,000	10,000	P10, P15, P17, M1
AND: 5K–	IND: 5A–	AND (AZAD), PAK	20,000	20,000	M1, P2
PAK: 1B–, 5G–, 5L–	IND: 5L–	PAK	6,800	6,800	M1, P1
AND: 10B–, 10N–	ZAI: 2B+	AND (SOELDNER)	1	25	P2
POR: 5F–, 4K–	ZAM: 2B–, 10R–	POR	1,000	30,000	M3
AND: 10N–, 10M–, 1D+	SAF: 10M–, 1D+	AND (SWAPO)	40,000	40 000	T2, P3, M2
AND: 10N–, 10M–	BOL: 10H–, 10M–, 10L–	AND	25	200	P11, M3, P1, P9
ISR: 2B+, 2B–, 5F–	EGY: 2B+, 2B–, 5F–	EGY, USR	5,968	5,968	M1, T5, P2
ISR: 2B+, 2B–, 5H–	EGY: 2B+, 2B–, 5H–	ISR	19,600	25,000	M2, T3, P2
AND: 10N–, 10A–, 10B–, 1D+	ETH: 10M–, 10H–	AND (EPLF)	36,000	2,000,000	T1, M5, P3, P6
AND: 10O–, 10M–	NIG: 5C–, 4B–, 10M–	AND (BIAFRA)	1,000,000	2 000,000	M3, P9, T4, P1
THI: 5A–, 5F–, 2B+	KHM: 5A–, 5F–	THI	500	500	M1, T5, P2
USR: 5G–	CZE: 1D+	USR	50	1,000	P7, P3, P15
AND: 10A–, 10B–, 10N–	UKI: 10E–, 10G–, 10M–	AND (IRA)	2,700	3,000	P2

Name	Start	End	Duration (years)	Items and issues	Intensity	Direct participants	External/ Indirect participants[1]
121. Philippines (uproar by National Front)	1968	1999	31	5	3	AND (NPA)// PHI, AND (CIVIL MILITIA)	//USA(2)
122. China–USSR (Ussuri conflict)	1969	1969	0	1	3	CHN//USR	—
123. Iran–Iraq (Shatt–al–Arab)	1969	1975		1	3	IRQ//IRN	//UKI (2), KURD (2)
124. Indonesia (West Irian III)	1969	1982	13	2, 3, 7	3	INS//AND (OPM)	//PNG (1), VAN (2)
125. Honduras–El Salvador (Soccer War I)	1969	1970	1	8,1	4	HON//SAL	—
126. Argentina (Montoneros)	1969	1977	8	4	4	AND (MONTON.)// ARG, AND (DEATH SQUADS)	//USA (2)
127. Portugal–Guinea (invasion of Conakry)	1970	1970	0	5	3	AND (G.s IN EXILE)//GUI	POR (3)//TAZ (2), ZAM (2), USA (2), EGY (2)
128. Jordan (Black September)	1970	1971	1	5	3	JOR//AND (PLO), SYR	USA (1)//USR (1), IRQ (3)
129. Philippines (Moros in Mindanao and Sulu)	1970	1999	29	2, 3	3	AND (MIM), AND (MNLF)//PHI	MAL (2)//USA (1)
130. Cambodia II	1970	1975	5	4, 5	4	AND (KHMER R.)// KHM	DRV (3)//USA (3), RVN (3)
131. Sri Lanka (Ceylon) (uproar)	1971	1971	0	5	3	AND (JVP)//SRI	//USA (2), UKI (2), AUL (2) ,USR (2), PAK (2), IND (3)
132. India XVII (Bangladesh III)	1971	1971	0	6, 2	4	AND (BNG), IND// PAK	USR (2)// CHN (1), USA (1),SAU (2), LIB (2), IRN(2)

Instruments of initiator	Instruments of affected party	Initiator	Minimum no. of victims/ casualties	Maximum no. of casualties	Outcome
AND: 10B–, 10F–, 10M–, 10O–	PHI: 10H–, 10E–, 10M–, 10R–, 10A+, 10C+, 10F+	AND (NPA)	10,000	12,000	M1, P2, P7
CHN: 5F–, 5G–, 1D+	USR:5F–,1D+	CHN	160	1,000	M1, T5, P1
IRQ: 1I–, 1C–, 5F–	IRN: 1H–, 5F–, 2B+, 1C–	IRQ	100	1,000	P17
INS: 5G–, 10M–, 10L–, 10S–	AND: 10F–, 10O–,10N–10B–	INS	50,000	100,000	T3, P2, M2, P11
HON: 1C–, 5F–, 1D+, 4I–	SAL: 5C–, 1C–, 5G–, 1D+	SAL	2,000	2,400	P17, M1
AND: 10B–, 10O–, 10N–	ARG: 10D–, 10H–, 10M–	AND (MONTON.)	10,000	10,000	P1, P11, M3
AND: 5F–, 5D–	GUI: 2B+	AND (EXILANTEN)	1	25	M3, P1
JOR: 10M–	SYR: 5F–, 5D–	JOR	2,000	2,440	M2, T3, P11
AND: 10N–, 10H+	PHI: 10M–, 10A+	AND (MIM)	100,000	150,000	P2, P17
AND: 10B–, 10N–	KHM: 10H–, 10M–	AND (R.KHMER)	150,000	1,000,000	M2, P5, P3
AND: 10D–, 10F–, 10M–	SRI: 10M–, 10G–, 10H–	AND (JVP)	1,200	2,000	M3, P1, P11
IND: 1E+, 5A–, 5F–, 5G–, 5L–	PAK: 5A–, 5F–, 5L–	IND	300,000	300,000	M2, T1, P3, P6

Name	Start	End	Duration (years)	Items and issues	Intensity	Direct participants	External/ Indirect participants[1]
133. Uganda–Tanzania (invasion)	1972	1972	0	5	3	AND (OBOTE)// UGA (AMIN)	TAZ (2)//
134. Rhodesia (civil war)	1972	1979	7	2	4	AND (ZANU), AND (ZAPU)//RHO	MZM (3), ZAM (2)// SAF (2)
135. Burundi I (genocide)	1972	1973	1	5	4	BUI//AND (HUTU), TAZ	ZAI (3)//
136. Iraq–Kuwait II (border)	1973	1973	0	3, 6, 7	3	IRQ//KUW	USR (1)//IRN (1)
137. Pakistan (Belochistan)	1973	1976	3	3	3	AND (BLF)//PAK	IRQ (2), AFG (1)//IRN (2)
138. Israel IV (Yom Kippur War)	1973	1973	0	1, 6	4	EGY, SYR, AND (PLO)//ISR	USR (2), IRQ (2), JOR (2), ALG (2), MOR (2)//USA (2)
139. Indochina II (cease-fire)	1973	1976	3	4, 5, 8	4	DRV, AND (FNL)// RVN	USR (2), CHN (2)//USA (2)
140. Iraq (Kurds II)	1974	1975	1	3	3	IRQ//AND (DPK–KURDS)	USR (2)//IRN (2)
141. Ethiopia (Red Terror)	1974	1978	4	4	3	AND (DERG), AND (MEISON), AND (EPRP)//ETH	—
142. Indonesia (East Timor [civil war I])	1974	1975	1	1,2	4	AND (UDT)//AND (FRETILIN)//INS// POR	—
143. Cyprus IV (Turkey invasion)	1974	1974	0	4	4	CYP, GRC, AND (NAT.GUARD), AND (EOKA–B)// CYP, AND (TMT)	GRC (2)//TUR (3)
144. Ethiopia (Tigray)	1974	1991	17	3, 5	4	AND (TPLF)//ETH	//CUB (3), USR (2)
145. Lebanon I	1975	1975	0	5	3	AND (FALANGE)// AND (PLO)	—
146. Rhodesia–Mozambique (attempt at destabilization)	1975	1979	4	2, 6	3	RHO, AND (RENAMO)//MZM	SAF (3)//

Instruments of initiator	Instruments of affected party	Initiator	Minimum no. of victims/ casualties	Maximum no. of casualties	Outcome
AND: 5F–, 1D+	UGA:5B–, 5F–, 1D+	AND (OBOTE)	300	300	P1, M3
AND: 10B–, 10F–, 10M–, 10N–	RHO:10G–, 10H–, 10I–, 10M–	AND (ZANU, ZAPU), MZM	12,000	20,000	M1, P3
BUI: 1D–, 10M–, 2B+	AND: 10B–	BUI	100,000	200,000	M2, P2
IRQ: 5F–, 5D+	KUW: 5A–, 5F–, 2B+	IRQ	25	100	M1, P2
AND: 10A–, 10B–, 10N–, 10O–	PAK: 10G–, 10M–, 10C+, 10F+	AND (BLF)	500	500	P17
EGY: 2A+, 2B+, 4B–, 5B+, 5G–	ISR: 2A+, 2B+, 4B–, 5G+	EGY, SYR	16 401	25,000	M1, T5, P2
"AND;DRV: 1D+, 5G–, 5F–"	RVN: 1D+, 5F–, 5G–	AND(FNL),DRV	100,000	100,000	M2, T3, P3, P5
IRQ: 10A+, 10M–	AND: 10A+, 10M–	IRQ	2,000	20,000	M2, P2, P11
AND: 10K–	ETH: 10E+	AND (DERG)	0	200,000	M2, P3
AND(UDT): 10K–, 10M–	AND (FRETILIN): 10M–, 10O–	AND (UDT)	2,000	2,000	M3, P16, P8
TUR: 5F–, 1D+, 5G–	CYP: 5A–, 1D+	GRC	1,500	5,000	M2, T1, P7
AND: 10M–	ETH: 10M–	AND (TPLF)	25,000	25,000	M2, P5, P7
AND (FALANGE): 10B–, 10K–	AND (PLO): 10K–	AND (FALANGE)	80	80	M1P2
RHO: 10M–	MZM:10M–	RHO, SAF (POR), AND	1,338	1,338	P5, M1

Name	Start	End	Duration (years)	Items and issues	Intensity	Direct participants	External/ Indirect participants[1]
147. Bangladesh (Chakma, Marma)	1975	1987	12	3	3	AND (SHANTI BAHINI)//BNG	—
148. Indonesia–FRETILIN (East Timor II)	1975	1976	1	1, 2, 5	3	AND (FRETILIN), POR//INS, AND (MRAC)	AUL (1)//
149. Lebanon II	1975	1976	1	5	4	LEB, AND (FALANGE)// AND (PLO), AND (MUSLIM.MILITIA NATIONAL MOVEMENT)	ISR (3)// SYR(3)
150. Rhodesia (Nagomia attack)	1976	1976	0	5, 6	3	RHO//MZM	USA (2)//USR (1)
151. Lebanon III	1976	1976	0	5	3	AND (ARMY), AND (NATIONALE BEW.)//SYR, AND (PLO)//LEB, AND (FALANGE)	—
152. Lebanon IV	1976	1979	3	57	3	AND (MOSLEM. MILIZEN), AND (PLO)//SYR// AND (CHRISTIAN MILITIA)	//ISR (3), SLA (3)
153. Morocco (Western Sahara II)	1976	1979	3	2, 6	3	AND (POLISARIO)// MOR, MAA	ALG (2), LIB (1)//FRA (3), USA (2)
154. South Africa (ANC, PAC)	1976	1994	18	4, 5	3	AND (ANC, PAC)// AND (INKATHA)// SAF	MZM (2), ANG (2)
155. Somalia–Ethiopia (Ogaden II)	1976	1978	2	6	4	AND (WSLF), SOM//ETH	//USR (2), CUB (3)
156. Angola (civil war II)	1976	1991	15	5, 7	4	AND (UNITA), AND (FNLA), SAF//ANG, AND (SWAPO)	USA (1)//USR (2), CUB (3)
157. Indonesia (East Timor III)	1976	1999	23	2, 5	4	INS//AND (FRETILIN)	//POR(1)

Instruments of initiator	Instruments of affected party	Initiator	Minimum no. of victims/ casualties	Maximum no. of casualties	Outcome
AND: 10B–, 10N–, 10O–	BNG: 10R–, 10G–, 10M–, 10A+	AND (SHANTI BAHINI)	200,000	200,000	P2, P11
INS: 5G–, 5K–, 5L–, 10R–, 1A–	AND: 5K–, 10N–, 1C–	AND (FRETELIN)	100,000	200,000	M4, P9, P11, T3
LEB: 10K–10M–10A+10F+	AND: 10K–, 10N–	AND (FALANGE)	1,000	1,000	M1, P2
RHO: 5F–	MZM: 5F–, 2B–	RHO	500	500	P1, M4
AND (ARMEE): 10K–, 10M–, 10o–	LEB: 10M–	AND	2,000	2,000	M1P2
AND: 10M–, 10o–	SYR: 10M–	AND (MOSLEM. MILIZEN)	1,500	1,500	M1, P2
AND: 10D–, 10O–, 10M–, 10N–	MOR:10R–,10M–,10G–	AND (POLISARIO), ALG	7,000	10,000	M4, P17, T5, P2
AND: 10A–, 10B–, 10N–, 10O–	SAF:10G–,10C+,10D+,10F+	AND (ANC, PAC)	15,000	15,000	P3, P5, P12
SOM: 5L–, 5D–, 1C–	ETH:1B–, 5L–, 1C–, 1E+	AND (WSLF), SOM	9,000	21,000	M3, P1, T5
AND: 10M–, 10N–;SAF: 5F–	ANG: 10M–	AND(UNITA), SAF	150,000	150,000	M1, M5, P12
INS: 10M–, 10E–, 10G–	AND: 10N–, 10O–, 10C+	INS	100,000	250,000	P2, M2, T5

Name	Start	End	Duration (years)	Items and issues	Intensity	Direct participants	External/ Indirect participants[1]
158. Pakistan (civil war in Karachi)	1977	1999	22	4	3	AND (MOHAJIR QAUMI MOVEMENT (MQM)//PAK	
159. Indochina IIIa	1977	1978	1	1,6	4	DRV, AND (EXILE KAMBODIANS)// KHM	USR(2)// CHN(2)
160. Nicaragua I (revolution)	1977	1979	2	4	4	AND (FSLN)//NIC	CUB(2), VEN(2)// USA(2), HON(3)
161. Tunisia (uprisings)	1978	1978	0	4, 5	3	AND (OPP)//TUN	—
162. Zaire (Shaba III)	1978	1978	0	5	3	AND (FLNC)//ZAI	ANG (2)//FRA (3), BEL (3), USA (2)
163. Ethiopia (Ogaden, WSLF)	1978	1988	10	3	3	AND (WSLF)//ETH	SOM (3)// USR (2)
164. Iran (Islamic Revolution I)	1978	1979	1	4, 5, 6	3	AND (AIRFORCE), AND (PASD.), AND (MUJAHED.), AND (FEDA.), AND (TUDEH)// IRN	//USA (2)
165. Afghanistan I (civil war I)	1978	1979	1	5, 4	3	AND (ISLAM. REBELS), AND (PARCHAHAM)// AFG (CHALQ)	//USR (3)
166. Uganda–Tanzania (border war)	1978	1979	1	1, 5, 6	4	UGA//TAZ, AND (UNLF)	LIB (3), USR (1)//USA (1)
167. Indochina IIIb	1978	1991	13	4, 5, 6	4	DRV, LAO, KHM// AND (KPNLF), AND (SIHANOUK), AND (POLPOT)	USR (2)//CHN (2), USA (2), THI (2)
168. Mozambique (civil war; RENAMO)	1978	1994	16	5, 6	4	SAF, AND (RENAMO)//MZM	//ZIM (3)

Instruments of initiator	Instruments of affected party	Initiator	Minimum no. of victims/ casualties	Maximum no. of casualties	Outcome
AND: 10A–, 10B–, 10M–, 10K–, 10O–	PAK: 10E–, 10G–, 10H–, 10M–	AND (MQM)	2,000	5,000	P2
DRV: 5F–	KHM: 5F–, 5D–	DRV	8,000	8,000	M3, P1
AND: 10O–, 10D–, 10N–, 10A–	NIC: 10E–, 10G–, 10H–, 10M–	AND (FSLN)	10,000	40,000	P5, M2, P3, P16
AND: 10A–, 10B–, 10F–, 10O–	TUN: 10G–, 10H–, 10M–, 10B+	AND (OPP)	51	51	P1, P11
AND: 10B–, 10N–, 1D+	ZAI: 10M–, 1D+	AND (FLNC)	700	700	P1, M3
SOM: 10N–, 10M–, 1E+, 5F–	ETH: 10M–, 5F–	AND (WSLF)	1,400	1,400	P1, M4
AND: 10A–, 10D–, 10F–, 10K–	IRN: 10H–, 10G–, 10L–	AND	0	4,000	P5
AND: 10M–, 10N–, 10O–, 10M–	AFG: 10K–, 10M–	AND	5,000	5,000	M4, P10, P15
UGA: 5F–, 5L–, 2B+, 10E+	TAZ: 5L–	UGA	4,000	4,000	M3, P1, T4
"DRV; LAO: 5A–, 5F–, 10H–, 5G–"	AND: 10D–, 10M–, 1D+, 10N–	DRV, LAO	25,000	150,000	M4, P5
"SAF: 5B+, 6A–;AND: 10B–, 10M–, 10N–"	MZM: 10M–, 10F+	SAF, AND (RENAMO)	100,000	400,000	M1, P17, P5

Name	Start	End	Duration (years)	Items and issues	Intensity	Direct participants	External/ Indirect participants[1]
169. Saudi Arabia (occupation of mosque)	1979	1979	0	5, 3	3	AND (HEDSHAS U.A.)//SAU	IRN (1)//FRA (3)
170. Iran (Kurds II)	1979	1988	9	3	3	AND (DPK–I KURDS)//AND (KOMULA–KURDS)//AND (PUK–KURDS)// IRN	—
171. Lebanon V	1979	1982	3	57	3	SYR, AND (MARADA)//LEB, AND (FALANGE), AND (FL)	//ISR (3)
172. Morocco (Western Sahara III)	1979	1991	12	2, 6	3	AND (POLISARIO)// MOR	ALG (2), LIB (1)//FRA (2), USA (2)
173. Iran (Islamic Revolution II)	1979	1981	2	4, 5	3	AND (MUJAH.), AND (FEDA.), AND (FORGHAN)//IRN (IRP UND PASD.)	—
174. Iraq (Kurds III)	1979	1986	7	3,5	3	IRQ//AND (KURD ORGANIZATION, SINCE 1989: KURDISTAN FRONT)	TUR (3)//SYR (2) , IRN (2)
175. China–Vietnam (war)	1979	1979	0	6,1	4	CHN//DRV	//USR (2)
176. Afghanistan II (Soviet intervention)	1979	1988	9	5, 4, 6	4	AND (ISLAM. ALLIANCE), AND (SHIITE RESISTANCE GROUP)//AFG, USR	USA (2), PAK (2), IRN (2), EGY (2), SAU (2)//
177. Tunisia (Gafsa)	1980	1987	7	5	3	AND (OPP), LIB// TUN	//FRA (2), USA (2)
178. Peru (Shining Path II)	1980	1996	16	4	4	AND (S.LUM)// AND (MRTA)//PER	

Instruments of initiator	Instruments of affected party	Initiator	Minimum no. of victims/ casualties	Maximum no. of casualties	Outcome
AND: 10B–	SAU: 10M–	AND	400	400	M3, T5, P1, P11
AND: 10O–, 10M–	IRN: 10A+, 10P–, 10M–	AND (KURDEN)	1,000	1,000	P2, P11
AND: 10M–10o–	LEB: 10M–	AND (MUSLIM MILIZEN)	1,500	1,500	M1, P2
AND: 10M–, 10N–, 10O–, 10D–	MOR: 10M–, 10G+, 10G–	AND (POLISARIO)	10,000	10,000	P2, M4, T5
AND: 10K–, 10B–, 10F–	IRN: 10M–, 10B–, 10E–	AND	4,000	20,000	P11
IRQ: 10A+, 10M–	AND: 10O–, 10M–	IRQ	182,000	200,000	M4, P2, P11
CHN: 5G–, 5D+, 1D+	DVR: 5A–, 5C–, 1D+	CHN	20,000	70,000	M1, M5, P2
USR: 1F+, 2B+, 5G–, 5A–, 3A–	AND:10O–, 10K–, 10M–	USR	14,454	1,200,000	M3, M5, P1, P11
LIB: 10N–, 10O–, 1C–, 1A+	TUN:10M–, 10G–, 1A–, 1C–, 1A+	AND (OPP)	42	42	P1, M1
AND: 10B–, 10D–, 10N–, 10O–, 10h+	PER: 10H–, 10M–, 10e–, 10g–, 10c+	AND (S. LUM)	20,000	30,000	P2, P11, P8

Name	Start	End	Duration (years)	Items and issues	Intensity	Direct participants	External/ Indirect participants[1]
179. Iran–Iraq I (Gulf War)	1980	1988	8	1, 4, 6	4	IRQ//IRN	FRA (2), EGY (2)//
180. Uganda (Obote)	1981	1986	5	5	3	AND (NRA–MUSEVENI)//UGA (OBOTE)	—
181. India XVIII (Khalistan/ Punjab)	1981	1999	18	3, 2	3	AND (SIKHS)//IND	—
182. Nicaragua II (Contras)	1981	1990	9	4, 3, 6	4	AND (KONTRAS), AND (ARDE), AND (MISKITO)//NIC	HON (2), USA (2)//CUB (2), USR (2)
183. El Salvador (civil war)	1981	1992	11	4	4	AND (FMLN)// SAL, AND (DEATH SQUADS)	NIC (2)//HON (3),USA (2)
184. Senegal (Casamance)	1982	1999	17	3	3	AND (MFDC)// SEN	GNB (2), GAM (1)
185. Syria (February—uproar in Hama)	1982	1982	0	5	3	AND (MUSLIM BROTHERS), AND (PARTS OF ARMY)//SYR	—
186. Israel–Lebanon III	1982	1985	3	8	3	ISR//AND (PLO), AND (AMAL)	USA (1)//SYR (3)
187. Argentina–United Kingdom (Falkland II)	1982	1982	0	1,7	4	ARG//UKI	//USA (2)
188. Lebanon VI	1982	1984	2	7, 5, 3	4	AND (AM AL), AND (PRO–SYR. PAL.), AND (DRUS. MILITIA)//SYR// PLO//ISR, LEB AND	//FRA (3), USA (3) ,ITA (3), UKI (3)
189. USA–Grenada	1983	1983	0	6	3	USA//GRN	BAR (3), JAM (3), DMA (3), ABA (3), STV (3)//CUB (3)
190. Zimbabwe (Matabele massacre)	1983	1983	0	5	3	ZIM//AND (NKOMO–GUERILLAS)	—

Instruments of initiator	Instruments of affected party	Initiator	Minimum no. of victims/ casualties	Maximum no. of casualties	Outcome
IRQ: 5F–, 5G–, 5H–	IRN:3A–,5A–	IRQ	400,000	450,000	P2, M1, T5
AND: 10N–, 10M–	UGA:10M–	AND (NRA–MUSEVENI)	90	90	M2, P3, P5
AND: 10B–, 10F–, 10N–, 10O–	IND: 10G–, 10M–, 10A+, 10F+	AND (SIKHS)	10,000	18,000	M1, P2, T5
AND: 10B–, 10M–, 10N–, 10O–	NIC:10H–, 10M–, 10C+, 10B+	AND (KONTRAS)	20,000	40,000	M1, P8, P10
AND: 10B–, 10D–, 10M–, 1D+	SAL: 10M–, 10A+, 1D+	AND (FMLN)	30,000	65,000	M1, P10, P17
AND: 10A–, 10B–, 10M–, 10N–	SEN: 10D+, 10H–, 10M–	AND (MFDC)	1,000	1,500	M4, T5, P2
AND (MUSLIM): 10K–, 10N–, 1L–	SYR: 10M–	AND (MUSLIMBRUEDER)	20,000	20,000	M3, T2, P9, P11
ISR: 5G–	AND (PLO):5F–	ISR	7,200	50,000	M2, P11
ARG: 5G–, 1G–, 5L–	UKI: 1C–, 1D–, 4B–, 5G–, 5L–	ARG	1,000	1,055	M3, P7, P1, T5
AND(MOSL): 10K–, 10M–, 10B–	ISR: 5F–, 5G–	AND	2,500	2,500	M5, P2, M1
USA: 1B+, 5F–, 4D+, 4C+	GRN: 5C–, 5L–	USA	78	78	M2, P5
ZIM: 10N–, 10R–	AND: 10N–	ZIM	500	500	P2

Name	Start	End	Duration (years)	Items and issues	Intensity	Direct participants	External/ Indirect participants[1]
191. India XIX (Assam I)	1983	1984	1	3, 5	3	AND (ASU), AND (AGSP), AND (OPP)//AND (BENGALIS), AND (LDA), IND	—
192. Sri Lanka (Tamils II)	1983	1987	4	3	4	AND (TIGER), AND (TULF)// SRI, AND (SINH. TERROR GROUP// AND (SLFP)	TAMILEN (2)//
193. Sudan (civil war II)	1983	1988	5	5, 3	4	AND (SPLA–SPLM), AND (SAP, 14 SUDANES)// SUD, AND (DUP), AND (UMMA)/NIF	AND (ANYA)//EGY, LIB, AND (ANYA)
194. Philippines (Aquino–Marcos)	1984	1986	2	4, 5, 6	3	AND (NAMFREL), AND (RAM), AND (CATH. CHURCH), AND (NDF), AND (BAJAN)//PHI	LIB (2), PLO (2)//USA (2)
195. Turkey (Kurds I)	1984	1989	5	3, 4, 2	3	AND (PKK)//TUR	LEB (1),SYR (1)//IRQ (1)
196. India XX (Ayodhya)	1984	1999	15	3	3	AND (RSS, BJP, VHP)//IND	
197. Burkina Faso–Mali (border II)	1985	1985	0	1, 7	3	MLI//UPP	—
198. Yemen PR (Aden—civil war)	1986	1986	0	5	4	AND (ISMAIL AND AL ATAS)//AND (MOHAMMAD)	—
199. Israel V (Intifada)	1987	1993	6	1, 3, 2	3	AND (PALES), AND (HAMAS), AND (ISLAM. JIHAD)/AND (PLO)//ISR	/SYR (2), IRQ (2), AND (PFLP) (3), AND (HEZBOLLAH) (3)//
200. Iran–Saudi Arabia (pilgrims I)	1987	1987	0	3, 4	3	AND (PILGRIMS)// SAU	IRN (1)//
201. Sri Lanka (Tamils III)	1987	1995	8	3, 5	4	AND (TIGER)// SRI//AND (JVP)// AND (MUSLIMS)	//IND (3), PAK (2), ISR (2)

Instruments of initiator	Instruments of affected party	Initiator	Minimum no. of victims/ casualties	Maximum no. of casualties	Outcome
AND: 10B–, 10F–, 10M–, 10O–	IND: 10G–, 10A+, 10D+	AND (ASU), AND (AGSP)	2,000	2,000	P10 (P12, P17)
AND (TIGER): 10M–, 10B–, 10N–	SRI: 10M–, 10G–, 10E–, 10B–	AND (TIGER)	1,000	25,000	M4, P2, P11, P15
AND: 10A+, 10N–	SUD: 10M–, 10F+, 10D+, 10K–	SUD	1,000 000	1,300,000	M1, P17, P10
AND: 10R–, 10D–, 10O–, 10Q–	PHI: 10S–, 10M–	AND (NAMFREL)	500	500	M3, P7, P10
AND: 10B–	TUR: 10H–, 10F–, 10G–	AND (PKK)	0	1,000	M4, T5, P2
AND: 10A–, 10B–, 10F–	IND:10E–, 10G–, 10H–	AND	1,200	2,500	T5, P2
UPP: 5F–, 5L–, 2D+	MLI: 5L–, 2D+	MLI	100	100	M1, M5, T5, P1
AND (ISM): 10O–	AND (MOH): 10K–	YPR	10,000	13,000	P5, M2
AND (PLO): 1D+, 10D+, 10F–	ISR:1D+, 10G+, 10G–, 10R–	AND (PALAESTINENSER)	1,700	0	P17, P6
AND: 3A–, 10A–, 10O–	SAU: 10M–	AND	447	447	P2
AND: 10O–, 10B–, 10M–	SRI: 10M–, 10P–, 10B–, 5L–	AND	30,000	100,000	M4, P15, P8, P11

Name	Start	End	Duration (years)	Items and issues	Intensity	Direct participants	External/ Indirect participants[1]
202. Algeria (October uprisings)	1988	1989	1	5	3	AND (OPPOSITION)// ALG	—
203. Papua–New Guinea (Bougainville II)	1988	1999	11	3, 7	3	AND (BRA)//PNG, AND (COPPER LTD.)	SLM
204. India XXII (Kashmir V)	1988	1999	11	1, 2	3	PAK, AND (MUSLIMS)//IND	
205. Egypt (Islamists vs. government)	1988	1999	11	5, 4	3	AND (GAMA'AT), AND (ICHWAN), AND (ISLAM. JIHAD)//EGY	IRN (1), SUD (2)// LYB (1)
206. Lebanon (Shiite militia)	1988	1990	2	5, 4	4	AND (AMAL)// AND (HEZBOLLAH)	SYR (3)//IRN (2)
207. Burundi II (Hutu)	1988	1988	0	5	4	AND (HUTU)//BUI	—
208. Somalia (civil war I)	1988	1991	3	3, 5	4	AND (SNM)// SOM//AND (SPM)	LIB (2)//
209. Mauritania–Senegal (tensions)	1989	1990	1	1, 7	3	MAA//SEN	—
210. China (student uprisings)	1989	1989	0	4	3	AND (STUDENTS)// CHN	—
211. Colombia (drug cartel)	1989	1999	10	8	3	AND (MEDELLIN CARTEL), AND (CALI CARTEL)// KOL	USA (2)
212. Lebanon VIII	1989	1990	1	5	4	AND (CHRISTIAN MILITIA–AOUN)// AND (MUSLIM MILITIA)	FRA (2), IRQ (2)//IRN (2), SYR (3)
213. Turkey (Kurds II)	1989	1999	10	3, 5	4	AND (PKK), AND (TKSP), AND (HEP), AND (DEP [HADEP])//TUR	//IRQ(1), AND(PUK)(3), AND (DVP)(3)

Instruments of initiator	Instruments of affected party	Initiator	Minimum no. of victims/ casualties	Maximum no. of casualties	Outcome
AND: 10A–, 10F–	ALG: 10E–, 10G–, 10F+	AND (OPPOSITION)	159	600	P17, P10
AND: 10B–, 10F–, 10N–	PNG:10M–, 10H–	AND (BRA)	1,000	5,000	T5, M1, P2
PAK, AND (Muslims): 1D+, 2B–, 5B+, 5L–, 5C–, 5K–	IND:1D+, 10H–, 10M–, 1C+, 5F–	AND (MUSLIMS)	15,000	20,000	T5, M4, P2
AND (GAMA'AT): 10B–, 10F–, 10M–, 10N–, 10O–	EGY:10E–, 10H–, 10M–, 10D–, 10G–	AND (GAMA'AT)	500	2,000	P1, P9, P11
AND: 10B–, 10K–	AND: 10B–, 10K–	AND	1,200	1,200	P17
AND: 10B–, 10O–	BUI: 10M–	BUI	5,000	50 000	P2, P11
AND: 10N–	SOM: 10M–	AND (SNM)	15,000	50,000	M2, P7, P3
MAA: 10G–, 4F–, 1D–, 5F–	SEN: 3A–, 1C–, 5F–	MAA	400	400	T5, P2
AND: 10A–	CHN: 10M–	AND(STUD.)	700	700	M3, P1
AND: 10B–,	COL: 10G–, 10H–, 10L–	AND (MEDELLIN)	3,000	3,000	P15, P2
AND: 10M–, 10K–	SYR: 5F–	AND (CHRISTL. MILIZEN)	1,500	1,500	M1, P7
AND: 10B–, 10N–	TUR: 10E–, 10G–, 10H–, 10M–	AND	5,250	9,440	P11, P2, M4, T5, P8

Name	Start	End	Duration (years)	Items and issues	Intensity	Direct participants	External/ Indirect participants[1]
214. Liberia (civil war)	1989	1995	6	5, 7	4	AND (NPFL)//LBR, AND (ULIMO), AND (LPC)	//IVO (2),UPP (2), NIG (ECOMOG) (3), SIE (2), GUI (2), FRA (1)
215. Nicaragua III (Recontras)	1990	1994	4	4, 5	3	AND (RECONTRAS)// AND (RECOMPAS)//NIC	USA (1)
216. Niger (Tuareg II)	1990	1995	5	1, 3	3	AND (FLAA, CRA)//NIR	ALG
217. Mali (Tuareg III)	1990	1999	9	1, 3	3	AND (MFUA)//MLI, AND (GANDA KOI)	ALG, LIB
218. Iraq–Kuwait V (annexing)	1990	1991	1	1, 7, 6	3	IRQ//KUW	SUD (2)//USA (3), UKI (3), FRA (3), MOR (3), USR (2)
219. South Africa (ANC—Inkatha)	1990	1994	4	3	3	AND (INKATHA)// [AND (ANC BIS 1994)] SAF (ANC–GOVERNMENT)	.
220. Indonesia (GAM movement in Aceh II)	1990	1999	9	2	3	AND (GAM, GERAKAN ACEH MERDEKA)//INS	MAL
221. Iraq–Kuwait VI (USA intervention)	1990	1991	1	1, 7, 6	4	IRQ//KUW, USA	JOR, YEM, PLO//UNO, EGY, SAU, MOR, IRN,U SR, FRA, UKI, ITA
222. Rwanda (civil war)	1990	1994	4	5, 7	4	AND (REBELS OF THE "FRONTE PATRIOTIQUE RWANDAIS")// RWA	UGA (2), FRA (3)
223. Yemen (unification) II	1991	1999	8	5, 4	3	AND (JSP)// AND (AVK), AND (ISLAH)	SAU (2?)//

Instruments of initiator	Instruments of affected party	Initiator	Minimum no. of victims/ casualties	Maximum no. of casualties	Outcome
AND: 10K–, 10M–, 10N–	LBR:10M–	AND (NPFL)	200,000	200,000	M1, P2
AND: 10B–, 10N–	NIC: 10A+, 10F+	AND (RECONTRAS), AND (RECOMPAS)	1	100	P17
AND: 10B–, 10N–, 1D+	NIR: 10A+, 10G–, 10H–, 10M–	AND (FLAA, CRA)	1,000	1,500	M1, P10, P17
AND: 10B–, 10N–, 1D+	MLI:10A+, 10F+, 10G–, 10H–, 10M–	AND (MFUA)	200	2,000	M1, P10, P17
IRQ: 1D+, 5G–	KUW:1D+	IRQ	4,200	4,200	M2, T3, P2
AND: 10B–, 10D–, 10F–, 10N–	SAF:10G– ,10H–,10M– ,10A+,10D+	AND (INKATHA FREEDOM PARTY, IFP)	12,000	12,000	P2
AND: 5K–, 10N–, 10O–	INS:10M–	AND (GAM)	2,000	20,000	P2
IRQ: 5B–, 4K–, 5G–	USA:5A–,2B–,4B– ,5B+,5H–	IRQ	100	100 000	T5, P2
AND (FRP): 10B–, 10M–	RWA:10G–,10H– ,10M–	AND (FRP)	500,000	1,000,000	M2, P7, P5
AND (JSP): 10A–	AND (AVK):10A+,10F+	AND	200	200	P17

Name	Start	End	Duration (years)	Items and issues	Intensity	Direct participants	External/ Indirect participants[1]
224. Somalia (Somaliland/ secession)	1991	1999	8	3	3	AND (SSDF, SNA, SSA)//SOM	
225. Kenya (Rift Valley)	1991	1995	4	5, 7	3	KEN//AND (KIKUYU)	
226. Togo (regime crisis)	1991	1994	3	4, 5	3	TOG//AND (COD)	GHA (1)
227. Haiti V (military government vs. President Aristide)	1991	1994	3	4, 5	3	AND (ARMY, PARAMILITARY)// HAI (LAVALLAS MOVEMENT)	USA (3)
228. Djibouti (Afar–Issas II)	1991	1994	3	3, 5	3	AND (FRUD)//DJI	FRA (3), ETH (1), ERI (1)
229. Sudan (SPLA split)	1991	1994	3	5	3	AND (SPLA–NASIR– RESP. –UNITED SINCE 93)/AND (SPLA–TORIT, GARANG)	—
230. Zaire (regime crisis)	1991	1999	8	4, 5	3	AND (UNION SACRÉE)//ZAI	BEL (3), FRA (3)
231. Iraq–Kuwait VII	1991	1994	3	1, 6, 7	3	IRQ//KUW, UNO	//USA (3)
232. Armenia–Azerbaijan (Nagorno–Karabakh II)	1991	1994	3	3, 2	4	ARM, AND (ARZACH)/GUS// AZI//RUS	ARM (2)/ // TUR (2)//
233. Sierra Leone (civil war)	1991	1999	8	5, 7	4	AND (RUF)//SIE	AND (NPFL) (3); AND (ULIMO) (3), GUI(3), NIG(3), GHA(2), GAM(2), UPP(1)
234. Russia (Chechnya)	1991	1999	8	1, 2	4	AND (CHECHEN LEADERSHIP)[2]// RUS	

Instruments of initiator	Instruments of affected party	Initiator	Minimum no. of victims/ casualties	Maximum no. of casualties	Outcome
AND: 1D+, 1F+, 1G–, 10H–, 10M–	SOM:1G–,5B+	AND (SSDF)	100	1,000	T1, P 6, P17
KEN: 10M–, 10H–	AND:10A–	KEN	1,500	1,500	P3,
TOG: 10G–, 10H–, 10M–	AND:10A–,10D–,10F–,10A+	TOG	1,000	2,000	P12, P17
AND: 10K–, 10I–, 10H–, 10E–	HAI:10D–,10F–,10O–	AND (ARMEE)	2,000	3,000	P5, P7
AND: 10B–, 10F–, 10N–	DJI:10H–,10M–,5L–	AND (FRUD)	1,000	5,000	M3, P12, P9
AND (SPLA, RIEK): 1D+, 5L–	AND (SPLA, GARANG): 1D+,5L–	AND	5,000	5,000	M4
AND: 10A–, 10B–, 10F–	ZAI: 10M–, 10H–, 10G–	AND (US)	500	10 000	P1
1F+, 3–, 5F–	1D+, 4B–, 4G–, 5A–	IRQ	0	200,000	T4, P14
AND (ARZACH): 3–, 10A+, 10M–	AZI: 2B+,3–, 10K–, 10M–, 10R–	AND (ARZACH)	15,000	15,000	M1, T5, P2
AND: 10M–, 10N–, 10D–, 10K–	SIE:10D+,10H–,10M–	AND (RUF)	50,000	50,000	M4, P7, P9, P2, P5
AND: 10A–, 10B–, 10M–, 10N–	USR:10E–,10H–,10M–	AND	72,000	72,000	T5, M4, P2

Name	Start	End	Duration (years)	Items and issues	Intensity	Direct participants	External/ Indirect participants[1]
235. Somalia (civil war II)	1991	1999	8	5	4	AND (SNA, AIDEED GROUP (SSA [ALI MAHDI]) // AND (SSDF)	UNOSOM (3), UNITAF (3)
236. Russian Federation (attempt at coup d'état)	1992	1993	1	5	3	AND (DUMA [RUSSIAN PARLIAMENT]// RUS (GOVERNMENT)	
237. Chad (autonomy of southern provinces)	1992	1999	7	3, 7	3	AND (CSNPD, FARF)//CHA	.
238. Algeria (Islamists vs. secularists II)	1992	1999	7	5, 4	3	AND (FIS), AND (GIA), AND (MIA)/ AND (MCB), AND (RCD), AND (FFS), AND (FLN)// ALG	LIB (1), IRN (2), SUD// FRA(1)
239. Tajikistan (civil war III)	1992	1999	7	3	3	AND (BADACHSCHAN)// TAJ (CHODSHENT)	RUS(3), IRN(1), USB(1), AFG(1)
240. Angola (civil war III)	1992	1994	2	5, 7	3	AND (UNITA)// ANG	2ZAI, 2UKI, 2FRA, 2URS, 2SAF, 1USA
241. Afghanistan IV (civil war III)	1992	1993	1	5, 3, 4	4	AND (HEKMATYAR)/ AND (SAYYAF)// AND (MASUD), AND (RABBANI)	IRN (2)/SAU (2)
242. Tajikistan (civil war II)	1992	1992	0	5, 3	4	AND (OPPOSITION)// TAJ (COMMUNISTS)	AND (AFG MUJ) (3)// RUS (3), USB (3), KYR (3), TKM (3), KZH
243. Bosnia–Herzegovina (Serbs–Croats)	1992	1994	2	1, 3, 5	4	AND (SERBS)// BOS	SER (2)//TUR (1), IRN (2)
244. Kurds–Kurds	1993	1999	6	4, 5	3	PUK//DPK	IRN (2), IRK (2)

Instruments of initiator	Instruments of affected party	Initiator	Minimum no. of victims/ casualties	Maximum no. of casualties	Outcome
AND (SNA): 10F–, 10N–, 10M–	AND (SSA): 10F–, 10M–, 10N–	AND (SNA)	300-000	300,000	M4, P2, M1
AND: 10A–, 10K–	USR: 10G–, 10M–	RUS	150	150	M3, P11
AND: 10A–, 10B–, 10M–, 1D+	CHA: 10H–, 10M–	AND (CSNPD, FARF)	500	500	T5, P2, P10
AND (FLN): 10O–, 10N–, 10L–; 10B–, 10M–	ALG: 10A+, E–, 10H–, 10I–, 10M–, 10C+	AND (FIS)	30,000	100,000	P3, P9
AND (BADACHSCH.): 10N–, 10O–, 10B–, 10D–, 10N+	TAJ: 10M–, 10E–, 10A+	AND (BADACHSCHAN)	0	100,000	P1, P2, P11
AND: 10M–, 10N–	ANG: 10D+, 10H–, 10M–	AND (UNITA)	500,000	500,000	M1, P12, P17
AND (HEKMATYAR): 10M–, 10O	AND (RABBANI): 10A+, 10M–	AND (HEKMATYAR)	4,000	4,000	P3, P5, M4
AND: 10M–, 10O–	TAJ: 10H–,10M–	AND (OPPOSITION)	20,000	20,000	P11, M3, P15
BOS: 10M–, 10H–	AND:10M–,10N– ,10O–	BOS (BOSN. ARMY)	200,000	200,000	M1, T4, P17
PUK: 1D+, 1D–, 2D+, 2E+, 5L	DPK: 1D+, 1D–, 2D+, 2E+, 5L	PUK	200	1,000	M1, P2

Name	Start	End	Duration (years)	Items and issues	Intensity	Direct participants	External/ Indirect participants[1]
245. Israel–Lebanon IV (Hezbollah vs. government)	1993	1999	6	1, 4	3	AND (ARAFAT–OPPONENTS), AND (HEZBOLLAH)/ AND (AMAL)// AND (SLA), ISR	//LEB (3)
246. Congo (regime crisis)	1993	1995	2	5	3	AND (URD–PCT)// CON	
247. Nigeria (Ogoni)	1993	1999	6	3, 7	3	AND (MOSOP)/ NIG	
248. Afghanistan V (civil war IV)	1993	1999	6	5, 3, 4	4	AND (HEKMATYAR), AND (DOSTUM)/ AND (WAHDAT)// AFG (RABBANI)/ AND (MASUD)	IRN (2)// SAU (2), AND (ISMAIL KHAN)//AND (TALEBAN 94)
249. Burundi III (civil war)	1993	1999	6	5	4	BUI (ARMY), AND (TUTSI–MILITIA)// AND (HUTU–MILITIA)	ZAI (1), TAN (1,2)
250. Surinam II (Toekayana)	1994	1999	5	3	3	AND (SURINAMESE LIBERATION ARMY), AND (TOEKAYANA)// SUR	
251. Mexico (Chiapas)	1994	1999	5	3	3	AND (EZPL)//MEX	
252. Ghana (Konkomba)	1994	1999	5	3,7	3	AND (CONCOMBA)// AND (NANUMBA)// GHA	
253. Rwanda (Hutu refugees)	1994	1999	5	4,5,7	3	AND (HUTU REFUGEES)//RWA	.BURI, ZAI

Instruments of initiator	Instruments of affected party	Initiator	Minimum no. of victims/ casualties	Maximum no. of casualties	Outcome
AND (HEZB.): 10B, 10M–, 10O–	ISR: 5L–, 1F+	AND (HEZBOLLAH)	200	1,000	P1, P2, M1, M4
AND: 10F–, 10N–, 10M–	CON: 10D+, 10F+, 10H–, 10M–	AND	2,000	2,000	P17
AND: 10A–, 10D–, 10F–, 10O–, 4G–, 2B–	NIG: 10H–, 10L–, 10M–	AND (MOSOP)	1,800	2,000	P2, T5
AND (HEKMATYAR): 10M–, 10o–	AND (RABBANI): 10M–	AND	6,000	6,000	M4, P2
BUI: 10K–, 10M–, 10H–, 10B–, 3A–	AND:10F–, 10N–,	BUI (ARMEE)	100,000	100,000	M4, P2, P12
AND: 10B–	SUR:10G–, 10H–	AND	200	500	P2
AND: 10N–, 10D–, 10A+	MEX:10A+,10M–, 10H–, 10D+, 10F+,10m–, 10A+,1G–	AND (EZPL)	150	550	M3, P17, P10;P2
AND: 10M–	GHA:10H–,10M–	AND (KONKOMBA)	2,000	5,000	P2
AND: 10K–, 10M–, 10O–	RWA:10M–, BUR: 10M–, ZAI: 10M–	AND (HUTU REFUGEES)	2,000	200,000	P2, P9, P11

Name	Start	End	Duration (years)	Items and issues	Intensity	Direct participants	External/ Indirect participants[1]
254. Yemen (70 Days War)	1994	1994	0	5,4	4	AND (ARMY–YPR), AND (JSP)//AND (ARMY–YAR), AND (AVK), AND (ISLAH)	SAU (2), GCC (O. QAT) (1), RUS (1)
255. Sri Lanka (Tamils IV)	1995	1999	4	1, 5, 7	4	AND (LTTE)//SRI// AND (MUSLIMS)	
256. Indonesia (Democratic Movement)	1997	1999	2	4, 5	3	AND//INS	
257. Yugoslavia (Serbia: Kosovo and Metohija)	1997	1999	2	5, 3, 2	3	YUG (SERBS)// AND (UCK, LDK (RUGOVA))	ALB (2)
258. Congo (Brazzaville, regime crisis)	1997	1997	0	5	4	CON (BRAZZAVILLE)// AND (DENIS SASSOU–NGESSO)	ANG (3), AND (UNITA) (3)
259. Eritrea–Ethiopia	1998	1999	1	1, 7	4	ERI//ETH	—

Note: Of the total of 259 armed conflicts with 25 minimum casualties selected from the COSIMO database of conflicts occurring in the years 1945–1999, 210 were asymmetric conflicts. Of these 210, 116 were asymmetric conflicts in which the great powers were involved.

Instruments of initiator	Instruments of affected party	Initiator	Minimum no. of victims/ casualties	Maximum no. of casualties	Outcome
AND (YPR): 10M–	AND (YAR):10M–	AND	10,000	15,000	P1, P11
AND: 10B–, 10M–, 10N–, 10H, 10E, 10G	SRI: 10H–, 10M–	AND (LTTE)	1,000	5,000	M1P2, P11
AND: 10A–, 10F–, 10N–	INS: 10D–, 10E–, 10G–, 10M–, 10C+, 10D+, 10E+	AND	1,000	2,000	P17, P7
YUG: 10G–, 10H–, 10A–, 10E–, 10M–	AND: 10A–, 10F–, 10N–, 10O–	YUG (SERBS)	2,200	2,200	P2, P10, P11
AND: 10H–, 10M–	CON (BRAZZAVILLE): 10I–	AND (DENIS SASSOU_ NGESSO)	10,000	10,000	M2, P5
ERI: 5G–, 5L–, 1I–, 10I–	ETH: 1I–, 5G–, 5L–, 10I–	ERI	4,100	4,100	T3, M4, P2

COUNTRY CODE LIST

Country code	Country or area name
ABA	Antigua and Barbuda
AFG	Afghanistan
ALB	Albania
ALG	Algeria
AND	Opposition, autonomy/liberation movements, etc.
ANG	Angola
ARG	Argentina
ARM	Armenia
AUL	Australia
AZI	Azerbaijan
BAR	Barbados
BEL	Belgium
BNG	Bangladesh
BOL	Bolivia
BOS	Bosnia and Herzegovina
BRA	Brazil
BUI	Burundi
BUL	Bulgaria
BUR	Burma
CAF	Central African Republic
CAN	Canada
CAO	Cameroon
CHA	Chad
CHI	Chile
CHN	China
COL	Colombia
CON	Congo, Republic of the (Congo-Brazzaville)
COS	Costa Rica
CUB	Cuba
CYP	Cyprus
CZE	Czechoslovakia
DDR (GDR)	German Democratic Republic
DJI	Djibouti

Country code	Country or area name
DMA	Dominica
DOM	Dominican Republic
DRV	Vietnam, Democratic Republic of (North Vietnam)
ECU	Ecuador
EGY	Egypt
ERI	Eritrea
ETH	Ethiopia
FRA	France
GAB	Gabon
GHA	Ghana
GNB	Guinea-Bissau
GRC	Greece
GRN	Grenada
GUA	Guatemala
GUI	Guinea
GUY	Guyana
HAI	Haiti
HON	Honduras
HUN	Hungary
IND	India
INS	Indonesia
IRN	Iran, Islamic Republic of
IRQ	Iraq
ISR	Israel
ITA	Italy
IVO	Côte d'Ivoire
JAM	Jamaica
JOR	Jordan
KEN	Kenya
KHM	Cambodia
KUW	Kuwait
KYR	Kyrgyzstan
KZH	Kazakhstan

Country code	Country or area name
LAO	Laos
LBR	Liberia
LEB	Lebanon
LIB	Libyan Arab Jamahiriya
MAA	Mauritania
MAG	Madagascar
MAL	Malaysia
MAW	Malawi
MEX	Mexico
MLI	Mali
MOR	Morocco
MZM	Mozambique
NEP	Nepal
NEW	New Zealand
NIC	Nicaragua
NIG	Nigeria
NIR	Niger
NTH	Netherlands
OMA	Oman
PAK	Pakistan
PAN	Panama
PAR	Paraguay
PER	Peru
PHI	Philippines
PLO	Palestine Liberation Organization
PNG	Papua New Guinea
POL	Poland
POR	Portugal
PRK	Korea, Democratic People's Republic of (North Korea)
QAT	Qatar
RHO	Rhodesia
ROK	Korea, Republic of (South Korea)
RUS	Russian Federation
RVN	Vietnam, Republic of (South Vietnam)
RWA	Rwanda

Country code	Country or area name
SAF	South Africa
SAL	El Salvador
SAU	Saudi Arabia
SEN	Senegal
SER	Serbia
SIE	Sierra Leone
SOM	Somalia
SPN	Spain
SRI	Sri Lanka
STV	Saint Vincent and the Grenadines
SUD	Sudan
SUR	Suriname
SYR	Syrian Arab Republic
TAJ	Tajikistan
TAW	Taiwan (Republic of China)
TAZ	Tanzania, Republic of
THI	Thailand
TIB	Tibet
TKM	Turkmenistan
TOG	Togo
TUN	Tunisia
TUR	Turkey
UGA	Uganda
UKI	United Kingdom of Great Britain and Northern Ireland
UNO	United Nations
UPP	Burkina Faso
USA	United States of America
USB	Uzbekistan
USR	Union of Soviet Socialist Republics
VAN	Vanuatu
VEN	Venezuela (Bolivarian Republic of)
YAR	Yemen Arab Republic
YPR	People's Democratic Republic of Yemen
YUG	Yugoslavia
ZAI	Zaire

Country code	Country or area name
ZAM	Zambia
ZIM	Zimbabwe

KEY TO SOME VARIABLES AS USED AND EXPLAINED IN THE COSIMO MANUAL, 1999[3]

Variable "Disputed issues in a conflict"

1. Territory, borders, sea borders
2. Decolonization, national independence
3. Ethnic, religious, or regional autonomy
4. Ideology, system
5. Internal power
6. International power
7. Resources
8. Others

Variable "Intensity of conflicts"

1. Latent conflict; completely nonviolent
2. Crisis; mostly nonviolent
3. Severe crisis; sporadic, irregular use of force, "war-in-sight" crisis
4. War; systematic, collective use of force by regular troops

Variable "Instruments of initiator / Instruments of affected party"

Ten types of instruments are specified that can be used positively or negatively in a conflict: (1) bilateral diplomacy; (2) multilateral diplomacy; (3) information, propaganda; (4) economic instruments; (5) military instruments; (6) secret agencies and services; (7) informal, subversive instruments; (8) alliances; (9) regional or universal integration or isolation; (10) internal instruments.

Several types of instruments were used for the purpose of this research: "military" and "internal" could have the following meaning:

5. Military instruments:

5A+ dispatching military observers
5B+ delivery of arms
5C+ peacekeeping forces
5D+ withdrawal of troops
5A– dispatching troops or vessels
5B– concentration of troops at borders
5C– mobilization
5D– alerts
5E– maneuvers
5F– sporadic military incidents
5G– intervention or invasion
5H– blockade
5I– ending military support
5K– terrorist attacks
5L– full-fledged war

10. Internal instruments:

10A+ government talks with opposition
10B+ ending of state of emergency
10C+ amnesty
10D+ recognition of opposition
10E+ change of government
10F+ fulfilling demands
10G+ agreement on settlement policy
10A– demonstration pro/contra the government
10B– street blockades, terrorist assaults

10C– bribery, corruption
10D– ideological mobilization, populism, charismatic leadership
10E– censorship
10F– unorganized resistance, street battles
10G– arrests, police action
10H– state of emergency, martial law
10I– expulsion of citizens, exile
10K– coup d'état
10L– liquidations
10M– military force
10N– organized resistance, rebellion
10O– antiregime demands
10P– boycott actions
10Q– nonviolent resistance
10R– ethnic cleansing, resettlement
10S– manipulation of elections

Variable "Outcomes, results and settlements"

Territorial outcome:

T1 separation of territory
T2 territorial loss
T3 annexation, unification, incorporation of territory
T4 denouncement of territorial claims
T5 status quo; initiator upholds territorial claims

Military outcome:

Ml stalemate, ceasefire, indecisive outcome
M2 victory of initiator
M3 defeat of initiator
M4 continuation of fighting
M5 withdrawal of troops

Political outcome:

P1 no agreement reached, status quo ante
P2 some issues still in dispute
P3, P4 conclusion of a consensual agreement

P5 change of regime
P6 formation of two different independent regimes
P7 fall of regime
P8 government position weakened
P9 government position strengthened
P10 strengthening of opposition
P11 suppression of opposition
P12 admission or inclusion of opposition into the government
[P13 this indicator is missing in the data set and its manual]
P14 denouncement of claims
P15 increased influence of an external power
P16 decreased influence of an external power
P17 compromise

Notes

CHAPTER 1

1 Raymond Aron, *Memoirs: Fifty Years of Political Reflection,* trans. George Holoch (New York: Holmes & Meier, 1990), 202, 206.
2 Raymond Aron, *Peace and War: A Theory of International Relations* (New Brunswick, NJ: Transaction Publishers, 2003), 5–6.
3 Aron, *Memoirs*, 211–212.
4 Raymond Aron, "A Half-Century of Limited War?," *Bulletin of the Atomic Scientists* 12, no. 4 (April 1956): 103.
5 Hans Morgenthau, *Politics among Nations: The Struggle for Power and Peace* (Boston: McGraw-Hill, 1993), 26.
6 Ibid., 183–216.
7 Ibid., 232–233.
8 Kenneth N. Waltz, "The Politics of Peace," *International Studies Quarterly* 11, no. 3 (September 1967): 199, 201.
9 Jack S. Levy, *War in the Modern Great Power System, 1495–1975* (Lexington: University Press of Kentucky, 1983); idem, "Theories of General War," *World Politics* 37, no. 3 (April 1985): 344–374; idem, "Domestic Politics and War," *Journal of Interdisciplinary History* 18, no. 4 (Spring 1988): 653–673.
10 Hedley Bull, *The Anarchical Society: A Study of Order in World Politics* (London: Macmillan, 1995 [1977]), 200.
11 Ivan Arreguín-Toft, *How the Weak Win Wars: A Theory of Asymmetric Conflict* (Cambridge: Cambridge University Press, 2005), 3–4.
12 William Zartman and Victor A. Kremenyuk, eds., *Cooperative Security: Reducing Third World Wars* (Syracuse, NY: Syracuse University Press, 1995); William Zartman and Guy Olivier Faure, eds., *Escalation and Negotiation in International conflicts* (Cambridge, UK: Cambridge University Press, 2005); Alan J. Kuperman, "The Stinger Missile and U.S. Intervention in Afghanistan," *Political Science Quarterly* 114, no. 2 (Summer 1999): 219–263.

13 John Ellis, *From the Barrel of a Gun: A History of Guerrilla, Revolutionary and Counter-insurgency Warfare, from the Romans to the Present* (London: Greenhill, 1995); Robert B. Asprey, *War in the Shadows: The Classic History of Guerrilla Warfare from Ancient Persia to the Present* (London: Little, Brown, 1994); Walter Laqueur, *Guerrilla Warfare: A Historical and Critical Study* (New Brunswick, NJ: Transaction Publishers, 1998); Ian Frederick and William Beckett, *Modern Insurgencies and Counter-insurgencies: Guerrillas and Their Opponents since 1750* (London: Routledge, 2001); Leroy Thompson, *Dirty Wars: Elite Forces vs. the Guerrillas* (Newton Abbot: David & Charles, 1990).

14 Joanna Wright, "PIRA Propaganda: The Constriction of Legitimacy," *Conflict Quarterly* 10, no. 3 (Summer 1990): 24–41; John A. Hannigan, "The Armalite and the Ballot Box: Dilemmas of Strategy and Ideology in the Provisional IRA," *Social Problems* 33, no. 1 (October 1985): 31–40.

15 Michael McKinley, "Lavish Generosity: The American Dimension of International Support for the Provisional Irish Republican Army, 1968–1983," *Conflict Quarterly* 7, no. 2 (Spring 1987): 20–42.

16 Waltz, "The Politics of Peace," 199.

17 Institute for National Security Studies, *Strategic Assessment: Engaging Power for Peace* (Washington, DC: National Defense University Press, 1998), esp. chap. 11; Robert D. Steele, "The Asymmetric Threat: Listening to the Debate," *Joint Forces Quarterly*, Autumn/Winter 1998–1999, 78–84.

18 Brantly Womack, *China and Vietnam: Politics of Asymmetry* (Cambridge: Cambridge University Press, 2006), 78.

19 Andrew Mack, "Why Big Nations Lose Small Wars: The Politics of Asymmetric Conflict," *World Politics* 27, no. 2 (January 1975): 175.

20 Ibid., 177–179.

21 Ibid., 188.

22 Andrew Mack's biography is available on the Human Security Report Project website (http://www.hsrgroup.org/about-hsrp/OurPeople/Andrew_Mack.aspx).

23 United Nations, "Press Conference to Launch Report on 'Shrinking Costs of War,'" press release, United Nations, Department of Public Information, News and Media Division, New York, January 20, 2010, http://www.un.org/News/briefings/docs/2010/100120_War.doc.htm.

24 Hans Morgenthau, "The Pathology of American Power," *International Security* 1, no. 3 (Winter 1977): 3–20; Michael Howard, "The Forgotten Dimensions of Strategy," *Foreign Affairs* 57, no. 5 (1979); J. Hart, "Three Approaches to the Measurement of Power in International Relations," *International Organization* 30, no. 2 (1976); James Lee Ray and Ayse Vural, "Power Disparities and Paradoxical Conflict Outcomes," *International Interactions* 12, no. 4 (1986); Zeev Maoz, "Power, Capabilities, and Paradoxical Conflict Outcomes," *World Politics* 41, no. 2 (1989); Seymour Melman, "Limits of Military Power: Economic and Other," *International Security* 11, no. 1 (Summer 1986): 72–87.

25 Michael Fischerkeller, "David versus Goliath: Cultural Judgments in Asymmetric Wars," *Security Studies* 7, no. 4 (Summer 1998): 1–43; A. F. K. Organski and Jacek Kugler, "Davids and Goliaths: Predicting the Outcomes of International Wars," *Comparative Political Studies* 11, no. 2 (July 1978):141–180; idem, *The War Ledger* (Chicago: University of Chicago Press, 1980), esp. chap. 11, "Davids and Goliaths: Predicting the Outcomes of International Wars."

26 M. Anderson, M. Arnsten, and H. Averch, *Insurgent Organization and Operations: A Case Study of the Viet Cong in the Delta, 1964–1966* (Santa Monica, CA: RAND Corp. for the Office of the Assistant Secretary of Defense and the Advanced Research Projects Agency, August 1967); Eliot A. Cohen, "Constraints on America's Conduct of Small Wars," *International Security* 9, no. 2 (1984):151–181; Frank C. Darling, "American Policy in Vietnam: Its Role in the Quakeland Theory and International Peace," *Asian Survey* 11, no. 8 (1971): 818–839; Michael J. Engelhardt, "America Can Win, Sometimes: U.S. Success and Failure in Small Wars," *Conflict Quarterly* (Winter 1989): 20–35; Laurence E. Grinter, "How They Lost: Doctrines, Strategies and Outcomes of the Vietnam War," *Asian Survey* 15, no.15 (1975): 1114–1132; Chang Jin Park, "American Foreign Policy in Korea and Vietnam: Comparative Case Studies," *Review of Politics* 37, no. 1 (1975): 20–47; J. A. Koch, *The Chieu Hoi Program in South Vietnam, 1962–1971: A Report Prepared for the Advanced Research Projects Agency* (Santa Monica, CA: RAND Corp., January 1973); Timothy Lomperis, "Vietnam's Offspring: The Lesson of Legitimacy," *Conflict Quarterly* (Winter 1986): 18–33; James McAllister and Ian Schulte, "The Limits of Influence in Vietnam: Britain, the United States and the Diem Regime, 1959–63," *Small Wars & Insurgencies* 17, no.1 (2006): 22–43; Mark Peceny, "Two Paths to the Promotion of Democracy during U.S. Military Interventions," *International Studies Quarterly* 39, no. 3 (1995): 371–401; G. D. T. Shaw, "Laotian 'Neutrality': A Fresh Look at a Key Vietnam War Blunder," *Small Wars & Insurgencies* 13, no. 1 (2002): 25–56; Willard J. Webb and Walter S. Poole, *History of the Joint Chiefs of Staff: The Joint Chiefs of Staff and the War in Vietnam, 1971–1973* (Washington, DC: Office of Joint History, Office of the Chairman of the Joint Chiefs of Staff, 2007).

27 Henry Kissinger, *Ending the Vietnam War: A History of America's Involvement in and Extrication from the Vietnam War* (New York: Simon & Schuster, 2003); Robert S. McNamara with Brian VanDeMark, *In Retrospect: The Tragedy and Lessons of Vietnam* (New York: Times Books, 1995).

28 C. D. Jones, "Just Wars and Limited Wars: Restraints on the Use of the Soviet Armed Forces," *World Politics* 28, no. 1 (1975): 44–68; S. P. Rosen, "Vietnam and the American Theory of Limited War," *International Security* 7, no. 2 (Fall 1982): 83–113; "Vietnam Reappraised: Debates," *International Security* 6, no. 1 (Summer 1981): 3–27; Guan-Fu Gu, "Soviet Aid to the Third World: An Analysis of Its Strategy," *Soviet Studies* 35, no.1 (January 1983): 71–89; W. Andrew Terrill, "Low Intensity Conflict in Southern Lebanon: Lessons and Dynamics of the Israeli-Shi'ite War," *Conflict Quarterly*, Summer 1987, 22–35; Errol A. Henderson and

David J. Singer, "Civil War in the Post-colonial World, 1946–92," *Journal of Peace Research* 37, no. 3 (May 2000): 275–299; Mark P. Lagon, "The International System and the Reagan Doctrine: Can Realism Explain Aid to 'Freedom Fighters'?," *British Journal of Political Science 22*, no.1 (January 1992): 39–70.

29 The September 1995 issue of *Annals of the American Academy of Political and Social Science*, titled "Small Wars," was devoted to small wars. In addition to the theoretical aspects, the articles analyzed specific armed conflicts, the role of the United States, the problem of the arms trade, international law application, experiences with postconflict transformation, and reconciliation. See, in this issue, Roger Beamount, "Small Wars: Definitions and Dimensions," 20–35; William V. O'Brien, "The Rule of Law in Small Wars," 36–46; Stephanie G. Neuman, "The Arms Trade, Military Assistance, and Recent Wars: Change and Continuity," 47–74; Richard H. Shultz Jr., "State Disintegration and Ethnic Conflict: A Framework for Analysis," 75–84; Carnes Lord, "The Role of the United States in Small Wars," 89–100; Michael Moodie, "The Balkan Tragedy," 101–115; James Brown, "The Turkish Imbroglio: Its Kurds," 116–129; Christine M. Knudsen and I. William Zartman, "The Large Small War in Angola," 130–143; Edwin G. Corr, "Societal Transformation for Peace in El Salvador," 144–156; Max G. Manwaring, "Peru's *Sendero Luminoso*: The Shining Path Beckons," 157–166; and Sumit Ganguly, "Wars without End: The Indo-Pakistan Conflict," 167–178.

30 Pierre Allen and Albert A. Stahel, "Tribal Guerrilla Warfare against a Colonial Power: Analyzing the War in Afghanistan," *Journal of Conflict Resolution* 27, no. 4 (1983): 590–617; Alexandre Bennigsen, *The Soviet Union and Muslim Guerrilla Wars, 1920–1981: Lessons for Afghanistan* (Santa Monica, CA: RAND Corp., 1981); David Charters, "Resistance to the Soviet Occupation of Afghanistan: Problems and Prospects," *Conflict Quarterly*, Summer 1980, 8–15; idem, "Coup and Consolidation: The Soviet Seizure of Power in Afghanistan," *Conflict Quarterly*, Spring 1981, 41–48; Lee Coldren, "Afghanistan in 1984: The Fifth Year of the Russo-Afghan War," *Asian Survey* 25, no. 2 (February 1985): 169–179; Joseph Collins, "Soviet Policy toward Afghanistan," *Proceedings of the Academy of Political Science: Soviet Foreign Policy* 36, no. 4 (1987): 198–210; idem, "The Use of Force in Soviet Foreign Policy: The Case of Afghanistan," *Conflict Quarterly*, Spring 1983, 20–47; Richard P. Cronin, "Afghanistan in 1988: Year of Decision," *Asian Survey* 29, no. 2 (February 1989): 207–215; Tad Daley, "Afghanistan and Gorbachev's Global Foreign Policy," *Asian Survey* 29, no. 5 (May 1989): 496–513; Erik P. Hoffman, "Soviet Foreign Policy from 1986 to 1991: Domestic and International Influences," *Proceedings of the Academy of Political Science* 36, no. 4 (1987): 254–272; Efraim Karsh, "Influence through Arms Supplies: The Soviet Experience in the Middle East," *Conflict Quarterly*, Winter 1989, 45–55; Mark N. Katz, "Why Does the Cold War Continue in the Third World?," *Journal of Peace Research* 27, no. 4 (November 1990): 353–357.

31 Melvin R. Laird, "Iraq: Learning the Lessons of Vietnam," *Foreign Affairs*, November/December 2005; Barnett R. Rubin, "Saving Afghanistan," *Foreign Affairs*, January/

February 2007; Robert M. Cassidy, *Russia in Afghanistan and Chechnya: Military Strategic Culture and the Paradoxes of Asymmetric Conflict* (Carlisle Barracks: PA: Strategic Studies Institute, 2003); Martin Ewans, *Conflict in Afghanistan: Studies in Asymmetric Warfare* (New York: Routledge, 2005); Mark Galeotti, *Afghanistan: The Soviet Union's Last War* (London: Cass, 2001); Eric Micheletti, *Special Forces: War against Terrorism in Afghanistan, 2001–2003*, trans. Cyril Lombardini (Paris: Histoire & Collections, 2003); Jules Stewart, *The Khyber Rifles: from the British Raj to Al Qaeda* (Sutton: Stroud, 2006).

32 Alexander Golts, "Military Reform in Russia and the Global War against Terrorism," *Journal of Slavic Military Studies* 17, no. 1 (2004): 29–41; Anatol Lieven, *Chechnya: Tombstone of Russian Power* (New Haven, CT: Yale University Press, 1999); John Russell, *Chechnya: Russia's "War on Terror"* (London: Routledge, 2007); David French, *The British Way in Warfare, 1688–2000* (London: Unwin Hyman, 1990); Daniel Isaac Helmer, "The Currency of Defeat: Asymmetric Warfare and Israel's Loss of the War in Lebanon," PhD diss., Oxford University, 2006; Adam Lowther, *Asymmetric Warfare and Military Thought* (London: Glen Segell, 2006); Rod Thornton, *Asymmetric Warfare: Threat and Response in the Twenty-first Century* (Cambridge: Polity, 2007); Womack, *China and Vietnam;* Ylva Isabelle Blondel, *The Power of Symbolic Power: An Application of O'Neill's Game of Honour to Asymmetric Internal Conflict* (Uppsala: Uppsala University, 2004); Jürgen Mittelstrass, ed., *Symmetry and Asymmetry* (Cambridge: Cambridge University Press, 2005).

33 Frank L. Jones, "*Blowtorch:* Robert Komer and the Making of the Vietnam Pacification Policy," *Parameters*, Autumn 2005, 103–118; George Quester, "The Guerrilla Problem in Retrospect," *Military Affairs* 39, no. 4 (December 1975): 192–196; D. Michael Shafer, "The Unlearned Lessons of Counterinsurgency," *Political Science Quarterly* 103, no. 1 (Spring 1988): 57–80; Benjamin C. Schwartz, *American Counterinsurgency Doctrine and El Salvador: The Frustrations of Reform and the Illusions of Nation Building. A Report for the National Defense Research Institute* (Santa Monica, CA: RAND Corp., 1991); Walter C. Soderlund, "An Analysis of the Guerrilla Insurgency and Coup D'État as Techniques of Indirect Aggression," *International Studies Quarterly* 14, no. 4 (December 1970): 335–360; Timothy P. Wickham-Crowley, "Terror and Guerrilla Warfare in Latin America, 1956–1970," *Comparative Studies in Society and History* 32, no. 2 (April 1990): 201–237; *Counterinsurgency: A Symposium*, April 16–20, 1962, ed. Stephen Hosmer (Santa Monica, CA: RAND Corp., 2006 [1962]); Paul Christopher, Colin P. Clarke and Beth Grill, *Victory Has a Thousand Fathers: Sources of Success in Counterinsurgency* (Santa Monica, CA: RAND Corp., 2010); idem, *Victory Has a Thousand Fathers: Detailed Counterinsurgency Case Studies* (Santa Monica, CA: RAND Corp., 2010).

34 L.V. Deriglazova, "K voprosu ob zvopliuttsii fenomena partizanskoi voiny" [On the evolution of the partisan war (guerrilla warfare) phenomenon], *Mirovaiia ekonomika i mezhdunarodnye otnosheniia* 4 (April 2009): 95–103.

35 Mao Zedong, "Voprosi strategii partizanskoi voiny protiv iaponskikh zakhvatchikov," "O zatiazhnoi voine," *Izbrannie proizvedeniia* [Strategy of guerrilla war against Japanese invaders; On protracted war, in selected writings] (Moscow: Izdatel'stvo inostrannykh literatur,1953), vol. 2.

36 Ernesto Guevara, *Guerrilla Warfare* (Harmondsworth: Penguin Books, 1969); Lewis H. Gann, *Guerrillas in History* (Stanford, CA: Hoover Institution Press, 1971); Walter Laquer, *Guerrilla: A Historical and Critical Study* (London: Weidenfeld & Nicolson. 1977); idem, ed., *The Guerrilla Reader: A Historical Anthology* (London: Wildwood House, 1978).

37 T. E. (Thomas Edward) Lawrence, *Secret Despatches from Arabia*, foreword by A. W. Lawrence (London: Golden Cockerel Press, 1939).

38 John E. Mack, *A Prince of Our Disorder: The Life of T.E. Lawrence* (Oxford: Oxford University Press, 1990); Jeffrey Meyers, *The Wounded Spirit: T.E. Lawrence's "Seven Pillars of Wisdom"* (Basingstoke: Macmillan, 1989); Malcolm Brown and Julia Cave. *A Touch of Genius: The Life of T.E. Lawrence* (London: Dent, 1988); James Lawrence, *The Golden Warrior: The Life and Legend of Lawrence of Arabia* (London: Weidenfeld & Nicolson, 1990); Hart Liddell and Henry Basil, *Lawrence of Arabia* (New York: Da Capo, 1989).

39 Herfried Münkler, "Terrorizm sevodnia: Voina stanovitsia assymetrichnoi" [Terrorism today: War is becoming asymmetric], *Internationale Politik* 1 (2004); N. Komleva and A. Borisov, "Assimetrichnye voini: Geopoliticheskaia tekhnologiia sovremennogo terrorizma" [Asymmetric wars: Geopolitical technology of contemporary terrorism], *Obozrevatel* 11-12 (2002), http://rau.su/observer/N11-12_02/index.htm; H. Richard Schultz, Jr., and J. Andrea Dew, *Insurgents, Terrorists, and Militants: The Warriors of Contemporary Combat* (New York: Columbia University Press, 2006); Robert M. Cassidy, *Counterinsurgency and the Global War on Terror: Military Culture and Irregular War* (Santa Barbara, CA: Praeger Security International, 2006); Hy S. Rothstein, *Afghanistan and the Troubled Future of Unconventional Warfare* (Annapolis, MD: Naval Institute Press, 2006).

40 See Ruth Wedgwood and Kenneth Roth, "Combatants or Criminals? How Washington Should Handle Terrorists," *Foreign Affairs*, May/June 2004.

41 G. V. Malinovskii, *Sovremennye pokal'nye voini imperializma protiv narodov, borioshchikhcia za natsional'nuiu nezavisimost: Uchebnoe posobie dlia slushatel'nei Voennoi Akademii khimicheskoi zashchity* [Contemporary local wars of imperialism against peoples fighting for national independence: Textbook for the students of the Military Academy of Chemical Protection] (Moscow, 1972); R. A. Savushkin, *Pokaln'ye voiny imperializma (1945–1978): Uchebnoe posobie* [Local imperialist wars (1945–1978): Textbook] (Moscow: BPA, 1979); I. I. Georgeadze, V. V. Larionov, and N. A. Antona, *Pokal'nie voiny: Istoriia i sovremennost*, pod red. I. E. Shavrov [Local wars: History and modernity, ed. I. E. Shavrov] (Moscow: Voenizdat, 1981).

42 *Malaiia voina: Opicannaia maiorom v sluzhbe korolia Prusskogo*, per. c frantsuskogo I. F. Bogdanovitchem [Small war: Described by a major in the Prussian king's service,

translated from French by I. F. Bogdanovich] (St. Petersburg, 1768); *Pravila maloii voiny i upotreblenie legkikh voisk: Ob'iasnenie primerami voiny maiorom Valentini,* pred. s nem Gen.-Maior Gogel [Rules of small war and use of light armies: Major Valentini's explanation by means of warfare examples, translated from German by Major General Gogel] (St. Petersburg, 1811); A. E. Engelgart, *Kratkoe nachermanie maloi voiny dlia vsekh rodov opuzhiia,* sost. General leitenant [A brief description of small warfare for all types of weapons, compiled by Lieutenant General] (St. Petersburg, 1850); A. E. Engelgart, *Srazheniia. Malaia voina. Kurs po voennomu dely* [Battles. Small War. Military Course. 1855] (B.M. Lit. Il'ina, 1855); V. Balk, *Malaia voina* (Samostoiatel'nyi vid voini, vedomoi slaboiu staronoiu protiv sil'nogo protivnika), iz vpechenie iz "Taktiki" Bapka, izd. Kak prakticheskoe rukovodstvo dlia komandnogo sostava [Small war (Separate type of war waged by a weak party against a strong one); extract from *Tactics*, by Balk. Published as a practical manual for officers, translated from German] (Moscow: Voennoe delo, 1919).

43 V. V. Kvachkov, retired colonel of the Soviet Main Intelligence Directorate, participated in combat operations in Afghanistan, Azerbaijan, and Tajikistan; was accused of organizing an assassination attempt on Anatoly Chubais in 2005; and was acquitted by a jury in June 2008. Kvachkov is the author of *Russian Special Task Force* (*Spetsnaz Rossii*), which was made available through the Internet portal "Military Business" (http://militera.lib.ru). In 2007, the book was published in Moscow by Panorama Publishing House.

44 Philip Towle, *Should the West Arm Guerrillas?* (London: Council for Arms Control, 1988); Franklin Mark Osanka, ed., *Modern Guerrilla Warfare: Fighting Communist Guerrilla Movements, 1941–1961,* introduction by Samuel P. Huntington (New York: Free Press of Glencoe, 1962); Daniel S. Papp, "Soviet Unconventional Conflict Policies and Strategies in the Third World," *Conflict Quarterly*, Fall 1988, 26–49.

45 "Lokal'nye voiny XX veka: Rol' SSSR" [Local wars of the twentieth century: The role of the USSR], *Otechestvennaia istoriia* 4 (992): 3–3; *Rossiia (SSSR) v lokal'nykh voinakh i vooryzhennykh konfliktakh vtoroi poloviny XX veka.* [Russia (USSR) in local wars and armed conflicts in the second half of the twentieth century], pod red. V. A. Zolotarev (Moscow: Kuchkovo pole, Politgraphresursy, 2000); S. Kolomnin, *Russkii spetsnaz v Afrike* [Russian Special Task Force in Africa] (Moscow: Eksmo, 2005).

46 *Ogranichennye voiny: Kratkii ukazatel literatury* [Limited wars. Brief literature index], Military Scientific Library of the General Staff (Moscow, 1965), 17.

47 *Afganski uroki: Vyvody dlia budushchevo v svete ideinogo naslediia A. E.* Snesareva [Afghan lessons: Conclusions for the future based on the legacy of A. E. Snesarev] (Moscow: Voennii Universitet, Russkoe put, 2003); *Problemy sotsial'noi reabilitatsii ychastnukov voiny v Afganistane (1979–1989 gg.)* [Problems of social rehabilitation of the Afghan war participants (1979–1989)] (Moscow, 1993); O. A. Grinevskiy, *Tainy sovietskoi diplomatii* [Mysteries of Soviet diplomacy] (Moscow: VAGRIUS, 2000); I. Rotar, *Pylaiushchie oblomki imperii: Zametki voennogo korrespondenta* [Empire shatters ablaze: Notes of a military correspondent] (Moscow: Novoe literaturnoe

obozrenie, 2001); 2001); M. Iu. Markelov, *Voina za kabrom: Professiia—russkii zhurnalist* [War offscreen: Profession— Russian journalist] (Moscow: Eksmo, Algoritm, 2004); *Afganistan bolit b moei dushe: Vospominaniia, dnevniki sovietskikh voinov, vypolniavshikh internatsional'nii dolg v Afganistane* [Afghanistan, a pain in my soul: Memoirs, diaries of Soviet soldiers fulfilling their international duty in Afghanistan] (Moscow: Molodaia gvardiia, 1990); A. D. Davydov, *Afganistan: Voiny moglo ne byt. Krest'iastvo i reform* [Afghanistan: War could be avoided. Peasantry and reforms (Moscow: Nauka, Vostochnaia literatura, 1993).

48 I. Kramnik, "Afganistan: Pokhod zapado po sovietskim grabliiam" [Afghanistan: The West repeating Soviet mistakes], *RIA Novosti*, February 15, 2009, http://www.rian.ru/analytics/20090215/162051686.html; "Russia Seeks Bigger Role in Afghanistan on Eve of BRIC Summit," *US Post Today*, June 15, 2009, http://www.usposttoday.com/russia-seeks-bigger-role-in-afghanistan-on-eve-of-bric-summit/.

49 I. Busygina, "Asymmetrichnaia integratsiia b Evropeisckom coiuze" [Asymmetric Integration in the European Union], *Mezhdunarodnye protsess* 5, no. 3 (15, September–December 2007); Iu. A. Bulanikova, "Strategiia integratsii Balkanskikh gosudarstv v EC i NATO: Sravnitel'niĭ analiz (1989–2007 gg.)" [Strategy of Balkan states' integration into the EU and NATO: Comparative analysis (1989–2007)], dissertation, MGIMO, Moscow, 2011.

50 A. R. Aklaev, *Etnopoliticheskaia konfliktologiia: Analiz i menedzhment: Uchebnoe pocobie* [Ethnopolitical conflict studies: Analysis and management: Textbook] (Moscow: Delo, 2005).

51 Ibid., 104.

52 Ibid., 107.

53 Robert Agranoff, *Accommodating Diversity: Asymmetry in Federal States* (Baden-Baden: Nomos, 1999).

54 Aklaev, *Etnopoliticheskaia konfliktologiia* [Ethnopolitical conflict studies], 31.

55 S. K. Oznobishchev, V. Ia. Potapov, and V. V. Skokov, *Kak gotovilcia "asimmetrichnii otvet" na "Strategicheskuiu oboronnuiu initsiativy" R. Reigana* [How the asymmetric response to Ronald Reagan's SDI was prepared] (Moscow: Lenand, 2008).

56 Iu. R. Khismatullina, "Simmetriia, asymmetrii i dissimmetriia v strukture I razvitii zhivoi materii" [Symmetry, asymmetry and dissymmetry in the structure and development of the living matter], dissertation, Saratov, 2005; A. A. Sanglibaev, "Sovremenii politicheskii protsess na Severnom Kavkaze: Proiavlenie etnostatusnoi asimmetrii" [Contemporary political process in the North Caucasus: Manifestation of ethno-status asymmetry], dissertation, Rostov-on-Don, 2001; E. A. Stepanova, "Terrorizm v asimmetrichnom konflikete na pokal'no-regional'nom global'nikh urovniakh (ideologicheskie I organizatsionnye aspekty" [Terrorism in asymmetric conflict at the local/regional and global levels (ideological and organizational aspects)], dissertation, Moscow, 2011; A. A. Sushentsov, "Politicheskaia strategiia SSHA v mezhdunarodnykh konfliktakh 2000-x godov (na primer situatsii v Afganistane i Irake)" [Political strategy of

the USA in international conflicts in the 2000s (case studies of Afghanistan and Iraq)], dissertation, Moscow, 2011.

57 S. G. Chekinov and S. A. Bogdanov, Asimmetrichye deistviia po obospecheniiu voennoi bezopasnosti Rossii" [Asymmetric actions to ensure Russian military security], *Voennaia mysl* 3 (2010):13–22; E. A. Stepanova, "Asymmetrichnyi konflikt kak cilovaia, statusnaia, idiologicheskaia i strukturnaia asymmetria" [Asymmetric conflict as power, status, ideological and structural asymmetry], *Voennaia mysl 5* (2010): 47–54; E. M. A. Khrustaev, "Diversionno–terroristicheskaia voina kak voenno-politicheskii feonomen" [Sabotage and terrorist war as a military-political phenomenon], *Mezhdunarodnye protsessy* 1, no. 2 (May–August 2003).

58 D. G. Baliuev, "Politika voine postindustrial'noi epokhi (Poniatie asimmetrii v voine I konflikte)" [Politics in war of the postindustrial era (The notion of asymmetry in war and conflict)], in *Sovremennaia mirovaia politika: Prikladnoi analiz,* otv. red. A. D. Bogutarov [Contemporary world politics: Applied analysis, ed. A. D. Bogaturov], sect. 1, chap. 11 (Moscow: Aspekt-Press, 2009); "Prikladnye aspektyanaliz global'noi bezopastnosti, gl. 5, v *Global'naia bezopasnost: Innovatsionnye metody analiza konfliktov*, pod. red. A. I. Smirnova [Applied aspects of global security analysis, chap. 5 in *Global Security: Innovative Methods of Conflict Analysis*, ed. A. I. Smirnov] (Moscow: Obschestvo "Znanie" Rossii, 2011), esp. sec. 5.3.

59 Owing to their close interconnection, asymmetric strategies can be separated from asymmetric tactics only tentatively. In general, however, strategy is the large-scale planning or organization of a military operation to achieve certain political and military objectives, while tactics have to do with the specific uses of troops and weaponry to achieve those objectives. As summed up in the frequently quoted words of Carl von Clausewitz, "Tactics is the art of using troops in battle; strategy is the art of using battles to win the war." An asymmetric strategy aims to compensate for resource and power inequality and to exhaust the stronger adversary's political ability to wage war, instead of unrealistically expecting military victory. Asymmetric tactics are actions designed to explore and attack a stronger opponent's vulnerabilities while sidestepping its power structures and military forces. Protracted guerrilla and terrorist operations meet the criteria of asymmetric strategies and tactics, as they are intended to inflict noticeable damage using limited means while avoiding direct clashes with a superior adversary.

60 Cassidy, *Russia in Afghanistan and Chechnya,* 5.

61 Robert H. Scales, "Adaptive Enemies: Dealing with the Strategic Threat after 2010," in *Future Warfare: Anthology*, rev. ed. (Carlisle Barracks, PA: US Army War College Press, 2001): 43.

62 General Colin L. Powell, "A Doctrinal Statement of Selected Joint Operational Concepts," November 10, 1992, 11, http://www.dtic.mil/doctrine/doctrine/research/p146.pdf.

63 General John M. Shalikashvili, "National Military Strategy, 1997," http://www.au.af.mil/au/awc/awcgate/nms/.

64 Joint Chiefs of Staff, *Joint Doctrine Encyclopedia* (Washington, DC: US Government Printing Office, 1997), 59, http://www.fas.org/man/dod-101/dod/docs/encya_b.pdf
65 Joint Chiefs of Staff, *Doctrine for Joint Interdiction Operations*, April 10, 1997, Joint Publication 3-03 (Washington, DC: US Government Printing Office, 1997; rev. ed., 2007): I-1–I-2, http://www.fas.org/irp/doddir/dod/jp3_03.pdf.
66 William S. Cohen, *Quadrennial Defense Review* (Washington, DC: Department of Defense, May 1997). See esp. "Section II. The Global Security Environment," http://www.fas.org/man/docs/qdr/sec2.html.
67 Chairman of the Joint Chiefs of Staff, *Joint Strategy Review* (Washington, DC, 1999), 2.
68 Robert Gates, *Quadrennial Defense Review* (Washington, DC: Department of Defense, February 2010), 80, 87, http://www.strategicstudiesinstitute.army.mil/pdffiles/qdr-2010.pdf.
69 See, for instance, documents on the website of the Strategic Studies Institute, US Army War College (http://www.strategicstudiesinstitute.army.mil/), and the website of the Center for Asymmetric Warfare, Naval Postgraduate School, Point Mugu, CA (http://www.cawnps.org).
70 See the serial publications *Parameters*, *Military Review*, *Joint Forces Quarterly*, and the like.
71 Lloyd J. Matthews, ed., *Challenging the United States Symmetrically and Asymmetrically: Can America Be Defeated?* (Carlisle Barracks, PA: Strategic Studies Institute, 1998).
72 Steven Metz and Douglas V. Johnson II, *Asymmetry and U.S. Military Strategy: Definition, Background, and Strategic Concepts* (Carlisle Barracks, PA: Strategic Studies Institute, 2001), http://www.strategicstudiesinstitute.army.mil/pdffiles/PUB223.pdf.
73 Cassidy, *Russia in Afghanistan and Chechnya,* 5-6; Steven Metz, "Strategic Asymmetry," *Military Review* 81 (July–August 2001): 24.
74 Joint Chiefs of Staff, *Doctrine for Joint Interdiction Operations,* Joint Publication 3-0, 17 September 2006, Incorporating Change 1, February 13, 2008 (JP 3-0 (CH 1), GL-16.
75 Joint Chiefs of Staff, *Doctrine for Join Interdiction Operations,* Joint Publication 3-0 (August 11, 2011), V-12, http://www.dtic.mil/doctrine/new_pubs/jp3_0.pdf.
76 J. Paget, *Counter-insurgency Campaigning* (London, 1967), 176.
77 "Holistic" is a term used in science to stress the necessity of considering all aspects or parts of an object under investigation, evaluating it as a whole rather than as its constituent parts. Holistic analysis focuses on complete systems; a holistic analysis of warfare looks beyond individual strategies, tactics, and battles to consider the entire conduct of a war.
78 Williamson Murray, ed., *A Nation at War in an Era of Strategic Change* (Carlisle Barracks, PA: Strategic Studies Institute, US Army War College, September 2004), 3.
79 Chris Field, *Asymmetric Warfare and Australian National Asymmetric Advantages: Taking the Fight to the Enemy,* Working Paper 136 (Duntroon, Australia: Land Warfare Studies Centre, November 2009); "Gaza Operation Investigations: An

Update" (Tel Aviv, Israel: Ministry of Foreign Affairs, January 2010), http://www.mfa.gov.il/NR/rdonlyres/8E841A98-1755-413D-A1D2-8B30F64022BE/0/GazaOperationInvestigationsUpdate.pdf.

80 Nicholas J. Newman, *Asymmetric Threats to British Military Intervention Operations* (London: Royal United Services Institute for Defence Studies, 2000).

81 See *Terrorism and Asymmetric Conflict in Southwest Asia, June 23–25, 2002: Conference Proceedings*, RAND's Center for Middle East Public Policy and the Geneva Center for Security Policy, ed. Shahram Chubin, Jerrold D. Green, and Andrew Rathmell (Santa Monica, CA: RAND Corp., 2002); *War and Morality: Proceedings of a RUSI Conference, "Morality in Asymmetric War and Intervention Operations": Held on 19–20 September 2002*, including a new section on the Iraq War and its aftermath, ed. Patrick Mileham (London: Royal United Services Institute for Defence Studies, 2004).

82 Montgomery C. Meigs, "Unorthodox Thoughts about Asymmetric Warfare," *Parameters,* Summer 2003, 4–18.

83 Ekaterina Stepanova, *Terrorism in Asymmetrical Conflict: Ideological and Structural Aspects*, SIPRI Research Report 23 (Oxford: Oxford University Press, 2008), 15–23.

84 J. G. Eaton, "The Beauty of Asymmetry: An Examination of the Context and Practice of Asymmetric and Unconventional Warfare from a Western/Centrist Perspective," *Defence Studies* 2, no. 1 (2002): 76-77.

85 General David H. Petraeus was director of the Central Intelligence Agency from September 2011 to November 2012. Prior to that, he served as Commander, International Security Assistance Force and US Forces Afghanistan (USFOR-A) (June 2010–July 2011); Commander, US Central Command (October 2008–June 2010); Commander, Multi-National Force-Iraq (May 2007–October 2008); Commander of the US Army's Combined Arms Center at Fort Leavenworth, Kansas (October 2005); and Commander of Multi-National Security Transition Command-Iraq (June 2004–September 2005). He also was the Assistant Chief of Staff for Operations of the NATO Stabilization Force and the Deputy Commander of the US Joint Counter-Terrorism Task Force-Bosnia. His biography is available on the Department of Defense website (http://www.defense.gov/bios/biographydetail.aspx?biographyid=166).

86 David H. Petraeus, "Lessons of History and Lessons of Vietnam," *Parameters* 16, no. 3 (1986): 45–46.

87 "Editors' Welcome to the Inaugural Issue of *Dynamics of Asymmetric Conflict* (DAC)," *Dynamics of Asymmetric Conflict* 1, no.1 (March 2008): 1.

88 Mittelstrass, *Symmetry and Asymmetry*; George Andrela and Anthony Dunning, eds., *The Age of Asymmetry and Paradox: Essays in Comparative Economics and Sociology* (London: Athena, 2007); Womack, *China and Vietnam;* Gary Witherspoon and Glen Peterson, *Dynamic Symmetry and Holistic Asymmetry in Navajo and Western Art and Cosmology* (New York: Peter Lang, 1995).

89 T. V. Paul, *Asymmetric Conflicts: War Initiation by Weaker Powers* (Cambridge: Cambridge University Press, 1994), 12, 16, 20, 23–35.

90 Ibid., 168–170, 174.

91 T. V. Paul, "Why Has the India-Pakistan Rivalry Been So Enduring? Power Asymmetry and an Intractable Conflict," *Security Studies* 15, no. 4 (October/December 2006): 600–630; idem, *The India-Pakistan Conflict: An Enduring Rivalry* (Cambridge: Cambridge University Press, 2005).

92 Fischerkeller, "David versus Goliath," 2–3.

93 Ibid., 10–13.

94 Ibid., 43.

95 Gil Merom, *How Democracies Lose Small Wars: State, Society, and the Failures of France in Algeria, Israel in Lebanon, and the United States in Vietnam* (Cambridge: Cambridge University Press, 2003), 3.

96 Ibid., 229–231.

97 See, for example, John Prados, *Vietnam: The History of an Unwinnable War, 1945–1975* (Lawrence: University Press of Kansas, 2009).

98 Ivan Arreguín-Toft, "How the Weak Win Wars: A Theory of Asymmetric Conflict," *International Security* 26, no. 1 (2001): 93–128; idem, *How the Weak Win Wars.*

99 Arreguín-Toft, *How the Weak Win Wars*, 31, 18.

100 Ibid., 222–223.

101 Ibid., 227.

102 Adam B. Lowther, *Americans and Asymmetric Conflict: Lebanon, Somalia, and Afghanistan* (Westport, CT: Praeger Security International, 2007), 14–15.

103 Ibid., 6.

104 Ibid., 143.

105 Ibid., 135–137.

106 Ekaterina Stepanova graduated from Moscow State University with a degree in history. She worked at a Moscow-based Russian research center that was established in the 1950s: the Institute of World Economy and International Relations. She led a research group in the study of military conflict at Stockholm International Peace Research Institute in 2006–2008. Her major research interest is international terrorism. The monograph primarily analyzes two main militant ideologies whose followers often rely on terrorism tactics: radical nationalism and religious extremism.

107 Stepanova, *Terrorism in Asymmetrical Conflict,* 153.

108 Ibid., 161–164.

109 William Zartman, "Dynamics and Constraints in Negotiations in Internal Conflicts," in *Elusive Peace: Negotiating an End to Civil War,* ed. William Zartman (Washington, DC: Brookings Institution Press, 1995), 3, 7–11.

110 Christopher Mitchell, "Asymmetry and Strategies of Regional Conflict Reduction," in Zartman and Kremenyuk, *Cooperative Security* 33–37.

111 Morris R. Cohen and Ernest Nagel, *An Introduction to Logic*, 2nd ed. (New York: Harcourt, Brace and World, 1962), 114.

112 Mitchell, "Asymmetry and Strategies of Regional Conflict Reduction," 26.

113 General conflict theory identifies structural (sustainable) and dynamic (changing) characteristics of conflict situations that influence the struggle results. Structural elements include the participants, the environment of their interaction, the causes of the conflict, the nature of the interaction, and the consequences of the conflict. Dynamic characteristics include the duration and intensity of the interaction, as well as the strategies and tactics of the struggle.

114 Mack, "Why Big Nations Lose Small Wars," 177–178.

115 Henry Kissinger, *Diplomacy* (New York: Simon & Schuster, 1995), 629.

116 Mack, "Why Big Nations Lose Small Wars," 178.

117 Aron, *Peace and War*, 34–35.

118 L. V. Deriglazova and S. Minasyan, *Nagorno-Karabakh: Paradoxes of Strength and Weakness in Asymmetric Conflict*, Analytical Reports of the Caucasus Institute 3 (Yerevan, Armenia: Caucasus Institute, June 2011), 10–24.

119 L. V. Deriglazova, "Ideal'ny proval: Voina SSHA v Irake cherez prizmu teorii asymmetrichnogo konflikta" [Ideal failure: US war in Iraq through the prism of asymmetric conflict theory], *Svobodnaia mysl* 3 (2010): 5–16.

CHAPTER 2

1 Andrew Mack, "Why Big Nations Lose Small Wars: The Politics of Asymmetric Conflict," *World Politics* 27, no. 2 (January 1975).

2 Lewis F. Richardson, *Statistics of Deadly Quarrels* (London: Stevens & Sons, 1960); Pitirim A. Sorokin, *Social and Cultural Dynamics: Fluctuation of Social Relationships, War and Revolution* (London: George Allen & Unwin, 1937), vol. 3; Quincy Wright, *A Study of War,* 2nd ed. (Chicago: University of Chicago Press, 1965).

3 Great Power Wars, 1495–1815, data set, Jack S. Levy and T. Clifton Morgan, producers, 1989 (Ann Arbor, MI: Inter-university Consortium for Political and Social Research [distributor], 1994), doi:10.3886/ICPSR09955.v1, http://www.icpsr.umich.edu; Jack S. Levy, *War in the Modern Great Power System, 1495–1975* (Lexington: University of Kentucky Press, 1985).

4 Dyadic Militarized Interstate Disputes (DYMID 2.0) Dataset—Version 2.0, developed by Zeev Maoz, University of California, Davis (last released June 3, 2008), http://psfaculty.ucdavis.edu/zmaoz/dyadmid.pdf (a dyadic version of the Militarized Interstate Dispute of Bremer, Jones, and Singer [1996]).

5 Correlates of War project, database, project founder J. David Singer, University of Michigan, Ann Arbor, http://www.correlatesofwar.org/datasets.htm

6 Minorities at Risk Project (2009): Minorities at Risk Dataset, developed by Ted Robert Gurr (College Park, MD: Center for International Development and Conflict Management, 2009), http://www.cidcm.umd.edu/mar/data.asp; Ted R. Gurr, *Minorities at Risk: A Global View of Ethnopolitical Conflicts* (Washington, DC: US Institute of Peace, 1993).
7 UN Collective Security System, a collective security database developed by Ernst Haas, University of California, Berkeley, http://www.usc.edu/dept/ancntr/Paris-in-LA/Database/haas.html/.
8 Major armed conflicts (1945–1995) data set, developed by Kalevi J. Holsti, University of British Columbia, Vancouver; Kalevi J. Holsti, *The State, War and the State of War* (Cambridge: Cambridge University Press, 1996).
9 André Miroir, Éric Remacle, and Olivier Paye, *Les conflits armés de 1945 à nos jours* (Paris: Services fédéraux des affaires scientifiques, techniques et culturelles, 1994).
10 Database on third-party interventions, developed by Patrick M. Regan, Binghamton University (SUNY), Binghamton, New York, http://bingweb.binghamton.edu/~pregan/; Patrick M. Regan, "Third Party Interventions and the Duration of Intrastate Conflict," *Journal of Conflict Resolution,* February 2002.
11 Uppsala Conflict Data Program (UCDP), project director Peter Wallensteen, Department of Peace and Conflict Research, Uppsala University, Sweden, http://www.prio.no/cwp/ArmedConflict/.
12 Kristine Eck, *A Beginner's Guide to Conflict Data: Finding and Using the Right Dataset*, UCDP Paper 1 Uppsala: Uppsala Conflict Data Program, Uppsala University, December 2005, 9, http://www.pcr.uu.se/digitalAssets/18/18128_UCDP_paper1.pdf.
13 Ibid., 64
14 Correlates of War project, National Material Capabilities (v4.0), http://www.correlatesofwar.org/.
15 UCDP/PRIO Armed Conflict Dataset, http://www.pcr.uu.se/research/ucdp/datasets/ucdp_prio_armed_conflict_dataset/
16 International Peace Research Institute, Oslo, (PRIO), Oslo, Norway, http://www.prio.no/.
17 Code Manual to Excel Data Bank Kosimo1 (COSIMO 1, database, National and International Conflicts,1945–1999, developed by Frank R. Pfetsch, University of Heidelberg, Germany),http://www.hiik.de/en/kosimo/data/codemanual_kosimo1b.pdf.
18 UCDP/PRIO Armed Conflict Dataset Codebook, Version 4–2007, 4, http://www.prio.no/sptrans/2119005713/UCDP_PRIO_Codebook_v4-2007.pdf.
19 Nils Petter Gleditsch, Peter Wallensteen, Mikael Eriksson, et al., "Armed Conflict 1946–2001: A New Dataset," *Journal of Peace Research* 39, no. 5 (2002): 617.
20 Nils Petter Gleditsch, Peter Wallensteen, Mikael Eriksson, et al., *Armed Conflict 1946–2000: A New Dataset. A Joint Report from the Conflict Data Project in the Department of Peace and Conflict Research at Uppsala University and the Conditions*

of War and Peace Program at the International Peace Research Institute, Oslo (PRIO) (Oslo: PRIO, 2001), 4.

21 UCDP/PRIO Armed Conflict Dataset Codebook, Version 4–2007, 3–4.

22 F. R. Pfetsch and C. Rohloff, *National and International Conflicts, 1945–1995* (London: Routledge, 2000), 27.

23 Ibid., 32.

24 Richardson, *Statistics of Deadly Quarrels*, 29–31, 73.

25 UCDP/PRIO Armed Conflict Dataset Codebook, Version 4–2007, 8-12.

26 Code Manual to Excel Data Bank Kosimo1b, 1–12, http://www.hiik.de/en/kosimo/data/codemanual_kosimo1b.pdf.

27 For an explanation of the research methodology, see Pfetsch and Rohloff, *National and International Conflicts, 1945–1995*, 49.

28 Ibid., 50.

29 Calculations throughout this chapter were made based on the UCDP data for 1946–2006 (http://www.pcr.uu.se/publications/UCDP_pub/Conflict_List_1946-2006.pdf).

30 Hans J. Morgenthau, *Politics among Nations: The Struggle for Power and Peace* (Boston: McGraw-Hill, 1993), 113–179.

31 Raymond Aron, *Peace and War: A Theory of International Relations* (Garden City, NY: Doubleday & Co., 1966), 98–99.

32 R. Kagan, *Of Paradise and Power: America and Europe in the New World Order* (New York: Alfred A. Knopf, 2003); I. Ia. Zebelev and M. A. Troitskii, Sila i vliianie v amerikano-rossiiskikh otnosheniiakh: Semioticheskii analiz, *Ocherki tekushchey politiki vypusk* 2 [I. A. Zevelev and, Mikhail Troitskiy. "Power and Influence in U.S.-Russian Relations: A Semiotic Analysis," *Essays on Current Politics 2*]. (Moscow: NOFMO [Scientific and International Forum on International Relations], 2006), 72.

33 Correlates of War project. National Material Capabilities Data Documentation. Version 4.0 (last update June 2010), 4, http://www.correlatesofwar.org/COW2%20Data/Capabilities/NMC_Codebook_4_0.pdf.

34 T. V. Paul, *Asymmetric Conflicts: War Initiation by Weaker Powers* (Cambridge, UK: Cambridge University Press, 1994).

35 Gil Merom, *How Democracies Lose Small Wars: State, Society, and the Failures of France in Algeria, Israel in Lebanon, and the United States in Vietnam* (Cambridge, UK: Cambridge University Press, 2003); Ivan Arreguín-Toft, *How the Weak Win Wars: A Theory of Asymmetric Conflict* (Cambridge, UK: Cambridge University Press, 2005); Ekaterina Stepanova, *Terrorism in Asymmetrical Conflict: Ideological and Structural Aspects,* SIPRI Research Report 23 (Oxford: Oxford University Press, 2008).

36 Both Jewish and Arab (Palestinian) organizations have used terrorist methods.

37 Indicator 13 is absent from the list and the database.

38 The United States, the USSR/Russia, China, France, and the United Kingdom are considered to be the great powers of the post–World War II era. The active participation of some regional leaders, such as Israel and India, in armed conflicts

also requires taking the indicators for those countries into account, but in the present research these countries are not included among the great powers.

39 George F. Kennan, *Memoirs, 1925–1950* (New York: Pantheon Books, 1967), 322.

40 I. Kende, "Wars of Ten Years (1967–1976)," *Journal of Peace Research* 15, no. 3 (1978): 232–234.

CHAPTER 3

1 *Raspad Britanskoi imperii*, pod. red. A. G. Mileikovskogo [The dissolution of the British Empire, ed. A. G. Mileikovsky] (Moscow: Nauka, 1964), 633.

2 Royal Institute of International Affairs, *The British Empire: A Report on Its Structure and Problems by a Study Group of Members of the Royal Institute of International Affairs* (London: Oxford University Press, 1939), 1, 13.

3 *Imperial Conference 1926: Inter-imperial Relations Committee, Report, Proceedings and Memoranda* (Balfour Declaration). Printed for the Imperial Conference, and transferred to the National Archives from the Commonwealth Department of Foreign Affairs, 1926, p. 2. Commonwealth documents, Museum of Australian Democracy, Canberra, http://foundingdocs.gov.au/scan-sid-13.html.

4 Royal Institute of International Affairs, *The British Empire*, 133–167.

5 G. S. Ostapenko, *Britanskie konservatory i dekolonizatsiia* [British Conservatives and decolonization] (Moscow: Institut vseobshchei istorii RAN [RAS Institute of World History], 1995), 73.

6 Royal Institute of International Affairs, *The British Empire*, 219, 236.

7 Ibid., 237–238.

8 Ibid., 249–252.

9 Royal Institute of International Affairs, *British Security: A Report by a Chatham House Study Group* (London: Royal Institute of International Affairs, 1946), 26, 29, 39.

10 Ibid., 47–49.

11 Ibid., 61–62.

12 Ibid., 82–89.

13 Ibid., 90–101.

14 Ibid., 129.

15 B. Porter, *The Lion's Share: A Short History of British Imperialism, 1850–2000* (London: Pearson Longman, 2004), 291.

16 K. Jeffery, "The Second World War," in *The Oxford History of the British Empire*, edited by W. Roger Louis, vol. 4: *The Twentieth Century* (Oxford: Oxford University Press, 1999), 326.

17 W. Roger Louis, *Imperialism at Bay, 1941–1945: The United States and the Decolonization of the British Empire* (Oxford: Clarendon Press, 1977), 134–136.

18 "Defence in the Mediterranean, Middle East and Indian Ocean: Memorandum by Mr. Bevin for Cabinet Defence Committee," March 13, 1946, CAB 131(2), DO (46)40, in *British Documents on the End of Empire* (*BDEE*), Ronald Hyam, general editor, ser. A, vol. 2: *The Labour Government and the End of Empire, 1945–1951*, pt. 3: *Strategy, Politics and Constitutional Change* (London: HMSO, 1992), 215–218; "Financing of Colonial Defence: Letter from Mr. Creech Jones to Sir S. Cripps (Exchequer)," January 10, 1949, DEFE 7/413, ibid., 365–367; "Strategy and Current Defence Policy in South-East Asia and the Far East: JPS Report (JP(50)47) to COS," April 6, 1950, DEFE 4/31, COS 70(50)4, May 2, 1950, ibid., 394–401.

19 I. I. Zhigalov, "Ob osobennostiakh britanskogo rabochego dvizhenie v period Vtoroii mirovoi voiny" [On the British Labour movement during World War II], in *Problemy Britanskoi istoriografii* (Moscow, 1982), 11].

20 Consul-General Wall to Sir R.W. Bullard (Tehran). Communicated by Tabriz dispatch no. 9, 17th Jan. Received in Foreign Office 30th Jan. "Soviet Policy in the Middle East: Analysis of the Effect on the Middle Eastern Problems of the New Soviet Approach Towards All Questions of Foreign Policy." In *British Documents on Foreign Affairs: Reports and Papers from the Foreign Office, Confidential Print*, pt. 4: *From 1946 through 1950*, ser. B: *Near and Middle East,* vol. 1: *Turkey, January 1946–December 1946* and *Eastern Affairs, January 1946–March 1946* (Bethesda, MD: University Publications of America, 1999), 1–9, 190–197.

21 "'Commonwealth Nomenclature': Cabinet Memorandum by Mr. Attlee," December 30, 1948, CAB 129/31, CP (48) 307, in *BDEE*, ser. A, pt. 2: *The Labour Government and the End of Empire, 1945–1951*, pt. 4 (London: HMSO, 1992), 179.

22 "Churchill Assails New Empire Title," *New York Times*, October 29, 1948, 18.

23 Louis, preface to *The Oxford History of the British Empire*, edited by W. Roger Louis, vol. 4: *The Twentieth Century* (Oxford: Oxford University Press, 1999), xi–xii.

24 David French, *The British Way in Warfare, 1688–2000* (London: Cambridge University Press, 1990), 202.

25 William Roger Louis, "The Dissolution of the British Empire in the Era of Vietnam: Presidential Address to the American Historical Association," *American Historical Review* 107, no. 1 (2002): 5–6.

26 Calculated based on Mileikovsky, *Raspad Britanskoi imperii* [The dissolution of the British Empire], 633–635.

27 Commonwealth Secretariat, Communication and Public Affairs Division, *Singapore Declaration of Commonwealth Principles* (London: Marlborough House, 2004), Arts. 1, 5, 6, 8, 9 (pp. 1–3).

28 In February 1942 the Japanese forces managed to inflict a devastating defeat on the British troops and conquered well-defended Singapore. British losses were around 25,000, versus 5,000 Japanese. The loss of Singapore was considered not only a military defeat but also a source of visible moral damage to the United Kingdom's standing in the world and in the region.

29 C. E. Carrington, "Decolonization: The Last Stages," *International Affairs* 38, no. 1 (January 1962) 29.

30 Ibid., 35–36, 39.

31 In 2006, the British press published excerpts from Prince Charles's diary. Prince Charles was present at the handover ceremony. On the flight back from China he discovered that he was the only one not seated in first class, unlike Foreign Secretary Robin Cook and other dignitaries. He then made the following entry: "Such is the end of Empire, I sighed to myself." "Charles' Diary Lays Thoughts Bare," *BBC News*, February 22, 2006, http://news.bbc.co.uk/2/hi/uk_news/4740684.stm.

32 *BDEE*, ser. A, vol. 2: *The Labour Government and the End of Empire, 1945–1951*, pt. 1: *High Policy and Administration*, D. J. Murray and S. R. Ashton, part editors (London: HMSO, 1992), vii.

33 Trevor Owen Lloyd, *The British Empire, 1558–1995* (Oxford: Oxford University Press, 2005).

34 *The Oxford History of the British Empire*, xii.

35 R. Hyam, "Introduction," in *BDEE*, ser. A, vol. 2: *The Labour Government and the End of Empire, 1945–1951*, pt. 1: *High Policy and Administration*, xxiii–lxxviii.

36 Niall Ferguson, *Empire: The Rise and Demise of the British World Order and the Lessons for Global Order* (New York: Basic Books, 2003), xxv.

37 W. J. Bryan, preface to *British Rule in India Condemned by the British Themselves* (London: Indian National Party, 1915), 14.

38 Mileikovsky, *Raspad Britanskoi imperii* [The dissolution of the British Empire], 13.

39 See K. B. Vinogradov, *Na oblomkakh imperii (Kolonial'naia politika Anglii na sovremennom etape)* [On the wrecks of empire (Colonial policy of England in its present stage)] (Leningrad, 1964); A. M. Glukhov, *Britanskii imperialism v Vostochnoi Afrike (1945–1960 gg.): Angliiskaia kolonial'naia politika* [British imperialism in East Africa (1945–1960): English colonial policy] (Moscow: Izdatel'stvo Instituta mezhdunarodnykh otnoshenii, 1962); *Mezhdunarodnye otnosheniia posle Vtoroi mirovoi voiny*, t. 3 (1956–1964 gg.), red. D.E. Mel'nikov i D. G. Tomashevskii [International relations after World War II, vol. 3 (1956–1964), ed. D. E. Melnikov and D. G. Tomashevski] (Moscow: Izdatel'stvo politicheskoi literaturi, 1965); *Politka Anglii v Iuzhnoi i Iugo-Vostochnoi Azii* [English policy in South and Southeast Asia] (Moscow: Nauka, 1966); *Kolonializm vchera i segodnia* [Colonialism yesterday and today] (Moscow: Nauka, 1964); Kolonializm I mezhimperialisticheskie protivorechiia v Afrike [Colonialism and inter-imperialist contradictions in Africa (Moscow: Izdatel'stvo vostochnoi literaturi, 1962); V. G. Trukhanovskii, *Vneshiaia politika Anglii posle Vtoroi mirovoi voiny* [English foreign policy after World War II] (Moscow: Gosudarstveno izdatel'stvo politicheskoi literaturi, 1957); A. G. Sudeikin, *Kolonial'naia politika leiboristskoi partii Anglii mezhdu dvumia mirovymi voinami* [Colonial policy of the English Labour Party between the two world wars] (Moscow: Nauka, 1976).

40 See M. D. Nikitin, *Chernaiia Afrika i britanskie kolonizatory: Stolknoveniie tsivilizatsii* [Black Africa and the British colonizers: Clash of civilizations] (Saratov: Nauchnaya kniga, 2005).
41 Ostapenko, *Britanskie konservatory i dekolonizatsiia* [British Conservatives and decolonization], 11.
42 W. Churchill, "Enclosure to R-483/2. Teheran, Iran. 21st December 1943," in *Churchill and Roosevelt: The Complete Correspondence*, vol. 3: *Alliance Declining: February 1944–April 1945*, edited by Warren F. Kimball (Princeton, NJ: Princeton University Press, 1984), 6–7.
43 D. Kaiser, "Churchill, Roosevelt, and the Limits of Power," *International Security* 10, no. 1 (1985): 218–219.
44 Louis, *Imperialism at Bay*, 513.
45 From the 1600s to the 1950s, more than 20 million people left the British Isles to start a new life overseas, and very few of them returned: Ferguson, *Empire*, xxv.
46 Alan Burns, *In Defence of Colonies: British Colonial Territories in International Affairs* (London: Allen & Unwin, 1957), 54–71.
47 "The Colonial Empire Today: 'Summary of Our Main Problems and Policies': CO International Relations Dept. paper. Annex: Some Facts Illustrating Progress to Date," May 1950, CO 537/5689, no. 69, in *BDEE*, ser. A, vol. 2: *The Labour Government and the End of Empire, 1945–1951*, pt. 1: *High Policy and Administration*, 335.
48 Ibid., ix, xxiii.
49 H. Macmillan, *Pointing the Way, 1959–1961* (New York: Harper & Row, 1972), 476.
50 A. Horne, *Harold Macmillan*, vol. 2: *1957–1986* (New York: Penguin Books, 1991), 195–196.
51 Quoted in M. Jones, *Conflict and Confrontation in South East Asia, 1961–1965: Britain, the United States and the Creation of Malaysia* (Cambridge: Cambridge University Press, 2002), 26.
52 Starting in the 1970s, the quantitative and qualitative parameters of war as defined by the Correlates of War project, led by J. David Singer of the United States, have become widely used. War is understood as an armed conflict between two or more political entities, one of which is represented by the state, in which at least 1,000 military troops sustain battle-related casualties in the course of a year or in the course of the whole conflict: D. Singer and M. Small, *The Wages of War, 1816–1965: A Statistical Handbook* (New York: John Wiley & Sons, 1972), 35.
53 Christopher Bayly and Tim Harper, *Forgotten Wars: The End of Britain's Asian Empire* (London: Penguin Books, 2008), xxviii.
54 J. Paget, *Counter-insurgency Campaigning* (London: Faber and Faber, 1967), 13, 180.
55 Ibid., 181.
56 Ivan Arréguin-Toft, *How the Weak Win Wars: A Theory of Asymmetric Conflict* (Cambridge: Cambridge University Press, 2005), 231–232.

57 "It was difficult to act using violent methods, when the Indian troops exceeded the British fivefold": Mileikovsky, *Raspad Britanskoi imperii* [The dissolution of the British Empire], 83–84.

58 Louis, *Imperialism at Bay, 1941–1945*, 149.

59 Ibid., 188.

60 Kaiser, "Churchill, Roosevelt, and the Limits of Power," 218.

61 Field Marshall Viscount Wavell to Lord Pethick-Lawrence, "Note on the Results to the British Commonwealth of the Transfer of Political Power in India, 13 July 1946," in Nicholas Mansergh and E. W. R. Lumly, series eds., *The Transfer of Power, 1942–7: Constitutional Relations between Britain and India*, vol. 8: *The Interim Government, 3 July–1 November 1946*, volume editor Penderel Moon, assisted by David M. Blake and Lionel Carter (London: HMSO, 1979), 49–52.

62 R. Hyam, "Introduction," in *BDEE*, ser. A, vol. 2: *The Labour Government and the End of Empire, 1945–1951*, pt. 1: *High Policy and Administration*, xxv.

63 Quoted in I. D. Parfenov, *Angliiskie leiboristi i kolonializm* [English Labour and colonialism] (Saratov: Izdatel'stvo Saratovskogo gosudarstvena universiteta, 1969), 45.

64 "India: Minutes of Cabinet Committee on Commonwealth Relations on India's Future Relations with the Commonwealth," January 7, 1949, CAB 134/119, CR 1 (49), in *BDEE*, ser. A, vol. 2: *The Labour Government and the End of Empire, 1945–1951*, pt. 4, 182.

65 S. G. Desiatskov and A. G. Sudeikin, "Angliiskaia politika v Palestine v 1917–1939 gg." [English policy in Palestine, 1917–1939], in *Problemy britanskoi istoriografii* (Moscow, 1982), 43.

66 Jakob Abadi, "The British Experience in Palestine: A Decade of Jewish Cooperation and Resistance, 1936–1946," *Conflict Quarterly*, Winter 1992, 63.

67 "Palestine: Military Implications of Future Policy; Political Implications of Future Policy. Cabinet Conclusions (Confidential Annexes)," January 15, 1947, CAB 128/11, CM 6(47) 3 & 4, in *BDEE*. ser. A, vol. 2: *The Labour Government and the End of Empire, 1945–1951*, pt. 1, 45, 47.

68 The Earl of Halifax to Mr. Bevin, "Annual Survey, 1944. Washington, 12th Feb. 1946." In *British Documents of Foreign Affairs: Reports and Papers from the Foreign Office, Confidential Print*, pt. 4, ser. C: *North America, 1946*, vol. 1 (Bethesda, MD: University Publications of America, 1999), 9.

69 A. Clayton, "'Deceptive Might': Imperial Defence and Security, 1900–1968," in *The Oxford History of the British Empire*, edited by W. Roger Louis, vol. 4: *The Twentieth Century* (Oxford: Oxford University Press, 1999),298.

70 "Memorandum of the Comparative Treatment of the Arabs during Disturbance of 1936–1939 and of the Jews during Disturbance of 1945 and Subsequent Years. Enclosed in 19 June 1947. High Commissioner for Palestine to Secretary of State for Colonies," in Bruce Hoffman, *The Failure of British Military Strategy within Palestine, 1939–1947* (Jerusalem: Daf-Chen Press, 1983), 76–81.

71 Hoffman, *The Failure of British Military Strategy within Palestine, 1939–1947*, 33–34.

72 D. Charters, *The British Army and Jewish Insurgency in Palestine, 1945–47* (Basingstoke: Macmillan, 1988).

73 The problem of separating these territories and creating ethnic states is implied.

74 "Palestine: Cabinet Conclusions on Relinquishing the Mandate and the Line to Be Taken at the UN," September 20, 1947, CAB 128/10, CM 76 (47)6, in *BDEE*, ser. A, vol. 2: *Labour Government and the End of Empire, 1945–1951*, pt. 1, 78.

75 "Palestine: Joint Cabinet Memorandum by Mr. Bevin and Mr. Creech Jones on Attitude to Be Taken at the UN, 3 December 1947." CAB 129/22, CP (47)320. In *BDEE*, ser. A, vol. 2: *Labour Government and the End of Empire, 1945–1951*, pt. 1, 79.

76 "Memorandum on the Present State of Jewish Affairs in the United States. Communicated in Washington dispatch no.344 of 25th February; received 12th March, 1946" (E 2198/14/31), in *British Documents on Foreign Affairs. Reports and Papers from the Foreign Office, Confidential Print*, pt. 4: *From 1946 through 1950*, ser. B: *Near and Middle East, 1946*, vol. 1: *Turkey, January 1946–December 1946*, and *Eastern Affairs, January 1946–March 1946*, 170.

77 Donald Mackay, *The Malayan Emergency, 1948–60: The Domino That Stood* (London: Brassey's, 1997).

78 "Strategy and Current Defence Policy in South-East Asia and the Far East. JPS Report (JP(50)47) to COS, 6 April 1950," DEFE 4/31, COS 70(50)4, May 2, 1950, in *BDDE*, ser. A, vol. 2: *The Labour Government and the End of Empire, 1945–1951*, pt. 3: *Strategy, Politics and Constitutional Change* (London: HMSO, 1992), 397.

79 "Policy in Regard to Malaya and Borneo: Cabinet Memorandum by Mr. Hall (Including Sarawak)," August 29, 1945, CAB 129/1, CP (45)133, in *BDEE*, ser. A, vol. 2: *The Labour Government and the End of Empire, 1945–1951*, pt. 3: *Strategy, Politics and Constitutional Change*, 153–156.

80 Clayton, "Deceptive Might," 296.

81 Memorandum, "Current Situation in Malaya." Central Intelligence Agency. ORE 33-49, published November 17, 1949, pp. 13, 6.

82 Mackay, *The Malayan Emergency, 1948–60*, 148.

83 Louis, "The Dissolution of the British Empire in the Era of Vietnam," 11, 13.

84 Jones, *Conflict and Confrontation in South East Asia, 1961–1965*.

85 Clayton, "Deceptive Might," 296–298.

86 Mackay, *The Malayan Emergency, 1948–60*, 148.

87 Ibid., 150.

88 "The English colonial policy in East Africa was distinguished by large fluctuations, from mass repressions and the use of armed force to constitutional maneuvers and flirtation with the national bourgeoisie, from pro forma, superficial concessions to real ones. But the most important was that British imperialism took a defensive position and gradually began giving ground under the pressure of the ever-growing national liberation movements": Glukhov, *Britanskii imperializm v Vostochnoi Afrike* [British imperialism in East Africa], 192–193.

89 French, *The British Way in Warfare, 1688–2000,* 216–217.
90 Charles W. Gwynn, *Imperial Policing*, 2nd ed. (London: Macmillan, 1939), 10–12.
91 Paget, *Counter-insurgency Campaigning*, 14–15.
92 Ibid., 16–19.
93 Ibid., 68, 176.
94 Ibid., 177–179.
95 Frank Kitson, *Low-Intensity Operations: Subversion, Insurgency and Peacekeeping* (London: Faber and Faber 1971), 15, 27.
96 Ibid., 49, 50.
97 Ibid., 144.
98 Ibid., 25.
99 Ibid., 152.
100 Ibid., 165–169.
101 Ibid., 200.
102 R. Stubbs, *Hearts and Minds in Guerrilla Warfare: The Malayan Emergency, 1948–1960* (Singapore: Oxford University Press, 1989); T. Mockaitis, *British Counterinsurgency, 1919–1960* (London: Macmillan, 1990); R. Thompson, *Defeating Communist Insurgency: Experiences from Malaya and Vietnam* (London: Chatto & Windus, 1966); Charters, *The British Army and Jewish Insurgency in Palestine, 1945–47*; I. Beckett, ed., *The Roots of Counter-insurgency: Armies and Guerilla Warfare, 1900–1945* (London: Blandford, 1988); idem, ed., *Modern Counter-insurgency* (Aldershot: Ashgate, 2007).
103 R. Smith, *The Utility of Force: The Art of War in the Modern World* (London: Penguin Books, 2005).
104 General Rupert Smith also participated in the military operation in Iraq in 1991, commanded UNPROFOR in Sarajevo in 1995, was General Officer Commanding Northern Ireland in 1996–1998, and was Deputy Supreme Allied Commander Europe in 1998–2001.
105 A. Bullock, *Ernest Bevin: Foreign Secretary, 1945–1951* (New York: W. W. Norton, 1983), 233.
106 Memorandum by Chancellor of Exchequer, "Defence Requirements and United States Assistance," July 31, 1950, C.P. (50) 181, pp. 2 (171)–3 (172), http://filestore.nationalarchives.gov.uk/pdfs/small/cab-129-41-cp-181.pdf.
107 "Conclusions of a Meeting of the Cabinet, 25 July 1950," C.M. (50) 50th Conclusions, pp. 173, 175, http://filestore.nationalarchives.gov.uk/pdfs/small/cab-128-18-cm-50-50-10.pdf.
108 "Conclusions of a Meeting of the Cabinet, 1 August 1950," C.M. (50), pp. 189, 190, http://filestore.nationalarchives.gov.uk/pdfs/small/cab-128-18-cm-50-52-12.pdf.
109 Most of these funds were allocated under the British loan. The Mutual Security Program, created by the Mutual Security Act of 1951, signed into law by President Harry S Truman, authorized nearly $7.5 billion to build military, economic, and political strength among nations that were American allies, mostly Western European nations, as a counterpoise to the perceived threat

posed by the Soviet Union during the Cold War (see http://www.archive.org/stream/reporttocongress1952unit/reporttocongress1952unit_djvu.txt).

110 Table 1114, "Mutual Security Program—ICA Aid Allotments, By Regions and By Country: 1949–1955," in *Statistical Abstract of the United States: 1956*, 77th annual ed. (Washington, DC: US Department of Commerce, Bureau of the Census, 1956), 896.

111 Ibid., 884.

112 Cabinet Conclusion 2, "Statement of Defence, 1957," March 18, 1957, C.C. (57), 21st Conclusions, 5 (165), http://filestore.nationalarchives.gov.uk/pdfs/small/cab-128-31-cc-57-21-21.pdf.

113 Cabinet Memorandum, "Statement of Defence, 1957," March 15, 1957, C. (57) 69. Note by the Minister of Defence, "Outline of Future Policy," 3rd proof (London: HMSO), 138, http://filestore.nationalarchives.gov.uk/pdfs/small/cab-129-86-c-57-69-19.pdf.

114 National Service, entailing a mandatory two-year military service for men starting from age 18, was canceled in 1963.

115 Cabinet Memorandum, "Statement of Defence, 1957." March 28, 1957, C. (57) 80. Note by the Minister of Defence, "Outline of Future Policy," 6th proof (London: HMSO), 212, 220, http://filestore.nationalarchives.gov.uk/pdfs/small/cab-129-86-c-57-80-30.pdf.

116 K. Hartley, "The British Experience with an All-Volunteer Force," in *Service to Country: Personnel Policy and the Transformation of Western Militaries*, ed. Curtis Gilroy and Cindy Williams (Cambridge, MA: MIT Press, 2007), 4, 6.

117 Cabinet Memorandum, "Statement of Defence, 1957," March 28, 1957, C.(57) 80, 214–218.

118 French, *The British Way of Warfare, 1688–2000*, 216.

119 "Conclusions of a Meeting of the Cabinet held at 10 Downing Street, S.W. 1, on Tuesday, 2nd April, 1957, at 11 a.m," C.C. 28 (57), National Archives, Catalogue reference CAB/128/31, p. 3, http://filestore.nationalarchives.gov.uk/pdfs/small/cab-128-31-cc-57-28-28.pdf.

120 According to the data in *The Oxford History of the British Empire*, India mobilized 2.4 million personnel for participation in World War II, Canada 1.06 million, Australia 995,000, South Africa 410,000, New Zealand 215,000, and other dependent territories combined 500,000 people: *The Oxford History of the British Empire*, Vol. 4, 308.

121 Clayton, "Deceptive Might," 281.

122 French, *The British Way in Warfare, 1688–2000*, 216.

123 Louis, *The Dissolution of the British Empire in the Era of Vietnam*, 8.

124 Memorandum by the Secretary of State for Defence, "Defence: Draft White Paper," July 4, 1967, C(67) 117. 3–8 (275–280), http://filestore.nationalarchives.gov.uk/pdfs/small/cab-129-131-c-117.pdf.

125 In 1968, the number of military and civilian personnel working for the British armed forces in the Far East or under contract was 80,000 according to the document. Ibid., 7 (279).

126 Ibid., 3–8 (275–280).

127 Cabinet Memorandum, "Defence Withdrawals. Memorandum by the Lord President of the Council," July 4, 1967, C(67) 116, 2 (270), http://filestore.nationalarchives.gov.uk/pdfs/small/cab-129-131-c-116.pdf.

128 French, *The British Way of Warfare, 1688–2000*, 220–221.

129 G. Peele, "The Changed Character of British Foreign and Security Policy," *International Security* 4, no. 4 (Spring 1980): 194.

130 Churchill to Roosevelt, "Enclosure to R-483/2. Teheran, Iran, 21 December 1943," in *Churchill and Roosevelt. The Complete Correspondence*, vol. 3: *Alliance Declining, February 1944–April 1945*, edited by Warren F. Kimball (Princeton, NJ: Princeton University Press, 1984), 10.

131 R. Ovendale, "Introduction," in *The Foreign Policy of the British Labour Governments, 1945–1951*, ed. Ritchie Ovendale (Leicester: Leicester University Press, 1984), 3–4, 16–17.

132 French, *The British Way in Warfare, 1688–2000*, 212.

133 Jeffery, "The Second World War," 326.

134 C. P. Kindleberger, *A Financial History of Western Europe* (London: George Allen & Unwin, 1984), 431–432.

135 Cited in Ferguson, *Empire*, 354.

136 Table no. 1111, "US Government Foreign Grants and Credits, by Program: 1945–1955," in *Statistical Abstract of the United States: 1956*, 77th annual ed. (Washington, DC: U.S. Department of Commerce, Bureau of the Census, 1956), 890.

137 "The British Loan: What It Means to Us," radio broadcast, *NBC University of the Air*, January 12, 1946, http://highered.mcgraw-hill.com/sites/dl/free/0072849037/35264/01_2_brit_loan.html.

138 "Relief Assistance. December 1, 1948," in *Treaties and Other International Agreements of the United States of America, 1776–1949*, vol. 12: *United Kingdom–Zanzibar* (Washington, DC: US State Department, 1974), 920–924.

139 V. G. Trukhanovskii, *Vneshniaia politka Anglii posle Btoroi mirovoi voiny* [English foreign policy after World War II], 69, 73–74.

140 A. Milward, *War, Economy, and Society, 1939–1945* (Berkeley: University of California Press, 1977), 351–352.

141 *Churchill and Roosevelt: The Complete Correspondence*, 3:7.

142 Kaiser, "Churchill, Roosevelt, and the Limits of Power," 205, 219.

143 Mr. Paul Mason to Mr. Wright (Washington), Foreign Office, December 22, 1945. In *British Documents of Foreign Affairs: Reports and Papers from the Foreign Office, Confidential Print*, pt. 4, Series C: *North America, 1946*, vol. 1 (Bethesda, MD: University Publications of America, 1999), 3.

144 "Report to the Secretary of State by the Under Secretary Stettinius on his mission to London, April 7–29, 1944," in *Foreign Relations of the United States, 1944*, vol. 3: *The British Commonwealth and Europe* (Washington, DC: US State Department, Office of the Historian, 1944), 2, 17.

145 Glukhov, *Britanskii imperialism v Vostochnoi Afrike* [British imperialism in East Africa], 79.

146 Kindleberger, *A Financial History of Western Europe*, 425, 443.

147 A. Williams, *Liberalism and War: The Victors and the Vanquished* (London: Routledge, 2006), 121.

148 S. Abouzahr, "The Tangled Web: America, France and Indochina, 1947–1950," *History Today* 54, no. 10 (October 2004): 50.

149 Data from table 1111, "U.S. Government Foreign Grants and Credits, by Program: 1945–1955," and table 1112, "U.S. Government Foreign Grants and Credits, by Country: 1945–1955," in *Statistical Abstract of the United States: 1956*, 77th annual ed. (Washington, DC: US Department of Commerce, Bureau of the Census, 1956), 890–894.

150 Table 373, "Military Assistance Program—Value of Grant Aid Deliveries: 1950–1969," in *Statistical Abstract of the United States: 1970,* 91st annual ed. (Washington, DC: US Department of Commerce, Bureau of the Census, 1970), 249.

151 P. Ovendale, "Macmillan and the Wind of Change in Africa, 1957–1960," *Historical Journal* 32, no. 2 (1995): 458.

152 I. D. Parfenov, *Angliiskie leiboristy i kolonializm* [English Labour and colonialism] (Saratov: Izdatel'stvo Saratovskogo gosuniversiteta, 1969); Sudeikin, *Kolonial'naia politika leiboristskoi partii Anglii* [Colonial policy of the English Labour Party]; Ryzhikov, *Britanskii leiborizm segodnia* [British Labour today]; Ostapenko, *Britanskie konservatory i dekolonizatsiia* [British Conservatives and decolonization].

153 Ryzhikov, *Britanskii leiborizm segodnia* [British Labour today], 41–42.

154 Ovendale, "Introduction," 1–2.

155 D. K. Fieldhouse, "The Labour Governments and the Empire-Commonwealth, 1945–1951," in *The Foreign Policy of the British Labour Governments, 1945–1951*, 83–84, 116–118.

156 Louis, *Imperialism at Bay*, 549.

157 Peele, "The Changed Character of British Foreign and Security Policy," 190–193.

158 Ibid., 197.

159 Ovendale, *Macmillan and the Wind of Change in Africa, 1957–1960*, 455–459.

160 Ferguson, *Empire*, 341.

161 "Joint Statement by President Roosevelt and Prime Minister Churchill, August 14, 1941," Avalon Project: Documents in Law, History and Diplomacy, Yale Law School, Lillian Goldman Law Library, http://avalon.law.yale.edu/wwii/at10.asp. "The Atlantic Conference: Memorandum of Conversation, by the Under Secretary of State (Welles), August 11, 1941," Avalon Project: Documents in Law, History and Diplomacy, Yale Law School, Lillian Goldman Law Library, http://avalon.law.yale.edu/wwii/at08.asp

162 "The Atlantic Conference: Resolution of September 24, 1941," Avalon Project: Documents in Law, History and Diplomacy. Yale Law School, Lillian Goldman Law Library, http://avalon.law.yale.edu/wwii/at17.asp.

163 Kaiser, "Churchill, Roosevelt, and the Limits of Power," 207.

164 Louis, *Imperialism at Bay*, 147.

165 "Restatement of Foreign Policy of the United States," address by President Harry S. Truman, press release, Office of the Press Secretary, The White House, released October 27, 1945, http://highered.mcgraw-hill.com/sites/dl/free/0072849037/35264/01_8_truman.html.

166 Louis, *Imperialism at Bay*, 567–568.

167 Burns, *In Defence of Colonies*, 5–6.

168 Ibid., 124.

169 Ibid., 297–298.

170 Ibid., 134–142.

171 Ibid., 6.

172 Ibid., 90–91.

173 "Macmillan Plea Made U.S. Abstain in U.N. Vote against Colonialism," *Washington Post*, December 22, 1960, A1.

174 Horne, *Harold Macmillan*, vol. 2: *1957–1986*, 273.

175 Williams, *Liberalism and War*, 55.

176 "A Report to the National Security Council. Memorandum for the Secretary of Defence. Implication of a Possible Chinese Communist Attack on Foreign Colonies in South China," July 26, 1949, p. 2, Digital National Security Archive, http://nsarchive.chadwyck.com/nsa/documents/PD/00150/all.pdf.

177 A Report to the National Security Council. Disposition of the Former Italian Colonies in Africa," August 4, 1948, NSC 19/2, p. 2, Digital National Security Archive, http://nsarchive.chadwyck.com/nsa/documents/PD/00061/all.pdf.

178 "A Report to the National Security Council by the Secretary of State on U.S. Position on the Disposition of the Former Italian Colonies," July 26, 1949, NSC 19/4, pp. 9, 3–4. Digital National Security Archive. http://nsarchive.chadwyck.com/nsa/documents/PD/00063/all.pdf.

179 "Current Situation in Malaya," Central Intelligence Agency. ORE 33-49, November 17, 1949, pp. 1, 2, 9.

180 Ferguson, *Empire*, 351.

181 Ibid., 346–354.

182 Ovendale, "Introduction," 17.

183 A. D. Bogaturov, *Velikie derzhavy na Tikhom okeane* [The great powers in the Pacific Ocean] (Moscow: Moskowskyj Obsshestvennyj Nauchyj Fond, 1997), 80–84, 117.

184 P. Kennedy, *The Rise and Fall of Great Powers: Economic Change and Military Conflict from 1500 to 2000* (New York: Random House, 1987), 393–394.

185 "British Military Intervention in Laos; For Foreign Secretary Douglas-Home from Prime Minister Macmillan," March 26, 1961, Digital National Security Archive, http://nsarchive.chadwyck.com/nsa/documents/VI/00383/all.pdf; "British Support in Laos; From the Prime Minister," March 26, 1961, Digital National Security Archive, http://nsarchive.chadwyck.com/nsa/documents/VI/00385/all.pdf; "SEATO Plan,"

March 28, 1961, Digital National Security Archive, http://nsarchive.chadwyck.com/nsa/documents/VI/00394/all.pdf.

186 Louis, *The Dissolution of the British Empire in the Era of Vietnam*, 7–8.

187 Mileikovsky, *Raspad Britanskoi imperii* [The dissolution of the British Empire], 3, 4.

188 A. Rothstein, *Vneshniaia politika Anglii i ee kritiki, 1830–1950* [British foreign policy and its critics, 1830-1950] (Moscow: Progress, 1973), 48–49.

189 A. M. Rodriguez, *Istoriia stran Azii i Afriki v noveishee vremia* [History of Asian and African countries in contemporary time] (Moscow: Prospect, 2006), 75–76.

190 Ferguson, *Empire*, 351–352.

CHAPTER 4

1 Barack Obama, "An America Built to Last," 2012 State of the Union address, press release, Office of the Press Secretary, The White House, Washington, D.C., January 1, 2012, http://www.whitehouse.gov/the-press-office/2012/01/24/remarks-president-state-union-address.

2 The United Nations Special Commission on Iraq, or UNSCOM, was established to inspect and verify the destruction of Iraqi chemical and biological weapons, as well as production and storage facilities for these weapons and for conventional ballistic missiles.

3 "Timeline: Iraq," BBC News, March 31, 2009, http://news.bbc.co.uk/2/hi/middle_east/country_profiles/737483.stm.

4 Iraq Liberation Act of 1998, Pub. L. No. 105-338, 112 Stat. 3178 (1998).

5 Bob Woodward, *Plan of Attack* (New York: Simon & Schuster, 2004), 9–11.

6 "Timeline: Iraq," March 31, 2009.

7 Woodward, *Plan of Attack*, 36–37.

8 "Planning Begins," in *Hard Lessons: The Iraq Reconstruction Experience*, draft of a federal report by the Office of the Special Inspector General for Iraq Reconstruction (SIGIR) (Washington, DC: Government Accountability Office, 2009), chap. 1, 3–17.

9 J. Kampfner, *Blair's Wars* (London: Free Press, 2003), 167–169.

10 George W. Bush, "President's Remarks at the United Nations General Assembly," September 12, 2002, http://merln.ndu.edu/merln/pfiraq/archive/wh/20020912-1.pdf.

11 UN Security Council meetings, no. 4707, February 14, 2003, S/PV.4707; UN Security Council meetings, no. 4717, March 7, 2003, S/PV4714; UN Security Council meetings, no. 4717, March 12, 2003; S/PV4717; UN Security Council meetings, no. 4721, March 19, 2003, S/PV4721.

12 UN Security Council meetings, no. 4717, March 12, 2003, S/PV4717, p. 18, http://daccess-dds-ny.un.org/doc/UNDOC/PRO/N03/276/62/PDF/N0327662.pdf.

13 UN Security Council Resolution, adopted by the Security Council at its 4644th meeting, on November 8, 2002, S/RES/1441 (2002), http://daccess-dds-ny.un.org/doc/UNDOC/GEN/N02/682/26/PDF/N0268226.pdf.

14 "President Pleased with U.N. Vote," remarks by the President on the United Nations Security Council resolution, November 8, 2002, press release, Office of the Press Secretary, The White House, November 8, 2002, http://merln.ndu.edu/merln/pfiraq/archive/wh/20021108-1.pdf.

15 "The National Security Strategy of the United States," The White House, September 2002, http://georgewbush-whitehouse.archives.gov/nsc/nss/2002/.

16 UN Security Council meetings, no. 4707, February 14, 2003, S/PV.4707, pp. 19, 21, http://daccess-dds-ny.un.org/doc/UNDOC/PRO/N03/248/19/PDF/N0324819.pdf.

17 *Would an Invasion of Iraq Be a "Just War"?*, Special Report 98 (Washington, DC: United States Institute of Peace, January 2003), 7, http://www.usip.org/files/resources/sr98.pdf.

18 "President Delivers 'State of the Union,'" press release, Office of the Press Secretary, The White House, January 28, 2003, http://georgewbush-whitehouse.archives.gov/news/releases/2003/01/20030128-19.html.

19 "Polling Report—Iraq," PollingReport.com, htttp://www.pollingreport.com/iraq.htm.

20 M. Gordon, "Bush Enlarges Case for War by Linking Iraq with Terrorists," *New York Times*, January 29, 2003, http://www.nytimes.com/2003/01/29/world/state-union-iraq-issue-bush-enlarges-case-for-war-linking-iraq-with-terrorists.html?pagewanted=2&src=pm.

21 "Debate over Iraq Fires Passions Not Seen since the Vietnam War," *USA Today*, March 6, 2003, http://www.usatoday.com/news/nation/2003-03-05-anti-war-usat_x.htm.

22 G. Husinger," Invading Iraq: Is It Justified?," in *Would an Invasion of Iraq Be a "Just War"?*, 11.

23 S. Kull, "Misperceptions, the Media and the Iraq War," Program on International Policy Attitudes (University of Maryland, Baltimore) / Knowledge Networks (Menlo Park, CA), October 2, 2003, 3, PIPA.org, http://www.pipa.org/OnlineReports/Iraq/IraqMedia_Oct03/IraqMedia_Oct03_rpt.pdf.

24 *Would an Invasion of Iraq Be a "Just War"?*, 1–13.

25 Joint Resolution Concerning the War Powers of Congress and the President, Public Law 93-148, 93rd Congress, H.J. Res. 542 (November 7, 1973).

26 US Senate Roll Call Votes, 107th Congress, 2nd session: Vote Summary on the Joint Resolution (H.J. Res. 114), http://thomas.loc.gov/cgi-bin/bdquery/z?d107:HJ00114.

27 Joint Resolution to Authorize the Use of United States Armed Forces against Iraq, October 2, 2002, H.J. Resolution 114, 107th Congress, 2nd Session (October 11, 2002).

28 Department of Defense briefing. Secretary Rumsfeld and General Myers, March 20, 2003, http://www.globalsecurity.org/wmd/library/news/iraq/2003/iraq-030320-dod01.htm

29 George W. Bush, "Presidential Letter," text of a letter from the President to the Speaker of the House of Representatives and President Pro Tempore of the Senate, press release, Office of the Press Secretary, The White House, March 21, 2003, http://georgewbush-whitehouse.archives.gov/news/releases/2003/03/20030321-5.html.

30 "The War Begins," *New York Times*, March 20, 2003, A32.

31 "U.S. Issues Most Wanted List," *World*, CNN.com, April 11, 2003, http://edition.cnn.com/2003/WORLD/meast/04/11/sprj.irq.wanted.cards/index.html

32 "The Long Way from Victory," *New York Times*, May 2, 2003, A32.

33 United Nations Assistance Mission for Iraq (UNAMI), Fact Sheet, August 7, 2007, 1, http://www.uniraq.org/documents/UNAMI_FactSheet-02Aug07_EN.pdf.

34 UN Security Council Resolution, adopted by the Security Council at its 4844th meeting, on October 16, 2003, S/RES/1511 (2003), p. 3, http://daccess-dds-ny.un.org/doc/UNDOC/GEN/N03/563/91/PDF/N0356391.pdf.

35 UN Security Council Resolution, adopted by the Security Council at its 4987th meeting, on June 8, 2004, S/RES/1546 (2004), http://daccess-dds-ny.un.org/doc/UNDOC/GEN/N04/381/16/PDF/N0438116.pdf.

36 UN Security Council meeting, no. 5189, May 31, 2005, S/PV.5189, pp. 3–4, http://daccess-dds-ny.un.org/doc/UNDOC/PRO/N05/365/52/PDF/N0536552.pdf.

37 *Report of the Secretary-General Pursuant to Paragraph 30 of Resolution 1546* (2004), Security Council, December 7, 2005, S/2005/766, p. 3, http://www.uniraq.org/FileLib/misc/SG_Report_S_2005_766_EN.pdf.

38 "'High Turnout' in Iraqi Election," BBC News, December 15, 2005, http://news.bbc.co.uk/2/hi/middle_east/4531904.stm.

39 *The Iraq Study Group Report*, James A. Baker III and Lee H. Hamilton, co-chairs (New York: Vintage Books, 2006), 35.

40 P. Baker, "Bush Veto Sets Up Clash on Budget. Democrats Make War-Funds Threat," *Washington Post*, November 14, 2007, A01.

41 Appendix II: "Levels of Violence and U.S. Force Levels in Iraq," in *Iraq: Key Issues for Congressional Oversight*, report no. GAO-09-294SP (Washington, DC: US Government Accountability Office, March 2009), 41.

42 Agreement between the United States of America and the Republic of Iraq on the Withdrawal of United States Forces from Iraq and the Organization of Their Activities during Their Temporal Presence in Iraq, November 17, 2008, 19–20, http://www.usf-iraq.com/images/CGs_Messages/security_agreement.pdf.

43 *Iraq: Key Issues for Congressional Oversight*, 9.

44 Strategic Framework Agreement for a Relationship of Friendship and Cooperation between the United States of America and the Republic of Iraq, § 8: Law Enforcement and Judicial Cooperation, November 17, 2008, 6, http://www.mnf-iraq.com/images/CGs_Messages/security_agreement.pdf.

45 *Iraq and Afghanistan: Security, Economics, and Governance Challenges to Rebuilding Efforts Should Be Addressed in U.S. Strategies, Before the Committee on Armed Services, H.R.* (testimony of Jacquelyn Williams-Bridges, Managing Director, International Affairs & Trade), March 25, 2009, 4 (GAO-09-476T); http://www.mnf-iraq.com/index.php?option=com_content&task=view&id=9202&Itemid=128.

46 "Obama Sets Iraq Deadline, Unveils New Strategy," Reuters, February 27, 2009, http://www.reuters.com/article/2009/02/27/us-obama-iraq-idUSTRE51P0AY20090227.

47 "Background Note: Iraq," US State Department, May 1, 2012, http://www.state.gov/r/pa/ei/bgn/6804.htm.

48 Michael E. O'Hanlon and Ian Livingston, "Iraq Index: Tracking Variables of Reconstruction & Security in Post-Saddam Iraq," Brookings Institution, Washington, DC, November 2011, 11, http://www.brookings.edu/iraqindex. According to the website, "the Iraq Index is a statistical compilation of economic, public opinion, and security data. This resource will provide updated information on various criteria, including crime, telephone and water service, troop fatalities, unemployment, Iraqi security forces, oil production, and coalition troop strength. The index is designed to quantify the rebuilding efforts and offer an objective set of criteria for benchmarking performance. It is the first in-depth, non-partisan assessment of American efforts in Iraq, and is based primarily on U.S. government information."

49 *5 years of the War in Iraq: Collection of Facts* [in Russian], Washington ProFile. March 13, 2008, 2, http://www.washprofile.org/?q=ru/node/7505 (22 May 2009) http://noravank.am/rus/articles/detail.php?ELEMENT_ID=2792.

50 *Stabilizing Iraq: DOD Cannot Ensure That U.S.-Funded Equipment Has Reached Iraqi Security Forces*. Report to Congressional Committees, report no. GAO-07-711 (Washington, DC: US Government Accountability Office, July 2007), 2, http://www.gao.gov/new.items/d07711.pdf.

51 *Iraq Study Group Report*, 6.

52 Ibid., ix, x, xiii.

53 Ibid., 12.

54 Ibid., 22.

55 Ibid., 27–32.

56 Ibid., 33–35.

57 Ibid., 40.

58 Ibid., 59–96.

59 "President's Address to the Nation, January 10, 2007, press release, Office of the Press Secretary, The White House, January 10, 2007, http://georgewbush-whitehouse.archives.gov/news/releases/2007/01/20070110-7.html.

60 "President Bush delivers State of the Union Address," January 23, 2007, press release, Office of the Press Secretary, The White House, http://georgewbush-whitehouse.archives.gov/news/releases/2007/01/print/20070123-2.html.

61 "General Petraeus's Opening Statement," transcript, *New York Times*, January 23, 2007, http://www.nytimes.com/2007/01/23/world/middleeast/24petraeustex tcnd.html?_r=1.

62 *Iraq: Key Issues for Congressional Oversight*, 9.

63 "Remarks of President Barack Obama: Responsibly Ending the War in Iraq," Camp Lejeune, NC, February 27, 2009, press release, White House Press Office, February 27, 2009, http://www.whitehouse.gov/the_press_office/Remarks-of -President-Barack-Obama-Responsibly-Ending-the-War-in-Iraq/.

64 S. Simon, "The Price of the Surge: How U.S. Strategy Is Hastening Iraq's Demise," *Foreign Affairs* 87, no. 3 (2008): 74.

65 "The Future of Global Engagement: A Discussion with Admiral Michael G. Mullen, Chairman of the Joint Chiefs of Staff, May 18, 2009," Brookings Institution, Washington, DC, URL: http://www.brookings.edu/events/2009/0518_global _engagement.aspx.

66 These problems were discussed in the UN Security Council, Congress, and the US State Department and were presented in reports of the Special Inspector General for Iraq Reconstruction. See UN Security Council meetings, no. 6087, February 26, 2009, S/PV.6087, http://daccess-dds-ny.un.org/doc/UNDOC /PRO/N09/249/91/PDF/N0924991.pdf; *Measuring Stability and Security in Iraq: Report to Congress in Accordance with the Department of Defense supplemental Appropriations Act 2008* (§ 9204, Public Law 110-252), March 2009, http:// www.defenselink.mil/pubs/pdfs/Measuring_Stability_and_Security_in_Iraq _March_2009.pdf; *Quarterly Report and Semiannual Report to the U.S. Congress, by the Special Inspector General for Iraq Reconstruction*, January 30, 2009, Arlington, VA; *Iraq Status Report* (Washington, DC: US Department of State, Bureau of Near Eastern Affairs, April 8, 2009), http://www.state.gov/documents /organization/121770.pdf.

67 According to data prepared by the UN High Commissioner for Refugees, in 2008 Iraq hosted some 2.8 million internally displaced persons, of whom 1.2 million had been displaced before February 2006 and an additional 1.6 million in 2006–2008 (http://www.unhcr.org/4919572d45.html). At the UN Security Council meeting on February 26, 2009, it was noted that in 2008, 220,000 Iraqis had returned to the country and another 500,000 more were expected in 2009, though "millions" of Iraqis are still beyond Iraqi borders. See UN Security Council meetings, no.6087, February 26, 2009, S/PV.6087, pp. 8, 14. http://daccess-dds-ny.un.org /doc/UNDOC/PRO/N09/249/91/PDF/N0924991.pdf.

68 According to UN data, 1 million Iraqis are not sufficiently provided with food and 6 million are provided through the state distribution of food; in 2008, breakouts of cholera were observed in Iraq. See UN Security Council meetings, no. 6087, February 26, 2009, S/PV.6087, p. 13. Moreover, according to UN data, around 15,500 physicians worked in Iraq, while around 100,000 physicians are needed for the population of 27.5 million; see "Iraqi Expatriate Professionals Back in Iraq for the First Conference on

Iraq's Capacities and Expertise," UNAMI Focus 29 (December 2008): 5, http://www.uniraq.org/FileLib/misc/Focus_December2008.pdf.

69 Table 2.1, "U.S. Appropriated Funds," in *Quarterly Report and Semiannual Report to the United States Congress, by the Special Inspector General for Iraq Reconstruction* (Washington, DC: US Government Accountability Office, January 30, 2009), 27.

70 Ibid., 14–15.

71 Ibid., 45.

72 *Measuring Stability and Security in Iraq*, 20.

73 G. Bruno, "Finding a Place for the 'Sons of Iraq,'" Washington, DC, Council on Foreign Relations, Backgrounder, January 9, 2009, 2, http://www.cfr.org/publication/16088/role_of_the_sons_of_iraq_in_improving_security.htm.

74 *Quarterly Report and Semiannual Report to the United States Congress, by the Special Inspector General for Iraq Reconstruction* (Washington, DC: US Government Printing Office, January 30, 2009), 7.

75 *Iraq: Key Issues for Congressional Oversight*, 25, 28.

76 UN Security Council meetings, no. 6087, February 26, 2009, S/PV.6087, p. 9, http://daccess-dds-ny.un.org/doc/UNDOC/PRO/N09/249/91/PDF/N0924991.pdf.

77 "Rossiia vzbolgaet irakskuiu neft" [Russia will stir up the Iraqi oil], *Kommersant* 64, no. 4119 (April 10, 2009), http://www.kommersant.ru/doc.aspx?fromsearch=07d03dc5-e2b0-4568-a640-dda6a1a0c417&docsid=1152768.

78 *Iraq: Key Issues for Congressional Oversight*, 21, 22.

79 *Effective Counterinsurgency: How the Use and Misuse of Reconstruction Funding Affects the War Effort in Iraq and Afghanistan, Hearing, Armed Services Committee, H.R.*, transcript, March 25, 2009, http://www.gpo.gov/fdsys/pkg/CHRG-111hhrg52944/html/CHRG-111hhrg52944.htm.

80 "*Effective Counterinsurgency: How the Use and Misuse of Reconstruction Funding Affects the War Effort in Iraq and Afghanistan, Before the Committee on Armed Services, H.R.* (testimony of Stuart W. Bowen Jr., Special Inspector General for Iraq Reconstruction), March 25, 2009, SIGIR 09-002T.

81 Ibid., 15.

82 UN Security Council meetings, no. 6087, February 26, 2009, S/PV.6087, p. 14.

83 "Iraq Index," November 30, 2011, 12.

84 J. Muir, "'No Delay' in US Withdrawal from Iraq," BBC News, April 27, 2009, http://news.bbc.co.uk/go/pr/fr/-/2/hi/middle_east/8020815.stm.

85 Andrew W. Terrill, *Regional Spillover Effect of the Iraq War* (Carlisle Barracks, PA: Strategic Studies Institute, December 2008), 7, 4, http://www.strategicstudiesinstitute.army.mil/pubs/display.cfm?pubID=901.

86 "Iraq Index," section titled "Estimated Availability of Essential Goods," November 30, 2011, p. 26.

87 "Business Environment Snapshot for Iraq," Washington, DC, World Bank 2012, http://rru.worldbank.org/besnapshots/BecpProfilePDF.aspx?economy=iraq.

88 "Global Peace Index Rankings, 2011," http://www.visionofhumanity.org/gpi-data/#/2011/scor/IQ.

89 Iraq Poll, February 2009. This survey was conducted for ABC News, the BBC, and NHK by D3 Systems of Vienna, VA, and KA Research Ltd. of Istanbul, Turkey (http://news.bbc.co.uk/2/shared/bsp/hi/pdfs/13_03_09_iraqpollfeb2009.pdf).

90 Brookings Institution, Saban Center for Middle East Policy, "Iraq Index," 47–50. http://www.brookings.edu/about/centers/saban/iraq-index.

91 "Iraqis 'More Upbeat about Future,'" BBC News, March 16, 2009, http://news.bbc.co.uk/2/hi/middle_east/7942974.stm.

92 International Republican Institute, "IRI Iraq Index: October 2010 Survey of Iraqi Public Opinion" (February 2, 2011) http://www.iri.org/sites/default/files/2011%20February%202%20IRI%20Index,%20October%2023-30,%202010.pdf.

93 International Republican Institute, "IRI Iraq Index" (June 16, 2011), http://www.iri.org/sites/default/files/2011%20June%2016%20IRI%20Iraqi%20Index,%20April%2013-18,%202011.pdf.

94 UN Security Council meetings, no. 6747, April 10, 2012, http://www.un.org/News/Press/docs//2012/sc10604.doc.htm.

95 "Operation Enduring Freedom: Coalition Deaths by Year," April 10, 2012, iCasualties.org, http://icasualties.org/OEF/index.aspx.

96 US Department of Defense, "Operation Iraqi Freedom (OIF), U.S. Casualty Status: Fatalities as of: September 11, 2013" (Washington, DC: US Department of Defense, September 11, 2013), http://www.defense.gov/news/casualty.pdf.

97 T. Shanker, "Army Is Worried by Rising Stress of Return Tours to Iraq," *New York Times*, April 6, 2008, http://www.nytimes.com/2008/04/06/washington/06military.html?_r=1&scp=1&sq=april%206%202008%20thom%20shanker&st=cse.

98 "Base Slayings Spur Probe of Mental Health Care," *Washington Post*, May 13, 2009, http://www.washingtonpost.com/wp-dyn/content/article/2009/05/12/AR2009051201127_pf.html.

99 *Current Status of Suicide Prevention Programs in the Military, Hearing, Committee on Armed Services, H.R.*, 212th Congress, September 9, 2011, http://www.gpo.gov/fdsys/pkg/CHRG-112hhrg68463/pdf/CHRG-112hhrg68463.pdf.

100 The Military Personnel Subcommittee will meet to receive testimony on psychological stress in the Military: What steps are leaders taking?, July 29, 2009, House of Representatives, 111th Congress, http://frwebgate.access.gpo.gov/cgi-bin/getdoc.cgi?dbname=111_house_hearings&docid=f:56936.pdf (July 8, 2009).

101 "Suicide Prevention Task Force Report. Executive Summary," 2010, http://www.stripes.com/polopoly_fs/1.115776.1282666756!/menu/standard/file/Suicide%20Prevention%20Task%20Force_EXEC%20SUM_08-20-10%20v6.pdf

102 Department of Defense Suicide Prevention Office, http://www.suicideoutreach.org/Default.aspx.

103 "Iraqi Deaths," iCasualties.org, May 28, 2009, http://icasualties.org/Iraq/IraqiDeaths.aspx.

104 *Iraq Family Health Survey Report*, IFHS 2006/7 (Washington, DC: World Health Organization, 2007), http://www.emro.who.int/iraq/pdf/ifhs_report_en.pdf.

105 S. Hurst, "Iraqi Official: 150,000 Civilians Dead," *Washington Post*, November 10, 2006, http://www.washingtonpost.com/wp-dyn/content/article/2006/11/10/AR2006111000164_pf.html.

106 "Iraqi Deaths from Violence 2003–2011," IraqBodyCount.org, January 2, 2012, http://www.iraqbodycount.org/analysis/numbers/2011/.

107 Gilbert Burnham, Riyadh Lafta, Shannon Doocy, and Les Roberts, "Mortality after the 2003 Invasion of Iraq: A Cross-Sectional Cluster Sample Survey," *Lancet* 368, no. 9545 (October 21, 2006): 1427.

108 "Iraqi Deaths due to U.S. Invasion," JustForeignPolicy.org, http://www.justforeignpolicy.org/iraq/counterexplanation.html.

109 A. Belasco, *The Cost of Iraq, Afghanistan, and Other Global War on Terror Operations since 9/11*, CRS Report for Congress, updated October 15, 2008 (Washington, DC: Congressional Research Service, 2008), CRS-7, 12, http://www.dtic.mil/cgi-bin/GetTRDoc?AD=ADA489440.

110 "'War on Terror' May Cost $2.4 Trillion: Congressional Budget Office Expects the Funds Would Keep 75,000 Troops Fighting in Iraq and Afghanistan for the Next 10 Years," CNN, October 24, 2007, http://money.cnn.com/2007/10/24/news/economy/cbo_testimony/index.htm?cnn=yes.

111 *Iraq: Key Issues for Congressional Oversight*, 1.

112 *Hard Lessons*, vii.

113 "Record Deficit Expected in 2009," *USA Today*, July 28, 2008, http://www.usatoday.com/news/washington/2008-07-27-deficit_N.htm.

114 *Fiscal Year 2010 Budget Overview Document: A New Era of Responsibility. Renewing America's Promise* (Washington, DC: Government Printing Office, 2009), Summary Tables, p. 114, http://www.gpoaccess.gov/usbudget/fy10/pdf/budget/summary.pdf.

115 Fiscal Year 2010 Budget Summary, 30–33, http://www.gpoaccess.gov/usbudget/fy10/pdf/fy10-newera.pdf.

116 *Analytical Perspectives, Budget of the U.S. Government, Fiscal Year 2010*, table 13–1, "Government Assets and Liabilities," 188, http://www.gpo.gov/fdsys/pkg/BUDGET-2010-PER/pdf/BUDGET-2010-PER.pdf.

117 A. A. Allawi, *The Occupation of Iraq: Winning the War, Losing the Peace* (North Yorkshire: Filey, 2007), 256, 257, 259, 260.

118 B. Knowlton, "It Rejects a House Inquiry That Found 'Too Many Uncertainties' on Iraqi Arms: White House Faces Tough Questioning," *New York Times/International Herald Tribune*, September 29, 2003, http://www.nytimes.com/2003/09/29/news/29iht-defend_ed3_.html?pagewanted=all.

119 NBC, *Meet the Press*, May 16, 2004, http://www.msnbc.msn.com/id/4992558/ns/meet_the_press/t/transcript-may/#.T6c7MTKhypc. The guests on the program included Secretary of State Colin Powell; Senator Joseph Biden (D-Del.),

ranking member, Senate Foreign Relations Committee; and Senator John McCain (R-Ariz.), Senate Armed Services Committee.

120 "10 Questions for Madeleine Albright," *Time*, September 22, 2003, http://www.time.com/time/magazine/article/0,9171,1101030922-485697,00.html.

121 CBS News poll, March 15–18, 2008, http://www.pollingreport.com/iraq2/htm.

122 CNN/Opinion Research Corporation poll, December 16-18, 2011, http://www.pollingreport.com/iraq.htm. N = 1,015 adults nationwide. Margin of error ±3%.

123 Figure 4.5: ABC News/*Washington Post* poll, May 29–June 1, 2007, http://www.pollingreport.com/iraq5.htm. N = 1,025 adults nationwide. Figure 4.6: Pew Research Center for the People & the Press, April 18–22, 2007, http://www.pollingreport.com/iraq6.htm. The survey was conducted by Princeton Survey Research Associates International.

124 CBS News/*New York Times* poll, April 25–29, 2008,14, http://www.pollingreport.com/iraq.htm.

125 Charles A. Kupchan and Peter L. Trubowitz, "Grand Strategy for a Divided America," *Foreign Affairs* 86, no. 4 (July/August 2007): 71.

126 *USA Today*/Gallup poll, July 6–8, 2007, 3–4, http://www.pollingreport.com.iraq3.htm; *USA Today*/Gallup poll, 2009. March 14–15, 2009; Gallup poll, August 5–8, 2010, http://www.pollingreport.com.iraq.htm.

127 CNN/Opinion Research Corporation poll, December 16–18, 2011, http://www.pollingreport.com/iraq.htm.

128 CBS News/*New York Times* poll, September 19–23, 2009, http://www.pollingreport.com/iraq.htm.

129 CNN/Opinion Research Corporation poll, December 16–18, 2011, http://www.pollingreport.com/iraq.htm.

130 CNN/Opinion Research Corporation poll, December 16–18, 2011, http://www.pollingreport.com/iraq.htm.

131 CBS News poll, August 20–24, 2010, http://www.pollingreport.com/iraq.htm.

132 CNN/Opinion Research Corporation poll, November 18–20, 2011, http://www.pollingreport.com/iraq.htm.

133 CBS News poll, August 20–24, 2010, http://www.pollingreport.com/iraq.htm.

134 ABC News/*Washington Post* poll, July 15–18, 2009, http://www.pollingreport.com/iraq2.htm.

135 Pew Research Center for the People & the Press survey, April 18–22, 2007, http://www.pollingreport.com/iraq5.htm; Pew Research Center for the People & the Press survey, June 18–29, 2008, http://www.pollingreport.com/iraq.htm. Both polls were conducted by Princeton Survey Research Associates International.

136 Gallup poll, August 5–8, 2010, http://www.pollingreport.com/iraq.htm.

137 *USA Today*/Gallup poll, February 21–24, 2008, http://www.pollingreport.com/iraq.htm.

138 NBC News/ *Wall Street Journal* poll, December 7–11, 2011, http://www.pollingreport.com/iraq.htm. Polling was conducted by the polling organizations of Peter Hart (D) and Bill McInturff (R).

139 D. Rumsfeld, *Known and Unknown: A Memoir* (New York: Penguin Books, 2011), 393.

140 H. Kissinger, *Diplomacy* (New York: Touchstone Books, 1996), 701.

141 M. Laird, "Iraq: Learning the Lessons of Vietnam," *Foreign Affairs* 84, no. 6 (November/December 2005): 24.

142 "Vietnam Reappraised," special issue, *International Security* 6, no. 1 (Summer 1981).

143 Ibid., 7.

144 "Wars of liberation, or guerilla wars, are always extraordinarily difficult for soldiers to fight; and they are particularly demoralizing for an army when the stakes are not very clear" (ibid., 10).

145 "Tendency of troops to become brutalized progressively by coping both with frustration and with an enemy itself acted in a way that was interpretable as brutal" (ibid., 12).

146 Ibid., 3–4.

147 Ibid., 9, 13.

148 Robert S. McNamara with Brian VanDeMark, *In Retrospect: The Tragedy and Lessons of Vietnam* (New York: Times Books, 1995), xvi.

149 McNamara wrote, "We saw in them a thirst for—and a determination to fight for—freedom and democracy. We totally misjudged the political forces within the country" (*In Retrospect*, 322).

150 Ibid., 321–322.

151 Ibid., 331.

152 "The Age Factor: Older Americans Most Negative about Iraq War," Gallup news service. May 11, 2007, http://www.gallup.com/poll/27562/Age-Factor-Older-Americans-Most-Negative-About-Iraq-War.aspx.

153 Kupchan and Trubowitz, "Grand Strategy for a Divided America," 82.

154 Jeffrey Record and Andrew W. Terrill, *Iraq and Vietnam: Differences, Similarities, and Insights* (Carlisle, PA: Strategic Studies Institute, May 2004).

155 Clark C. Smith, *Vietnam ... in Iraq: Reflections on the New Quagmire* (Berkley, CA: Winter Soldier Archive, 2004), 42–43, 48–55.

156 William G. Howell and Jon C. Pevehouse, "When Congress Stops Wars: Partisan Politics and Presidential Power," *Foreign Affairs* 86, no. 5 (September/October 2007), 107.

157 Laird, "Iraq," 39–40.

158 C. Gelpi and P. Feaver, *Choosing Your Battle: American Civil-Military Relations and the Use of Force* (Princeton, NJ: Princeton University Press, 2004); C. Gelpi, "The Cost of War: How Many Casualties Will Americans Tolerate?," *Foreign Affairs* 85, no. 1 (Jan./Feb. 2006); J. Mueller, "Response to C. Gelpi," *Foreign Affairs* 85, no. 1 (January/February 2006).

159 J. Mueller, "The Iraq Syndrome," *Foreign Affairs* 84, no. 6 (November/December 2005): 45.

160 McNamara, *In Retrospect*, 323, 333.

161 L. Diamond, *Squandered Victory: The American Occupation and the Bungled Effort to Bring Democracy to Iraq* (New York: Times Books, 2005), 292.

162 D. Phillips, *Losing Iraq: Inside the Postwar Reconstruction Fiasco* (New York: Basic Books, 2005).

163 J. Dobbins, "Who Lost Iraq?," *Foreign Affairs* 86, no. 5 (September/October 2007): 74.

164 A. Cordesman, *The Iraq War: Strategy, Tactics, and Military Lessons* (Westport, CT: Praeger, 2003).

165 "What to Do in Iraq: A Roundtable," Larry Diamond, James Dobbins, Chaim Kaufmann, Leslie H. Gelb, and Stephen Biddle, participants, *Foreign Affairs* 85, no. 4 (July/August 2006), 150–169.

166 S. Biddle, "Seeing Baghdad, Thinking Saigon," *Foreign Affairs* 85, no. 2 (March/April 2006): 2–14.

167 M. Laird, "Iraq: Learning the Lessons of Vietnam," *Foreign Affairs* 84, no. 6 (November/December 2005): 22–43.

168 A. Krepinevich, "How to Win in Iraq?," *Foreign Affairs* 84, no. 5 (September/October 2005): 87–104.

169 C. Kahl, "How We Fight," *Foreign Affairs* 85, no. 6 (November/December 2006): 83–101.

170 John Dumbrell and David Ryan, eds., *Vietnam in Iraq: Tactics, Lessons, Legacies, and Ghosts* (London: Routledge, 2007).

171 "Spillover" is a term that is used in integration theory to describe the expansion of a phenomenon and its transition from one sphere of relations to other ones. For instance, the analysis of European integration is often regarded as the expansion of integration from economic to political and social spheres, as well as the geographic expansion of integration.

172 Andrew W. Terrill, *Regional Spillover Effect of the Iraq War* (Carlisle Barracks, PA: Strategic Studies Institute, December 2008), http://www.strategicstudiesinstitute.army.mil/pubs/display.cfm?pubID=901.

173 Thomas E. Ricks, *Fiasco: The American Military Adventure in Iraq* (New York: Penguin Press, 2006).

174 The Situation Room, Wolf Blitzer, interview with Thomas Ricks, CNN.com, March 7, 2009, http://transcripts.cnn.com/TRANSCRIPTS/0903/07/sitroom.01.html.

175 Thomas E. Ricks, *The Gamble* (New York: Penguin Press, 2009), 312.

176 "Afghan Taliban Spokesman: We Will Win the War," Nic Robertson , interview with Taliban spokesman Zabiullah Mujahid, CNN.com, May 5, 2009, http://edition.cnn.com/2009/WORLD/asiapcf/05/04/robertson.interview.zabiullah.mujahid/index.html#cnnSTCText.

177 R. N. Haass, *War of Necessity, War of Choice: A Memoir of Two Iraq Wars* (New York: Simon & Schuster, 2009).

178 Brookings Institution, "War of Necessity, War of Choice," transcript of a conference held on June 1, 2009, 10, http://www.brookings.edu/~/media/events/2009/6/01%20iraq%20wars/20090601_iraq_war.pdf.

179 Ibid., 44–45.

180 Barack Obama, "Remarks by the President on a New Beginning," speech delivered at Cairo University, Cairo, Egypt, June 4, 2009, Office of the Press Secretary, The White House, http://i2.cdn.turner.com/cnn/2009/images/06/04/obama.anewbeginning.pd.f

181 Z. Brzezinski, "A Tale of Two Wars," *Foreign Affairs* 88, no. 3 (May/June 2009).

182 "Obama Reverses Course on Alleged Prison Abuse Photos," CNN, May 13, 2009, http://edition.cnn.com/2009/POLITICS/05/12/prisoner.photos/index.html.

183 An especially vocal critic of the Democratic administration and its policies in Iraq and Afghanistan was Dick Cheney, US vice president during the Bush administration, who was a "shadow" figure for a long time, as he avoided interviews and open debates.

184 "UK Combat Operations End in Iraq," BBC News, April 30, 2009, http://news.bbc.co.uk/go/pr/fr/-/2/hi/uk_news/8026136.stm.

185 Gavin Hewitt, "Uncertainty over UK Iraq Legacy," BBC News, April 29, 2009, http://news.bbc.co.uk/go/pr/fr/-/2/hi/uk_news/8023876.stm.

186 "Terrorism Statistics: Terrorists Acts, 1968–2006: Fatalities (Most Recent) by Country," graph, NationMaster.com, http://www.nationmaster.com/graph/ter_ter_act_196_fat-terrorist-acts-1968-2006-fatalities.

APPENDIX

1 The COSIMO manual defines the variable "Indirect or external participants" as "parties that become involved during a conflict" and labels the involvement as (1) diplomatic, political, and/or economic support; (2) weapons sales; or (3) military intervention. For example, the first of the Indochina wars was fought between FRN,RVN,LAO//DRV,AND(PATHET LAO),AND(KHMER ISSARAK), and the external parties were USA(2),UKI(2)//CHN(2).

2 The COSIMO data set spells this information as "Russia (Czechnia)" and "CZECH LEADERSHIP." The spelling has been revised to fit the common romanization.

3 *Code Manual to Excel Data Bank Kosimo1b,* Heidelberg Institute for International Conflict Research, http://www.hiik.de/en/kosimo/kosimo1.html.

Index

Figures, notes, and tables are indicated by f, n, and t following page numbers. Personal surnames starting with "al-" are alphabetized by the next part of the name (e.g., "al-Maliki" is in the Ms).